Applied and Numerical Harmonic Analysis

Series Editor
John J. Benedetto
University of Maryland

Editorial Advisory Board

Akram Aldroubi
Vanderbilt University

Douglas Cochran
Arizona State University

Ingrid Daubechies
Princeton University

Hans G. Feichtinger
University of Vienna

Christopher Heil
Georgia Institute of Technology

Murat Kunt
Swiss Federal Institute of Technology, Lausanne

James McClellan
Georgia Institute of Technology

Wim Sweldens
Lucent Technologies, Bell Laboratories

Michael Unser
Swiss Federal Institute
of Technology, Lausanne

Martin Vetterli
Swiss Federal Institute
of Technology, Lausanne

M. Victor Wickerhauser
Washington University

Advances in Mathematical Finance

Michael C. Fu
Robert A. Jarrow
Ju-Yi J. Yen
Robert J. Elliott
Editors

Birkhäuser
Boston • Basel • Berlin

Michael C. Fu
Robert H. Smith School of Business
Van Munching Hall
University of Maryland
College Park, MD 20742
USA

Ju-Yi J. Yen
Department of Mathematics
1326 Stevenson Center
Vanderbilt University
Nashville, TN 37240
USA

Robert A. Jarrow
Johnson Graduate School of Management
451 Sage Hall
Cornell University
Ithaca, NY 14853
USA

Robert J. Elliott
Haskayne School of Business
Scurfield Hall
University of Calgary
Calgary, AB T2N 1N4
Canada

Cover design by Joseph Sherman.

Mathematics Subject Classification (2000): 91B28

Library of Congress Control Number: 2007924837

ISBN-13: 978-0-8176-4544-1   e-ISBN-13: 978-0-8176-4545-8

Printed on acid-free paper.

©2007 Birkhäuser Boston

9 8 7 6 5 4 3 2 1

www.birkhauser.com   (KeS/EB)

In honor of Dilip B. Madan on the occasion of his 60th birthday

ANHA Series Preface

The *Applied and Numerical Harmonic Analysis (ANHA)* book series aims to provide the engineering, mathematical, and scientific communities with significant developments in harmonic analysis, ranging from abstract harmonic analysis to basic applications. The title of the series reflects the importance of applications and numerical implementation, but richness and relevance of applications and implementation depend fundamentally on the structure and depth of theoretical underpinnings. Thus, from our point of view, the interleaving of theory and applications and their creative symbiotic evolution is axiomatic.

Harmonic analysis is a wellspring of ideas and applicability that has flourished, developed, and deepened over time within many disciplines and by means of creative cross-fertilization with diverse areas. The intricate and fundamental relationship between harmonic analysis and fields such as signal processing, partial differential equations (PDEs), and image processing is reflected in our state-of-the-art *ANHA* series.

Our vision of modern harmonic analysis includes mathematical areas such as wavelet theory, Banach algebras, classical Fourier analysis, time-frequency analysis, and fractal geometry, as well as the diverse topics that impinge on them.

For example, wavelet theory can be considered an appropriate tool to deal with some basic problems in digital signal processing, speech and image processing, geophysics, pattern recognition, biomedical engineering, and turbulence. These areas implement the latest technology from sampling methods on surfaces to fast algorithms and computer vision methods. The underlying mathematics of wavelet theory depends not only on classical Fourier analysis, but also on ideas from abstract harmonic analysis, including von Neumann algebras and the affine group. This leads to a study of the Heisenberg group and its relationship to Gabor systems, and of the metaplectic group for a meaningful interaction of signal decomposition methods. The unifying influence of wavelet theory in the aforementioned topics illustrates the justification for providing a means for centralizing and disseminating information from the broader, but still focused, area of harmonic analysis. This will be a key role of *ANHA*. We intend to publish with the scope and interaction that such a host of issues demands.

Along with our commitment to publish mathematically significant works at the frontiers of harmonic analysis, we have a comparably strong commitment to publish major advances in the following applicable topics in which harmonic analysis plays a substantial role:

Antenna theory　　　　　*Prediction theory*
Biomedical signal processing　　*Radar applications*
Digital signal processing　　*Sampling theory*
Fast algorithms　　*Spectral estimation*
Gabor theory and applications　　*Speech processing*
Image processing　　*Time-frequency and*
Numerical partial differential equations　*time-scale analysis*
Wavelet theory

The above point of view for the *ANHA* book series is inspired by the history of Fourier analysis itself, whose tentacles reach into so many fields.

In the last two centuries Fourier analysis has had a major impact on the development of mathematics, on the understanding of many engineering and scientific phenomena, and on the solution of some of the most important problems in mathematics and the sciences. Historically, Fourier series were developed in the analysis of some of the classical PDEs of mathematical physics; these series were used to solve such equations. In order to understand Fourier series and the kinds of solutions they could represent, some of the most basic notions of analysis were defined, e.g., the concept of "function." Since the coefficients of Fourier series are integrals, it is no surprise that Riemann integrals were conceived to deal with uniqueness properties of trigonometric series. Cantor's set theory was also developed because of such uniqueness questions.

A basic problem in Fourier analysis is to show how complicated phenomena, such as sound waves, can be described in terms of elementary harmonics. There are two aspects of this problem: first, to find, or even define properly, the harmonics or spectrum of a given phenomenon, e.g., the spectroscopy problem in optics; second, to determine which phenomena can be constructed from given classes of harmonics, as done, for example, by the mechanical synthesizers in tidal analysis.

Fourier analysis is also the natural setting for many other problems in engineering, mathematics, and the sciences. For example, Wiener's Tauberian theorem in Fourier analysis not only characterizes the behavior of the prime numbers, but also provides the proper notion of spectrum for phenomena such as white light; this latter process leads to the Fourier analysis associated with correlation functions in filtering and prediction problems, and these problems, in turn, deal naturally with Hardy spaces in the theory of complex variables.

Nowadays, some of the theory of PDEs has given way to the study of Fourier integral operators. Problems in antenna theory are studied in terms of unimodular trigonometric polynomials. Applications of Fourier analysis abound in signal processing, whether with the fast Fourier transform (FFT), or filter design, or the adaptive modeling inherent in time-frequency-scale methods such as wavelet theory. The coherent states of mathematical physics are translated and modulated Fourier transforms, and these are used, in conjunction with the uncertainty principle, for dealing with signal reconstruction in communications theory. We are back to the raison d'être of the *ANHA* series!

John J. Benedetto
Series Editor
University of Maryland
College Park

Preface

The "Mathematical Finance Conference in Honor of the 60th Birthday of Dilip B. Madan" was held at the Norbert Wiener Center of the University of Maryland, College Park, from September 29 – October 1, 2006, and this volume is a Festschrift in honor of Dilip that includes articles from most of the conference's speakers. Among his former students contributing to this volume are Ju-Yi Yen as one of the co-editors, along with Ali Hirsa and Xing Jin as co-authors of three of the articles.

Dilip Balkrishna Madan was born on December 12, 1946, in Washington, DC, but was raised in Bombay, India, and received his bachelor's degree in Commerce at the University of Bombay. He received two Ph.D.s at the University of Maryland, one in economics and the other in pure mathematics. What is all the more amazing is that prior to entering graduate school he had never had a formal university-level mathematics course! The first section of the book summarizes Dilip's career highlights, including distinguished awards and editorial appointments, followed by his list of publications.

The technical contributions in the book are divided into three parts. The first part deals with stochastic processes used in mathematical finance, primarily the Lévy processes most associated with Dilip, who has been a fervent advocate of this class of processes for addressing the well-known flaws of geometric Brownian motion for asset price modeling. The primary focus is on the Variance-Gamma (VG) process that Dilip and Eugene Seneta introduced to the finance community, and the lead article provides an historical review from the unique vantage point of Dilip's co-author, starting from the initiation of the collaboration at the University of Sydney. Techniques for simulating the Variance-Gamma process are surveyed in the article by Michael Fu, Dilip's longtime colleague at Maryland, moving from a review of basic Monte Carlo simulation for the VG process to more advanced topics in variation reduction and efficient estimation of the "Greeks" such as the option delta. The next two pieces by Marc Yor, a longtime close collaborator and the keynote speaker at the birthday conference, provide some mathematical properties and identities for gamma processes and beta and gamma random variables. The final article in the first part of the volume, written by frequent collaborator Robert Elliott and his co-author John van der Hoek, reviews the theory of fractional Brownian motion in the white noise framework and provides a new approach for deriving the associated Itô-type stochastic calculus formulas.

The second part of the volume treats various aspects of mathematical finance related to asset pricing and the valuation and hedging of derivatives. The article by Bob Jarrow, a longtime collaborator and colleague of Dilip in the mathematical finance community, provides a tutorial on zero volatility spreads and option adjusted spreads for fixed income securities – specifically bonds with embedded options – using the framework of the Heath-Jarrow-Morton model for the term structure of interest rates, and highlights the characteristics of zero volatility spreads capturing *both* embedded options and mispricings due to model or market errors, whereas option adjusted spreads measure only the mispricings. The phenomenon of market bubbles is addressed in the piece by Bob Jarrow, Phillip Protter, and Kazuhiro Shimbo, who provide new results on characterizing asset price bubbles in terms of their martingale properties under the standard no-arbitrage complete market framework. General equilibrium asset pricing models in incomplete markets that result from taxation and transaction costs are treated in the article by Xing Jin – who received his Ph.D. from Maryland's Business School co-supervised by Dilip – and Frank Milne – one of Dilip's early collaborators on the VG model. Recent work on applying Lévy processes to interest rate modeling, with a focus on real-world calibration issues, is reviewed in the article by Wolfgang Kluge and Ernst Eberlein, who nominated Dilip for the prestigious Humboldt Research Award in Mathematics. The next two articles, both co-authored by Ali Hirsa, who received his Ph.D. from the math department at Maryland co-supervised by Dilip, focus on derivatives pricing; the sole article in the volume on which Dilip is a co-author, with Massoud Heidari as the other co-author, prices swaptions using the fast Fourier transform under an affine term structure of interest rates incorporating stochastic volatility, whereas the article co-authored by Peter Carr – another of Dilip's most frequent collaborators – derives forward partial integro-differential equations for pricing knock-out call options when the underlying asset price follows a jump-diffusion model. The final article in the second part of the volume is by Hélyette Geman, Dilip's longtime collaborator from France who was responsible for introducing him to Marc Yor, and she treats energy commodity price modeling using real historical data, testing the hypothesis of mean reversion for oil and natural gas prices.

The third part of the volume includes several contributions in one of the most rapidly growing fields in mathematical finance and financial engineering: credit risk. A new class of reduced-form credit risk models that associates default events directly with market information processes driving cash flows is introduced in the piece by Dorje Brody, Lane Hughston, and Andrea Macrina. A generic one-factor Lévy model for pricing collateralized debt obligations that unifies a number of recently proposed one-factor models is presented in the article by Hansjörg Albrecher, Sophie Ladoucette, and Wim Schoutens. An intensity-based default model that prices credit derivatives using utility functions rather than arbitrage-free measures is proposed in the article by Ronnie Sircar and Thaleia Zariphopoulou. Also using the utility-based pricing approach is the final article in the volume by Marek Musiela and Thaleia Zariphopoulou, and they address the integrated portfolio management optimal investment problem in incomplete markets stemming from stochastic factors in the underlying risky securities.

Besides being a distinguished researcher, Dilip is a dear friend, an esteemed colleague, and a caring mentor and teacher. During his professional career, Dilip was one of the early pioneers in mathematical finance, so it is only fitting that the title of this Festschrift documents his past and continuing love for the field that he helped develop.

Michael Fu
Bob Jarrow
Ju-Yi Yen
Robert Elliott
December 2006

MATHEMATICAL FINANCE
CONFERENCE in honor of
the 60th birthday of Dilip B. Madan

KEYNOTE SPEAKER:
Marc Yor (Université Paris VI, France)

INVITED SPEAKERS:

Peter Carr (Bloomberg LP and Courant Institute)
Freddy Delbaen (ETH-Zurich, Switzerland)
Bruno Dupire (Bloomberg LP)
Ernst Eberlein (University of Freiburg, Germany)
Robert Elliott (University of Calgary, Canada)
Hélyette Geman (Université Paris IX Dauphine and ESSEC, France)
Ali Hirsa (Caspian Capital)
Lane Hughston (King's College London)
Robert Jarrow (Cornell University)
Ajay Khanna
Andreas Kyprianou (University of Bath, United Kingdom)
Frank Milne (Queen's University, Canada)
Marek Musiela (BNP Paribas, UK)
Philip Protter (Cornell University)
Wim Schoutens (Katholieke Universiteit Leuven, Belgium)
Eugene Seneta (University of Sydney, Australia)
Thaleia Zariphopoulou (University of Texas at Austin)

Organizers:
John J. Benedetto (University of Maryland)
Peter Carr (Bloomberg LP and Courant Institute)
Michael C. Fu (University of Maryland)
Ioannis Konstantinidis (University of Maryland)
Ju-Yi Yen (Vanderbilt University and University of Maryland)

Hosted by the Norbert Wiener Center

Sponsored in part by the Department of Finance,
Robert H. Smith School of Business,
University of Maryland

SEPTEMBER 29 - OCTOBER 1, 2006
www.norbertwiener.umd.edu/Madan/

Conference poster (designed by Jonathan Sears).

Photo Highlights (September 29, 2006)

Dilip delivering his lecture.

Dilip with many of his Ph.D. students.

Norbert Wiener Center director John Benedetto and Robert Elliott.

Left to right: CGMY (Carr, Geman, Madan, Yor).

VG inventors (Dilip and Eugene Seneta) with the Madan family.

Dilip's wife Vimla cutting the birthday cake.

Career Highlights and List of Publications

Dilip B. Madan

Robert H. Smith School of Business
Department of Finance
University of Maryland
College Park, MD 20742, USA
dmadan@rhsmith.umd.edu

Career Highlights

1971 Ph.D. Economics, University of Maryland
1975 Ph.D. Mathematics, University of Maryland

2006 recipient of Humboldt Research Award in Mathematics
President of Bachelier Finance Society 2002–2003
Managing Editor of *Mathematic Finance, Review of Derivatives Research*
Series Editor on Financial Mathematics for CRC, Chapman and Hall
Associate Editor for *Quantitative Finance, Journal of Credit Risk*

1971–1975: Assistant Professor of Economics, University of Maryland
1976–1979: Lecturer in Economic Statistics, University of Sydney
1980–1988: Senior Lecturer in Econometrics, University of Sydney
1981–1982: Acting Head, Department of Econometrics, Sydney
1989–1992: Assistant Professor of Finance, University of Maryland
1992–1997: Associate Professor of Finance, University of Maryland
1997–present: Professor of Finance, University of Maryland

Visiting Positions:
La Trobe University, Cambridge University (Isaac Newton Institute),
Cornell University, University Paris,VI, University of Paris IX at Dauphine

Consulting:
Morgan Stanley, Bloomberg, Wachovia Securities, Caspian Capital, FDIC

Publications (as of December 2006 (60th birthday))

1. The relevance of a probabilistic form of invertibility. *Biometrika*, 67(3):704–5, 1980 (with G. Babich).
2. Monotone and 1-1 sets. *Journal of the Australian Mathematical Society, Series A*, 33:62–75, 1982 (with R.W. Robinson).
3. Resurrecting the discounted cash equivalent flow. *Abacus*, 18-1:83–90, 1982.
4. Differentiating a determinant. *The American Statistician*, 36(3):178–179, 1982.
5. Measures of risk aversion with many commodities. *Economics Letters*, 11:93–100, 1983.
6. Inconsistent theories as scientific objectives. *Journal of the Philosophy of Science*, 50(3):453–470, 1983.
7. Testing for random pairing. *Journal of the American Statistical Association*, 78(382):332–336, 1983 (with Piet de Jong and Malcolm Greig).
8. Compound Poisson models for economic variable movements. *Sankhya Series B*, 46(2):174–187, 1984 (with E. Seneta).
9. The measurement of capital utilization rates. *Communications in Statistics: Theory and Methods*, A14(6):1301–1314, 1985.
10. Project evaluations and accounting income forecasts. *Abacus*, 21(2):197–202, 1985.
11. Utility correlations in probabilistic choice modeling. *Economics Letters*, 20:241–245, 1986.
12. Mode choice for urban travelers in Sydney. *Proceedings of the 13th ARRB and 5th REAAA Conference*, 13(8):52–62, 1986 (with R. Groenhout and M. Ranjbar).
13. Simulation of estimates using the empirical characteristic function. *International Statistical Review*, 55(2):153–161, 1987 (with E. Seneta).
14. Chebyshev polynomial approximations for characteristic function estimation. *Journal of the Royal Statistical Society, Series B*, 49(2):163–169, 1987 (with E. Seneta).
15. Modeling Sydney work trip travel mode choices. *Journal of Transportation Economics and Policy*, XXI(2):135–150, 1987 (with R. Groenhout).
16. Optimal duration and speed in the long run. *Review of Economic Studies*, 54a(4a):695–700, 1987.
17. Decision theory with complex uncertainties. *Synthese*, 75:25–44, 1988 (with J.C. Owings).
18. Risk measurement in semimartingale models with multiple consumption goods. *Journal of Economic Theory*, 44(2):398–412, 1988.
19. Stochastic stability in a rational expectations model of a small open economy. *Economica*, 56(221):97–108, 1989 (with E. Kiernan).
20. Dynamic factor demands with some immediately productive quasi fixed factors. *Journal of Econometrics*, 42:275–283, 1989 (with I. Prucha).

21. Characteristic function estimation using maximum likelihood on transformed variables. *Journal of the Royal Statistical Society, Series B*, 51(2):281–285, 1989 (with E. Seneta).

22. The multinomial option pricing model and its Brownian and Poisson limits. *Review of Financial Studies*, 2(2):251–265, 1989 (with F. Milne and H. Shefrin).

23. On the monotonicity of the labour-capital ratio in Sraffa's model. *Journal of Economics*, 51(1):101–107, 1989 (with E. Seneta).

24. The Variance-Gamma (V.G.) model for share market returns. *Journal of Business*, 63(4):511–52,1990 (with E. Seneta).

25. Design and marketing of financial products. *Review of Financial Studies*, 4(2):361–384, 1991 (with B. Soubra).

26. A characterization of complete security markets on a Brownian filtration. *Mathematical Finance*, 1(3):31–43, 1991 (with R.A. Jarrow).

27. Option pricing with VG Martingale components. *Mathematical Finance*, 1(4):39–56, 1991 (with F. Milne).

28. Informational content in interest rate term structures. *Review of Economics and Statistics*, 75(4):695–699, 1993 (with R.O. Edmister).

29. Diffusion coefficient estimation and asset pricing when risk premia and sensitivities are time varying. *Mathematical Finance*, 3(2):85–99, 1993 (with M. Chesney, R.J. Elliott, and H. Yang).

30. Contingent claims valued and hedged by pricing and investing in a basis. *Mathematical Finance*, 4(3):223–245, 1994 (with F. Milne).

31. Option pricing using the term structure of interest rates to hedge systematic discontinuities in asset returns. *Mathematical Finance*, 5(4):311–336, 1995 (with R.A. Jarrow).

32. Approaches to the solution of stochastic intertemporal consumption models. *Australian Economic Papers*, 34:86–103, 1995 (with R.J. Cooper and K. McLaren).

33. Pricing via multiplicative price decomposition. *Journal of Financial Engineering*, 4:247–262, 1995 (with R.J. Elliott, W. Hunter, and P. Ekkehard Kopp).

34. Filtering derivative security valuations from market prices. *Mathematics of Derivative Securities*, eds. M.A.H. Dempster and S.R. Pliska, Cambridge University Press, 1997 (with R.J. Elliott and C. Lahaie)

35. Is mean-variance theory vacuous: Or was beta stillborn. *European Finance Review*, 1:15–30, 1997 (with R.A. Jarrow).

36. Default risk. *Statistics in Finance*, eds. D. Hand and S.D. Jacka, Arnold Applications in Statistics, 239–260, 1998.

37. Pricing the risks of default. *Review of Derivatives Research*, 2:121–160, 1998 (with H. Unal).

38. The discrete time equivalent martingale measure. *Mathematical Finance*, 8(2):127–152, 1998 (with R.J. Elliott).

39. The variance gamma process and option pricing. *European Finance Review*, 2:79–105, 1998 (with P. Carr and E. Chang).

40. Towards a theory of volatility trading. *Volatility*, ed. R.A. Jarrow, Risk Books, 417–427, 1998 (with P. Carr).
41. Valuing and hedging contingent claims on semimartingales. *Finance and Stochastics*, 3:111–134, 1999 (with R.A. Jarrow).
42. The second fundamental theorem of asset pricing theory. *Mathematical Finance*, 9(3):255–273, 1999 (with R.A. Jarrow and X. Jin).
43. Pricing continuous time Asian options: A comparison of Monte Carlo and Laplace transform inversion methods. *Journal of Computational Finance*, 2:49–74, 1999 (with M.C. Fu and T. Wang).
44. Introducing the covariance swap. *Risk*, 47–51, February 1999 (with P. Carr).
45. Option valuation using the fast Fourier transform. *Journal of Computational Finance*, 2:61–73, 1999.
46. Spanning and derivative security valuation. *Journal of Financial Economics*, 55:205–238, 2000 (with G. Bakshi).
47. A two factor hazard rate model for pricing risky debt and the term structure of credit spreads. *Journal of Financial and Quantitative Analysis*, 35:43–65, 2000 (with H. Unal).
48. Arbitrage, martingales and private monetary value. *Journal of Risk*, 3(1):73–90, 2000 (with R.A. Jarrow).
49. Investing in skews. *Journal of Risk Finance*, 2(1):10–18, 2000 (with G. McPhail).
50. Going with the flow. *Risk*, 85–89, August 2000 (with P. Carr and A. Lipton).
51. Optimal investment in derivative securities. *Finance and Stochastics*, 5(1):33–59, 2001 (with P. Carr and X. Jin).
52. Time changes for Lévy processes. *Mathematical Finance*, 11(1):79–96, 2001 (with H. Geman and M. Yor).
53. Optimal positioning in derivatives. *Quantitative Finance*, 1(1):19–37, 2001 (with P. Carr).
54. Pricing and hedging in incomplete markets. *Journal of Financial Economics*, 62:131–167, 2001 (with P. Carr and H. Geman).
55. Pricing the risks of default. *Mastering Risk Volume 2: Applications*, ed. C. Alexander, Financial Times Press, Chapter 9, 2001.
56. Purely discontinuous asset price processes. *Handbooks in Mathematical Finance: Option Pricing, Interest Rates and Risk Management*, eds. J. Cvitanic, E. Jouini, and M. Musiela, Cambridge University Press, 105–153, 2001.
57. Asset prices are Brownian motion: Only in business time. *Quantitative Analysis of Financial Markets*, vol. 2, ed. M. Avellanada, World Scientific Press, 103–146, 2001 (with H. Geman and M. Yor).
58. Determining volatility surfaces and option values from an implied volatility smile. *Quantitative Analysis of Financial Markets*, vol. 2. ed. M. Avellanada, World Scientific Press, 163–191, 2001 (with P. Carr).

59. Towards a theory of volatility trading. *Handbooks in Mathematical Finance: Option Pricing, Interest Rates and Risk Management*, eds. J. Cvitanic, E. Jouini and M. Musiela, Cambridge University Press, 458–476, 2001 (with P. Carr).

60. Pricing American options: A comparison of Monte Carlo simulation approaches. *Journal of Computational Finance*, 2:61–73, 2001 (with M.C. Fu, S.B. Laprise, Y. Su, and R. Wu).

61. Stochastic volatility, jumps and hidden time changes. *Finance and Stochastics*, 6(1):63–90, 2002 (with H. Geman and M. Yor).

62. The fine structure of asset returns: An empirical investigation. *Journal of Business*, 75:305–332, 2002 (with P. Carr, H. Geman, and M. Yor).

63. Pricing average rate contingent claims. *Journal of Financial and Quantitative Analysis*, 37(1):93–115, 2002 (with G. Bakshi).

64. Option pricing using variance gamma Markov chains. *Review of Derivatives Research*, 5:81–115, 2002 (with M. Konikov).

65. Pricing the risk of recovery in default with APR violation. *Journal of Banking and Finance*, 27(6):1001–1025, June 2003 (with H. Unal and L. Guntay).

66. Incomplete diversification and asset pricing. *Advances in Finance and Stochastics: Essays in Honor of Dieter Sondermann*, eds. K. Sandmann and P. Schonbucher, Springer-Verlag, 101–124, 2002 (with F. Milne and R. Elliott).

67. Making Markov martingales meet marginals: With explicit constructions. *Bernoulli*, 8:509–536, 2002 (with M. Yor).

68. Stock return characteristics, skew laws, and the differential pricing of individual stock options. *Review of Financial Studies*, 16:101–143, 2003 (with G. Bakshi and N. Kapadia).

69. The effect of model risk on the valuation of barrier options. *Journal of Risk Finance*, 4:47–55, 2003 (with G. Courtadon and A. Hirsa).

70. Stochastic volatility for Lévy processes. *Mathematical Finance*, 13(3):345–382, 2003 (with P. Carr, H. Geman, and M. Yor).

71. Pricing American options under variance gamma. *Journal of Computational Finance*, 7(2):63–80, 2003 (with A. Hirsa).

72. Monitored financial equilibria. *Journal of Banking and Finance*, 28:2213–2235, 2004.

73. Understanding option prices. *Quantitative Finance*, 4:55–63, 2004 (with A. Khanna).

74. Risks in returns: A pure jump perspective. *Exotic Options and Advanced Levy Models*, eds. A. Kyprianou, W. Schoutens, and P. Willmott, Wiley, 51–66, 2005 (with H. Geman).

75. From local volatility to local Lévy models. *Quantitative Finance*, 4:581–588, 2005 (with P. Carr, H. Geman and M. Yor).

76. Empirical examination of the variance gamma model for foreign exchange currency options. *Journal of Business*, 75:2121–2152, 2005 (with E. Daal).

77. Pricing options on realized variance. *Finance and Stochastics*, 9:453–475, 2005 (with P. Carr, H. Geman, and M. Yor).
78. A note on sufficient conditions for no arbitrage. *Finance Research Letters*, 2:125–130, 2005 (with P. Carr).
79. Investigating the role of systematic and firm-specific factors in default risk: Lessons from empirically evaluating credit risk models. *Journal of Business*, 79(4):1955–1988, July 2006 (with G. Bakshi and F. Zhang).
80. Credit default and basket default swaps. *Journal of Credit Risk*, 2:67–87, 2006 (with M. Konikov).
81. Itô's integrated formula for strict local martingales. *In Memoriam Paul-André Meyer – Séminaire de Probabilités XXXIX*, eds. M. Émery and M. Yor, Lecture Notes in Mathematics 1874, Springer, 2006 (with M. Yor).
82. Equilibrium asset pricing with non-Gaussian returns and exponential utility. *Quantitative Finance*, 6(6):455–463, 2006.
83. A theory of volatility spreads. *Management Science*, 52(12):1945–56, 2006 (with G. Bakshi).
84. Asset allocation for CARA utility with multivariate Lévy returns. forthcoming in *Handbook of Financial Engineering* (with J.-Y. Yen).
85. Self-decomposability and option pricing. *Mathematical Finance*, 17(1):31–57, 2007 (with P. Carr, H. Geman, and M. Yor).
86. Probing options markets for information. *Methodology and Computing in Applied Probability*, 9:115–131, 2007 (with H. Geman and M. Yor).
87. Correlation and the pricing of risks. forthcoming in *Annals of Finance* (with M. Atlan, H. Geman, and M. Yor).

Completed Papers

88. Asset pricing in an incomplete market with a locally risky discount factor, 1995 (with S. Acharya).
89. Estimation of statistical and risk-neutral densities by hermite polynomial approximation: With an application to Eurodollar Futures options, 1996 (with P. Abken and S. Ramamurtie).
90. Crash discovery in options markets, 1999 (with G. Bakshi).
91. Risk aversion, physical skew and kurtosis, and the dichotomy between risk-neutral and physical index volatility, 2001 (with G. Bakshi and I. Kirgiz).
92. Factor models for option pricing, 2001 (with P. Carr).
93. On the nature of options, 2001 (with P. Carr).
94. Recovery in default risk modeling: Theoretical foundations and empirical applications, 2001 (with G. Bakshi and F. Zhang).
95. Reduction method for valuing derivative securities, 2001 (with P. Carr and A. Lipton).
96. Option pricing and heat transfer, 2002 (with P. Carr and A. Lipton).
97. Multiple prior asset pricing models, 2003 (with R.J. Elliott).

98. Absence of arbitrage and local Lévy models, 2003 (with P. Carr, H. Geman and M. Yor).
99. Bell shaped returns, 2003 (with A. Khanna, H. Geman and M. Yor).
100. Pricing the risks of deposit insurance, 2004 (with H. Unal).
101. Representing the CGMY and Meixner processes as time changed Brownian motions, 2006 (with M. Yor).
102. Pricing equity default swaps under the CGMY Lévy model, 2005 (with S. Asmussen and M. Pistorious).
103. Coherent measurement of factor risks, 2005 (with A. Cherny).
104. Pricing and hedging in incomplete markets with coherent risk, 2005 (with A. Cherny).
105. CAPM, rewards and empirical asset pricing with coherent risk, 2005 (with A. Cherny).
106. The distribution of risk aversion, 2006 (with G. Bakshi).
107. Designing countercyclical and risk based aggregate deposit insurance premia, 2006 (with H. Unal).
108. Sato processes and the valuation of structured products, 2006 (with E. Eberlein).
109. Measuring the degree of market efficiency, 2006 (with A. Cherny).

Contents

Part I

Variance-Gamma and Related
Stochastic Processes

The Early Years of the Variance-Gamma Process

Eugene Seneta

School of Mathematics and Statistics FO7
University of Sydney
Sydney, New South Wales 2006, Australia
eseneta@maths.usyd.edu.au

Summary. Dilip Madan and I worked on stochastic process models with stationary independent increments for the movement of log-prices at the University of Sydney in the period 1980–1990, and completed the 1990 paper [21] while respectively at the University of Maryland and the University of Virginia. The (symmetric) Variance-Gamma (VG) distribution for log-price increments and the VG stochastic process first appear in an Econometrics Discussion Paper in 1985 and two journal papers of 1987. The theme of the pre-1990 papers is estimation of parameters of log-price increment distributions that have real simple closed-form characteristic function, using this characteristic function directly on simulated data and Sydney Stock Exchange data. The present paper reviews the evolution of this theme, leading to the definitive theoretical study of the symmetric VG process in the 1990 paper.

Key words: Log-price increments; independent stationary increments; Brownian motion; characteristic function estimation; normal law; symmetric stable law; compound Poisson; Variance-Gamma; Praetz t.

1 Qualitative History of the Collaboration

Our collaboration began shortly after my arrival at the University of Sydney in June 1979, after a long spell (from 1965) at the Australian National University, Canberra. After arrival I was in the Department of Mathematical Statistics, and Dilip a young lecturer in the Department of Economic Statistics (later renamed the Department of Econometrics). These two sister departments, neither of which now exists as a distinct entity, were in different Faculties (Colleges) and in buildings separated by City Road, which divides the main campus.

Having heard that I was an applied probabilist with focus on stochastic processes, and wanting someone to talk with on such topics, he simply walked into my office one day, and our collaboration began. We used to meet about once a week in my office in the Carslaw Building at around midday, and our meetings were accompanied by lunch and a long walk around the campus. This routine at Sydney continued until he left the University in 1988 for the University of Maryland at College Park, his alma mater. I remember telling him on several of these walks that Sydney was too small a pond for his talents.

I was to be on study leave and teaching stochastic processes and time series analysis at the Mathematics Department of the University of Virginia, Charlottesville, during the 1988–1989 academic year. This came about through the efforts of Steve Evans, whom Dilip and I had taught in a joint course, partly on the Poisson process and its variants, at the University of Sydney. The collaboration on what became the foundation paper on the Variance-Gamma (VG) process (sometimes now called the Madan–Seneta process) and distribution [21] was thus able to continue, through several of my visits to his new home. The VG process had appeared in minor roles in two earlier joint papers [16] and [18].

There was a last-to-be published joint paper from the 1988 and 1989 years, [22], which, however, did not have the same econometric theme as all the others, being related to my interest in the theory of nonnegative matrices.

After 1989 there was sporadic contact, mainly by e-mail, between Dilip and myself, some of which is mentioned in the sequel, until I became aware of work by another long-term friend and colleague, Chris Heyde, who divides his time between the Australian National University and Columbia University. His seminal paper [8] was about to appear when he presented his work at a seminar at Sydney University in March 1999. Among his various themes, he advocates the t distribution for returns (increments in log-price) for financial asset movements. This idea, as shown in the sequel, had been one of the motivations for the work on the VG by Dilip and myself, on account of a 1972 paper of Praetz [25].

I was to spend the fall semester of the 1999–2000 academic year again at the University of Virginia, where Wake (T.W.) Epps asked me to be present at the thesis defence of one of his students, and surprised me by saying to one of the other committee members that as one of the creators of the VG process, I "was there to defend my turf." Wake had learned about the VG process during my 1988–1989 sojourn, and a footnote in [21] acknowledges his help, and that of Steve Evans amongst others. Wake's book [2] was the first to give prominence to VG structure. More recently, the books of Schoutens [27] and Applebaum [1] give it exposure.

At about this time I had also had several e-mail inquiries from outside Australia about the fitting of the VG model from financial data, and there was demand for supervision of mathematical statistics students at Sydney University on financial topics. I was supervising one student by early 2002, so I asked Dilip for some offprints of his VG work since our collaboration, and then turned to him by e-mail about the problem of statistical estimation of parameters of the VG. I still have his response of May 20, 2002, generously telling me something of what he had been doing on this.

My personal VG story restarts with the paper [28], produced for a Festschrift to celebrate Chris Heyde's 65th birthday. In this I tried to synthesize the various ideas of Dilip and his colleagues on the VG process with those of Chris and his students on the t process. The themes are: subordinated Brownian motion, skewness of the distribution of returns, returns over unit time forming a strictly stationary time series, statistical estimation, long-range dependence and self-similarity, and duality between the VG and t processes. Some of the joint work with M.Sc. dissertation graduate students [30; 5; 6] stemming from that paper has been, or is about to be, published.

I now pass on to the history of the technical development of my collaboration with Dilip.

2 The First Discussion Paper

The classical model (Bachelier) in continuous time $t \geq 0$ for movement of prices $\{P(t)\}$, modified to allow for drift, is

$$Z(t) = \log\{P(t)/P(0)\} = \mu t + \theta^{1/2} b(t), \tag{1}$$

where $\{b(t)\}$ is standard Brownian motion (the Wiener process); μ is a real number, the drift parameter; and θ is a positive scale constant, the diffusion parameter. Thus the process $\{Z(t)\}$ has stationary independent increments in continuous time, which over unit time are:

$$X(t) = \log\{P(t)/P(t-1)\} \sim \mathcal{N}(\mu, \theta), \tag{2}$$

where $\mathcal{N}(\mu, \theta)$ represents the normal (Gaussian) distribution with mean μ and variance θ. When we began our collaboration, it had been observed for some time that although the assumed common distribution of the $\{X(t)\}$ given by the left-hand side of (2) for historical data indeed seemed symmetric about some mean μ, the tails of the distribution were heavier than the normal. The assumption of independently and identically distributed (i.i.d.) increments was pervasive, making the process $\{Z(t), t = 1, 2, \ldots\}$ a random walk.

Having made these points, our first working paper [12], which has never been published, retained all the assumptions of the classical Bachelier model, with drift parameter given by $\mu = r - \theta/2$. This arose out of a model in continuous time, in which the instantaneous rate of return on a stock is assumed normally distributed, with mean r and variance θ. Consequently, $\{e^{-rt}P(t)\}$ is seen to be a martingale, consistent with option pricing measure for European options under the Merton–Black–Scholes assumptions, where r is the interest rate. The abstract to [12] reads:

> Formulae are developed for computing the expected profitability of market strategies that involve the purchase or sale of a stock on the same day, with the transaction to be reversed when the price reaches either of two prespecified limits or a fixed time has elapsed.

This paper already displayed Dilip's computational skills (not to mention his analytical skills) by including many tables with the headings: Probabilities of hitting the upper [resp., lower] barrier. The profit rate r is characteristically taken as 0.002.

Dilip had had a number of items presented in the Economic Statistics Papers series before number 45 appeared in February 1981. These were of both individual and joint authorship. Several, from their titles, seem of a philosophical nature, for example:

- No. 34. D. Madan. Economics: Its Questions and Answers
- No. 37. D. Madan. A New View of Science or at Least Social Science
- No. 38. D. Madan. An Alternative to Econometrics in Economic Data Analyses

There were also technical papers foreshadowing things to come on his return to the University of Maryland, for example:

- No. 23. P. de Jong and D. Madan. The Fast Fourier Transform in Applied Spectral Inference

Dilip was clearly interested in, and very capable in, an extraordinarily wide range of topics. Our collaboration seems to have marked a narrowing of focus, and continued production on more specific themes. The incoming Head of the Department of Econometrics, Professor Alan Woodland, also encouraged him in this.

3 The Normal Compound Poisson (NCP) Process

Our first published paper [14], based on the the Economics Discussion Paper [13] dated February 1982, focuses on modelling the *second* differences in log-price: $\log\{P(t)/P(0)\}$ by the first difference of the continuous-time stochastic process $\{Z(t), t \geq 0\}$, where

$$Z(t) = \mu t + \sum_{i=1}^{N(t)} \xi_i + \theta^{1/2} b(t). \tag{3}$$

Here $\{N(t)\}$ is the ordinary Poisson process with arrival rate λ, and $\{b(t)\}$ is standard Brownian motion (the Wiener process), μ is a real constant, and $\theta > 0$ is a scale parameter. The ξ_i, $i = 1, 2, \ldots$, form a sequence of i.i.d. $\mathcal{N}(0, \sigma^2)$ rv's, probabilistically independent of the process $\{b(t)\}$. The process $\{Z(t), t \geq 0\}$, called in [14] the Normal Compound Poisson (NCP) process, is therefore a process with independent stationary increments, whose distribution (the NCP distribution) over unit time interval, is given by:

$$X|V \sim \mathcal{N}(\mu, \theta + \sigma^2 V), \text{where } V \sim \text{Poisson}(\lambda). \tag{4}$$

Hence the cumulative distribution function (c.d.f.) and the characteristic function (c.f.) of X are given by

$$F(x) = \sum_{n=0}^{\infty} \frac{\lambda^n e^{-\lambda}}{n!} \Phi\left(\frac{x - \mu}{\theta + \sigma^2 n}\right), \tag{5}$$

$$\phi_X(u) = \exp\left(i\mu u - u^2\theta/2 + \lambda(e^{-u^2\sigma^2/2} - 1)\right). \tag{6}$$

Here $\Phi(\cdot)$ is the c.d.f. of a standard normal distribution.

The distribution of X is thus a normal with mixing on the variance, is symmetric about μ, and has the same form irrespective of the size of time increment t. It is long-tailed relative to the normal in the sense that its kurtosis value

$$3 + \frac{3\lambda\sigma^4}{(\theta + \sigma^2\lambda)^2}$$

exceeds that of the normal (whose kurtosis value is 3). When the NCP distribution is symmetrized about the origin by putting $\mu = 0$, it has a simple real characteristic function of closed form.

The NCP process from the structure (3) clearly has jump components (the ξ_is are regarded as "shocks" arriving at Poisson rate), and through the Brownian process add-on $\theta^{1/2}b(t)$ in (3), has obviously a Gaussian component. The NCP distribution and NCP process, and the above formulae, are due to Press [26]; in fact, our NCP structure is a simplification of his model (where $\xi_i \sim \mathcal{N}(\nu, \sigma^2)$) to symmetry by taking $\nu = 0$. Note that the nonsimplified model of Press is normal with mixing on both the mean and variance:

$$X|V \sim \mathcal{N}(\mu + \nu V, \theta + \sigma^2 V),$$

where as before $V \sim \text{Poisson}(\lambda)$. Normal variance–mean mixture models have been studied more recently; see [30] for some cases and earlier references.

The c.f.s of the symmetric stable laws,

$$e^{iu\mu - \gamma|u|^\beta}, \ \gamma > 0, \ 0 < \beta < 2, \tag{7}$$

where γ is a scale parameter, were also fitted in this way, as was the c.f. of the normal (the case $\beta = 2$). The distributions fitted were thus symmetric about a central point μ, and the c.f.s, apart from allowing for the shift to μ, were consequently real-valued, and of simple closed form. Like the NCP, the process of independent stable increments has both continuous and jump components.

The continuous-time strictly stationary process of i.i.d. increments with stable law also has the advantage of having laws of the same form for an increment over a time interval of any length, and heavy tails, but such laws have infinite variance.

The data for the statistical analysis consisted of five series of share prices and six series on economic variables. The share prices taken were daily last prices on the Sydney Stock Exchange.

Estimation in [14] was undertaken by Press's minimum distance method, which amounts to minimizing the L_1-norm of the difference of the empirical and actual log c.f. evaluated at a set of arguments $u_1, u_2, \ldots, u_n$. Each fitted c.f. was then inverted using the fast Fourier transform, and comparison made between the fitted and empirical c.f.s, using Kolmogorov's test. A variant of this estimation procedure was to be investigated in the next two papers [16], [18]. This is anticipated in [14] by mentioning the paper [4].

It was remarked that the estimate of θ for the NCP was generally small, thus arguing for a purely compound Poisson process, and against the stable laws.

The VG distribution and process were yet to make their appearance. The VG was to share equal standing with the NCP in the next two papers [16] and [18]. An important difference between the VG and NCP models, however, is that the VG turns out to be a pure jump process, and the limit of compound Poisson processes.

The paper [14] displays to a remarkable degree Dilip's knowledge and proficiency in statistical computing methodology, and his to-be-ongoing focus on the c.f.

4 The Praetz t and VG Distributions

The symmetric VG distribution (and corresponding VG process) first occur in our writings in [15] as the fourth of five parametric classes of distribution with real c.f. of simple closed form. The first three of these, including the origin-centered NCP, had been considered in [14]. The fifth parametric class is related to the NCP but was constructed to generate continuous sample paths. A revised version of [15] was eventually published, two years later, in the *International Statistical Review* [ISR] [16] after trials and tribulations with another journal. It was received by *ISR* in May 1986 and revised November 1986.

Because a random variable X with real c.f. is symmetrically distributed about zero, c.f. $E[e^{iuX}] = E[\cos uX]$, so for a set of i.i.d. observations $X_1, X_2, \ldots, X_n$, with characteristic function $\phi(u; \alpha)$, where α denotes an m-dimensional vector of parameters, the empirical c.f. is

$$\hat{\phi}(u) = \frac{\sum_{i=1}^{n}\cos(uX_i)}{n}.$$

Selecting a set of p values $u_1, u_2, \ldots, u_p$ of u for $p \geq m$, construct the p-dimensional vector $\mathbf{z}(\alpha) = \{z_j(\alpha)\}$ where $z_j(\alpha) = \sqrt{n}(\hat{\phi}(u_j) - \phi(u_j; \alpha))$. The vector $\mathbf{z}(\alpha)$ is asymptotically normally distributed with covariance matrix $\Sigma(\alpha) = \{\sigma_{jk}(\alpha)\}$, where the individual covariances may be expressed explicitly in terms of $\phi(\cdot; \alpha)$ evaluated at various combinations of the u_j, u_k. In the simulation study, specific true values α^* in the five distributions studied were used to construct $\Sigma(\alpha^*)$, and then estimates of α were obtained by minimizing $\mathbf{z}^T(\alpha)\Sigma^+\mathbf{z}(\alpha)$, where Σ^+ is a certain generalized inverse of $\Sigma(\alpha^*)$.

There is obviously some motivation from the estimation procedure of Press [26], inasmuch as what is being minimized is a generalized distance, but the main motivation was the work of Feuerverger and McDunnough [4]. The estimation was successful for just the first two classes, the normal ($m = 1$), the symmetric stable ($m = 2$), and the fourth class, the VG (with $m = 2$). The remaining classes, including the NCP, each had three parameters.

On the whole, the paper motivated further consideration of more effective c.f. estimation methods.

But the main feature of this paper, as regards history, was the appearance of the c.f. of the symmetric VG and the associated stochastic process. This introduction of the VG material is expressed, verbatim, as follows in both [15] and [16]:

> The fourth parametric class is motivated by the derivation of the t distribution proposed by Praetz (1972). Praetz took the variance of the normal to be uncertain with reciprocal of the variance distributed as a gamma variable. The characteristic function of this distribution is not known in closed form, nor is it known what continuous time stochastic process gives rise to such a period-one distribution. We will, in contrast, take the variance itself to be distributed as a gamma variable. Letting $Y(t)$ be the continuous time process of independent gamma increments with mean $m\tau$ and variance $v\tau$ over nonoverlapping increments of length τ and $X(t) = b(Y(t))$, where b is again standard Brownian motion, yields a continuous time stochastic process $X(t)$ with $X(1)$ being a normal variable with a gamma variance. Furthermore, the characteristic function of $X(1)$ is easily evaluated by conditioning on $Y(1)$ as
>
> $$\phi_4(u; m, v) = [(m/v)/(m/v + u^2/2)]^{m^2/v}.$$

Thus the distribution of an increment over unit time is specified by mixing a normal variable on the variance:

$$X|V \sim \mathcal{N}(0, V), \tag{8}$$

where $V \sim \Gamma(\gamma, c)$. By this notation we mean that the probability density function (p.d.f.) of V is

$$g_V(w) = c^\gamma w^{\gamma-1} e^{-cw}/\Gamma(\gamma), \ w > 0; \ \ 0 \text{ otherwise.} \tag{9}$$

Praetz's paper [25], published in 1972 and in the *Journal of Business* as was that of Press [26], had come to our attention sometime after February 1982, the date of [13]. He focuses on the issue that although evidence of independence of returns (log-price increments of shares) had been widely accepted, the actual distribution of returns seemed to be highly nonnormal.

He describes this distribution as typically symmetric, with fat tails, a high peaked center, and hollow in between, and proposes a general mixing distribution on the variance of the normal.

Then he settles on the inverse gamma distribution for the variance V, described by the p.d.f.:

$$g_I(w) = \sigma_0^{2m}(m-1)^m w^{-(m+1)} e^{(m-1)\sigma_0^2/w}/\Gamma(m), \ w > 0; \ 0 \text{ otherwise, } (10)$$

so that

$$E[V] = \sigma_0^2, \qquad \text{Var } V = \sigma_0^4/(m-2).$$

The reason for the choice is doubtless because this p.d.f. for V produces for the p.d.f. of X:

$$f_T(x; \nu, \delta) = \frac{\Gamma(\frac{\nu+1}{2})}{\delta\sqrt{\pi}\Gamma(\frac{\nu}{2})} \cdot \frac{1}{(1 + (\frac{x}{\delta})^2)^{(\nu+1)/2}}, \tag{11}$$

where $-\infty < x < \infty, \nu = 2m, \delta = \sigma_0(\nu-2)^{1/2}$. This expression (11) corresponds to (6) in [25], which is misprinted there. Equation (11) is the p.d.f. of a t distribution with scaling parameter δ. The classical form Student-t distribution with n degrees of freedom has $\delta = \sqrt{n}$, with n a positive integer.

An appealing feature of the t p.d.f. is that it has at least fatter tails than the normal, actually of Pareto (power-law) type, and that it has a well-known closed form. Whether it is sufficiently peaked at the origin and hollow in the intermediate range, however, is still a matter of debate [6].

The simple form of the p.d.f. (11) makes estimation from i.i.d. readings straightforward. Praetz notes that there had been difficulties in estimating the parameters of the symmetric stable laws (for which the p.d.f. was not known, although the form of c.f. is simple, (7)), and the parameters of the compound events distribution of Press [26].

In his Section 4, Praetz fits the scaled t, the normal, the compound events distribution, and the symmetric stable laws (he calls these stable Paretian) to 17 share-price index series from weekly observations from the Sydney Stock Exchange for the nine years 1958–1966: a total of 462 observations. He says of an earlier paper of his [24] that none of the series gave normally distributed increments.

Using a χ^2 statistic for goodness of fit, he concludes that the scaled t distribution gives superior fit in all cases. The compound events model causes larger χ^2 values through inability as in the past to provide suitable estimates of the parameters.

He also makes the interesting comparison that the symmetric stable laws (c.f.s given by (7) with $1 < \beta < 2$) used by Mandelbrot to represent share price changes are intermediate between a Cauchy and a normal distribution (the respective cases with $\beta = 1, 2$), and that the scaled t distribution also lies between these two extremes (resp., $\nu = 1, \nu \to \infty$).

These various issues served as stimulus for the creation of the VG distribution and empirical characteristic function estimation methods in [15] and [16]. Praetz [25] had shifted the argument into p.d.f. domain, whereas in [14] we had stayed with simple closed-form c.f.s, and persisted with this.

My recollection is that Dilip began by attempting to find the c.f. for mixing on the variance V as in (8), and thus integrating out the conditional expectation expression:

$$E\left[E[e^{iXu}|V]\right], \tag{12}$$

taking V to have the p.d.f. of an inverse gamma distribution, and trying to find the c.f. corresponding to t. I think that when I looked at the calculation, he had in fact calculated the c.f. of X, where

$$X|V \sim \mathcal{N}(0, 1/V),$$

so that in effect

$$X|V \sim \mathcal{N}(0, V),$$

where V has an ordinary gamma distribution. There was in any case some interaction at this point, and the result was the beautifully simple c.f. of the VG distribution.

At the time, the corresponding c.f. of the t distribution was not known.

It may be interesting to Dilip and other readers if I break in the technical history of our collaboration to reflect on background concerning Praetz and myself.

5 The Praetz Confluence

The VG distribution is a direct competitor to the Praetz t, and is in fact dual to it [28, Section 6]; [7]. Our paper [21] was deliberately published in the same journal.

Praetz was an Australian econometrician, and focused on Sydney Stock Exchange data, as did all our collaborative papers on the VG before [21].

Peter David Praetz was born February 7, 1940. His university training was B.A. (Hons) 1961, M.A. 1963, both at University of Melbourne. His Ph.D from the University of Adelaide, South Australia, in 1971, was titled *A Statistical Study of Australian Share Prices*. From 1966 to 1970 he was Lecturer then Senior Lecturer in the Faculty of Commerce, University of Adelaide. For 1971–1975 he was Senior Lecturer, Faculty of Economics and Politics at Monash University in Melbourne, and in 1975–1989 Associate Professor (this was equivalent to full Professor in the U.S. system) jointly in the Department of Econometrics and Operations Research and in the Department of Accounting and Finance, Monash University. He took early retirement in 1989 on account of his health, and died October 6, 1997.

As a young academic at the Australian National University, Canberra, from the beginning of 1965, and interested primarily in stochastic processes, I had noticed in the *Australian Journal of Statistics* the well-documented paper of Praetz [24] investigating thoroughly the adequacy of various properties of the simple Brownian model for returns (Bachelier), in particular independence of nonoverlapping increments and the tail weight of the common distribution. Praetz's address is given as the University of Adelaide. The paper was received August 20, 1968 and revised December 11, 1968. There are no earlier papers of Praetz cited in it, although he did have several papers published earlier in *Transactions of the Institute of Actuaries of Australia,* so it is likely associated with work for his Ph.D. dissertation in a somewhat new direction.

In the paper [25] there is a footnote which reads:

> Department of Economics, University of Adelaide, Adelaide, South Australia. I am grateful to Professor J. N. Darroch, who suggested the possibility of this approach to me.

John N. Darroch is a mathematical statistician who was a Senior Lecturer at the University of Adelaide's then-Department of Mathematics, from about August 1962 to August 1964. In 1963 I took his courses in Mathematical Statistics (with Hogg and Craig's first edition as back-up text), and in Markov chains (with Kemeny and Snell's 1960 edition of *Finite Markov Chains* as back-up text). These courses largely determined my future research and teaching directions. In 1964 until his departure he supervised my work on absorbing Markov chains and nonnegative matrices for my M.Sc. dissertation. After his departure from Adelaide, he spent two years at the University of Michigan, Ann Arbor, and then returned to Adelaide as Professor and Head of Statistics at the newly created Flinders University of South Australia. One of his prevailing interests was in contingency tables, and he was on friendly terms with Professor H. O. Lancaster, another leader in that area who was at the University of Sydney, and at the time editor of the *Australian Journal of Statistics*, which Lancaster had founded. It is possible Darroch refereed [24]. In any case, the contact mentioned by Praetz in 1972 took place on Darroch's return to Adelaide. Regrettably in retrospect, I never had any contact with Praetz.

The step in Praetz's paper for computing the p.d.f. of X, where

$$X|V \sim \mathcal{N}(\mu, V),$$

where V has inverse gamma distribution described by (10), with its Bayesian and decision-theoretic interpretation, leading to a t distribution, which is of fundamental importance in elementary statistical inference theory and practice, has strong resonance with my memory of John Darroch's teaching, and Hogg and Craig's book.

John Darroch is now happily retired in Adelaide, still in very good health, and pursuing interests largely other than in mathematical statistics, not the least of which are the works of Shakespeare, and just a little in option pricing. We have kept in touch over the years, and last met in Adelaide in January of this year (2006). I wrote to him a few weeks later relating to the Praetz footnote, but his memory of the contact is very vague. But, albeit in an indirect way through Praetz, he nevertheless influenced the genesis of the VG distribution.

6 Chebyshev Polynomial Approximations and Characteristic Function Estimation

6.1 The Theme

The work leading to [18] (in its first incarnation [17]) overlapped with that leading to [16]. The paper [18] was received August 1986 and revised December 1986. It had very similar motivation, and very similar introduction, and was partly a result of our dissatisfaction with the effectiveness of empirical characteristic estimation procedures in [16], and the delays in getting that material published. The paper [18] cites only [15] as motivation, and focuses again on the simple c.f. structure of the symmetric stable laws, the VG and NCP, using the same parametrization for the VG as in [15] and [16]. Much is made of the point that the VG and NCP distributions for i.i.d. returns are consistent with an underlying continuous-time stochastic process for log-prices, in contrast to the Praetz t distribution.

The essential idea is that if one has a random variable with real c.f. $\phi_X(u)$ of simple closed form, it is possible to express the p.d.f. of a transformed variable T in terms of $\phi_X(u)$. The transforming function $\psi(\cdot)$ is taken as a simple bounded periodic function of period 2π, which maps the interval $[-\pi,\ \pi)$ onto the interval $[-b,b)$, for fixed b. The transformation of the whole sample space of X to that of T is in general not one-to-one, thus involving some loss of information. A sample $X_1, X_2, \ldots, X_n$ of i.i.d. random variables is transformed via

$$T_i = \psi(\omega X_i), \qquad i = 1, 2, \ldots, n,$$

where $\omega > 0$ is a parameter chosen at will to control the loss of information in going to the transformed sample. The random variable $T = \psi(\omega X)$ is bounded on the interval $(-b/\omega, b/\omega]$, and because the original random variable is symmetrically distributed about 0 (it has real c.f.), if the function ψ is an even (or odd) function, the random variable T will be symmetrically distributed on $(-b/\omega, b/\omega]$.

Because the random variable T is bounded and its p.d.f. depends on the same parameters as the distribution of X, if it is explicitly available, these parameters may be estimated by maximum likelihood procedures from the transformed observations $T_1, T_2, \ldots, T_n$.

In Madan and Seneta [18] the choice $\psi(v) = \cos v$, $-\infty < v < \infty$ is made, so $T_i = \cos \omega X_i$, $i = 1, \ldots, n$. There is an intimate relation between trigonometric functions and Chebyshev polynomials, and in fact the p.d.f. of T is given by

$$g(t) = \sum_{n=0}^{\infty} 2^n \phi_X(n\omega) q_n(t)(\pi(1-t^2)^{1/2})^{-1}, \tag{13}$$

where $q_0(y) = 1$ and $q_n(y)$, $n \geq 1$, is the nth Chebyshev polynomial. Here $(\pi(1-y^2)^{1/2})^{-1}$ is the weight function with respect to which the Chebyshev polynomials form an orthogonal family. There is a simple recurrence relation between Chebyshev polynomials by which the successive polynomials may be calculated. The value $\omega = 1$ was used in [18]. After some simulation investigation, parameters were estimated for daily returns from 19 large company stocks on the Sydney Stock Exchange, after standardizing the returns to unit sample mean and unit sample variance, using the symmetric stable law (two parameters, γ, β); the VG (two parameters, m, v); the NCP (three parameters $\theta, \lambda, \sigma^2$); and the normal (one parameter, σ^2). The χ^2 goodness-of-fit test indicated that the VG and NCP models for returns were superior to the symmetric stable law model. The following passage from [17; 18, p.167] verbatim, indicates that the deeper quantitative structure of the VG stochastic process in continuous time had already been explored in preparation for [21]:

> We observe . . . that the shares with low β's . . . also have large estimates for v and this is consistent as high v's and low β's generate long tailedness, the Kurtosis of the v.g. being $3+(3v/m^2)$, ... The low values of θ in the n.c.p. model are suggestive of the most appropriate model being a pure jump process. The v.g. model can be shown to be a pure jump process while the process of independent stable increments has both continuous and jump components. Judging on the basis of the chi-squared statistics, it would appear that the stable model generally overstates the longtailedness.

6.2 Variations on the Theme

There was one last Econometric Discussion Paper [19] in our collaboration. The paper following on from it, [20], was received December 1987, and revised October 1988. It was thus completed in revised form in the Fall of 1988, when Dilip was already at Maryland and I at Virginia. It addresses a number of practical numerical issues relating to the implementation of the procedure of [18], including the truncation point of the expansion (13) of the p.d.f., and the choice of ω.

The paper also addresses the possibility of using the simpler transformation $\psi(v) = v, \mod 2\pi$; and also problems of parameter estimation by this method for a nonsymmetrically distributed random variable. Section 4 of [19] on simulation results was never published. It relates to applying the estimation procedure to an asymmetric stable law with c.f. of general form; and consequently there are still occasional requests for the Discussion Paper, because the p.d.f. for the stable case is unavailable.

However, inasmuch as the p.d.f. of the symmetric VG distribution was explicitly stated in [21], there was no need for estimation methodology using the c.f. as soon as MATLAB became available because that could handle the special function that the p.d.f. involves; and likewise for the asymmetric VG distribution, given later in [10].

7 The VG paper of 1990

The *Journal of Business* does not show received/revised dates for [21], but the original submission, printed on a dot-matrix printer, carries the date February 1988, and I have handwritten notes dated 23 March 1988 relating to things that need to be addressed in a possible revision.

To give an idea of the evolution to [21], the following presents some of the flow of the original submission, which begins by taking the distribution of the return R to be given by

$$R|V \sim \mathcal{N}(\mu, \sigma^2 V),$$

where $V \sim \Gamma(\gamma, c)$. Taking $X = R - \mu$ for the centered return, the c.f. is

$$\phi_X(u) = [1 + (\sigma^2 v/m)(u^2/2)]^{-m^2/v}, \tag{14}$$

where $m = \gamma/c$ is the mean of the gamma distribution and $v = \gamma/c^2$ is its variance. Taking $m = E[V] = 1$ to correspond to unit time change in the gamma process $\{Y(t)\}, t \geq 0$, by which the Brownian motion process is subordinated to give the VG process $\{Z(t)\}$, where

$$Z(t) = \mu t + b(Y(t)), \tag{15}$$

the c.f. (14) becomes

$$\phi_X(u) = [1 + (\sigma^2 v u^2/2)]^{-1/v}. \tag{16}$$

From the c.f. (13), it is observed that X may be written as

$$X = Y - Z, \tag{17}$$

where Y, Z are i.i.d. gamma random variables because $(1 + a^2 u^2) = (1 - iau)(1 + iau)$. Kullback [9] had investigated the p.d.f. of such an X, and this gives the p.d.f. symmetric about the origin corresponding to (16) for $x > 0$ as

$$f(x) = \frac{\sqrt{2/v}}{\sigma} \frac{(x\sqrt{2/v}/\sigma)^{(2/v-1)/2}}{2^{(2/v-1)/2}\Gamma(1/v)\sqrt{\pi}} K_{(2/v-1)/2}(x\sqrt{2/v}/\sigma), \tag{18}$$

where $K_\eta(x)$ is a Bessel function of the second kind of order η and of imaginary argument. At the time it was thought of as a power series.

Usage of (17) leads to the process $\{X(t)\}, t \geq 0$, where

$$X(t) = Z(t) - \mu t = b(Y(t)), \tag{19}$$

to be written in terms of

$$X(t) = U(t) - W(t), \tag{20}$$

where $\{U(t), t \geq 0\}$ and $\{W(t), t \geq 0\}$ are i.i.d. gamma processes.

The preceding more or less persists in the published version [21], along with the possibility of estimation by using the p.d.f. of $\Theta = \omega X$, mod 2π, the submission citing [19]. The published paper replaces this with the by-then published [20]. (See our Section 6.2.)

In the original submission, the fact that $\{Z(t), t \geq 0\}$ is a pure-jump process is argued in a similar way to how Dilip had first obtained the result in early 1986 ([17], in which the fact is first mentioned, actually carries the date July 1986). I still see him walking into my office one day and saying, "The VG process is pure-jump!"

I still think that this is one of the most striking features arguing for use of the VG as a feasible financial model. Another is that the VG distribution of an increment over a time interval of any length is still VG, so the form persists over any time interval. This is a feature in common with the NCP and stable forms, is aesthetically pleasing, and not shared by the t process. Other positive attributes are examined in [6].

Here, verbatim, is the leadin to Dilip's argument in the original submission:

> We now show that the process of i.i.d. gamma increments is purely discontinuous and so $X(t)$ is a pure jump process. Let $Y(t)$ be the process with i.i.d. gamma increments. Consider any interval of time, $[t, t+h]$, and define ΔY to be $Y(t+h) - Y(t)$. Also define y^k to be the k-th largest jump of process $Y(t)$ in the time interval $[t, t+h]$. We shall derive the exact density of y^k and show that

$$E[\Delta Y] = E\left[\Sigma_{k=1}^{\infty} y^k\right]. \tag{3.3}$$

Equation 3.3 then implies that

$$H = \Delta Y - \Sigma_{k=1}^{\infty} y^k$$

has zero expectation.

The derivation (over pp. 7–10) of the p.d.f. of the size of the kth jump in a gamma process $\{Y(t)\}$, which allows $E[y^k]$ to be calculated, is followed on pp. 11–13 by an argument which shows that the gamma process can be approximated by a nondecreasing compound Poisson process $\{Y^n(t)\}$ with mean Poisson arrival rate of jumps β_n/ν and density of jumps given by

$$\beta_n^{-1}[e^{-x/\nu}/x]\mathbf{1}_{\{x>1/n\}}, \ x > 0,$$

where the normalization constant is given by

$$\beta_n = \int_0^{\infty} [e^{-x/\nu}/x]\mathbf{1}_{\{x>1/n\}}dx.$$

Because in this approach nonnegative random variables are being considered, it is possible to work with Laplace transforms of real nonnegative argument s rather than c.f.s; this is done using a spectral approach extracted from

[29], Theorem A, which gives the representation of a Laplace transform of a nonnegative random variable as

$$\Psi(s) = \exp\left\{ -\int_{0-}^{\infty} \frac{1 - e^{-sx}}{1 - e^{-x}} \mu(dx) \right\}, \qquad s > 0,$$

for a measure $\mu(\cdot)$ on the positive half line.

Thus the VG process as the difference of two i.i.d. gamma processes can be approximated by the difference of two i.i.d. nondecreasing compound Poisson processes. This conclusion is obtained more compactly in the published version [21].

That the subordinated process is a pure jump process is argued there directly from the Lévy representation of the log-c.f. of $Z(t)$, which shows that there is no Gaussian component. This argument is in a short Appendix in line with the editor's instruction to compress and then move any mathematical proofs to the back of the paper, to accord with the requirement that papers in the *Journal of Business* should be accessible to a wide audience. In the discussion of the Lévy representation, it is shown that the process $\{Z(t)\}$ can be viewed as the limit of compound Poisson processes.

A direct approach through the c.f. of this kind is obviously more appropriate when considering the VG process allowing for skewness, as is done in what in my perception is the next major breakthrough paper for VG theory [10], where a commensurate form of the VG p.d.f. generalizing (18) is also given.

However, the fact that the skewed VG process is a difference of two independent gamma processes is still a key player in that exposition, even though the processes are no longer identically distributed.

I remember an e-mail from Dilip, telling me he had achieved such more general results during a conference at the Newton Institute, Cambridge, U.K. I believe [10] also took much too long to find a publisher. It now enjoys a well-deserved status.

The photograph shown in Figure 1 taken in my Sydney office shows Dilip and me finalizing a first revision of what became [21] just before he left Sydney for the University of Maryland, and so dates to about July 1988. There were to be several further revisions during the 1988–1989 U.S. academic year, and I recall last touches on a brief conference visit which I made to the United States in January 1990.

Acknowledgments

I am indebted to Keith McLaren, Professor of Econometrics at Monash University, and Emeritus Professor Helen Praetz, Peter Praetz's widow, for biographical and academic information about him.

Fig. 1. Dilip Madan (left) in my Sydney office circa July 1988, finalizing a first revision of what became the original VG paper [21].

References

1. D. Applebaum. *Lévy Processes and Stochastic Calculus.* Cambridge University Press, 2003.
2. T.W. Epps. *Pricing Derivative Securities.* World Scientific, 2000.
3. E.F. Fama. The behaviour of stock-market prices. *J. Business*, 38:34–105,1965.
4. A. Feuerverger and P. McDunnough. On the efficiency of empirical characteristic function procedures. *J. R. Statist. Soc., Ser. B*, 43:20–27, 1981.
5. R. Finlay and E. Seneta. Stationary-increment Student and Variance-Gamma processes. *J. Appl. Prob.*, 43:441–453, 2006.
6. T. Fung and E. Seneta. Tailweight, quantiles and kurtosis: A study of competing distributions. *Operations Research Letters*, in press, 2006.
7. S.W. Harrar, E. Seneta, and A.K. Gupta. Duality between matrix variate t and matrix variate V.G. distributions. *J. Multivariate Analysis*, 97:1467–1475,2006.
8. C.C. Heyde. A risky asset model with strong dependence through fractal activity time. *J. Appl. Prob.*, 36: 1234–1239, 1999.
9. S. Kullback. The distribution laws of the difference and quotient of variables independently distributed in Pearson type III laws. *Ann. Math. Statist.*, 7:51–53,1936.
10. D.B. Madan, P. Carr, and E.C. Chang. The variance gamma process and option pricing. *European Finance Review*, 2:79–105, 1998.

11. D.B. Madan and F. Milne. Option pricing with VG martingale components. *Mathematical Finance*, 1(4): 19–55,1991.

12. D.B. Madan and E. Seneta. The profitability of barrier strategies for the stock market. *Economic Statistics Papers*, No.45, 39 pp, University of Sydney, 1981.

13. D.B. Madan and E. Seneta. Residuals and the compound Poisson process. *Economic Statistics Papers*, No. 48, 17 pp, University of Sydney, 1982.

14. D.B. Madan and E. Seneta. Compound Poisson models for economic variable movements. *Sankhya, Ser.B*, 46:174–187, 1984.

15. D.B. Madan and E. Seneta. Simulation of estimates using the empirical characteristic function. *Econometric Discussion Papers*, No. 85-02, 18 pp, University of Sydney, 1985.

16. D.B. Madan and E. Seneta. Simulation of estimates using the empirical characteristic function. *International Statistical Review*, 55:153–161, 1987.

17. D.B. Madan and E. Seneta. Chebyshev polynomial approximations and characteristic function estimation. *Econometric Discussion Papers*, No. 87-04, 13 pp, University of Sydney, 1987.

18. D.B. Madan and E. Seneta. Chebyshev polynomial approximations and characteristic function estimation. *J.R. Statist. Soc., Ser. B*, 49:163–169, 1987.

19. D.B. Madan and E. Seneta. Characteristic function estimation using maximum likelihood on transformed variables. *Econometric Discussion Papers*, No. 87-08, 9 pp, University of Sydney, 1987.

20. D.B. Madan and E. Seneta. Chebyshev polynomial approximations for characteristic function estimation: Some theoretical supplements. *J.R. Statist. Soc., Ser. B*, 51:281–285, 1989.

21. D.B. Madan and E. Seneta. The Variance-Gamma (V.G.) model for share market returns. *J. Business*, 63: 511–524, 1990.

22. D.B. Madan and E. Seneta. On the monotonicity of the labour-capital ratio in Sraffa's model. *Journal of Economics [Zeitschrift für Nationalökonomie]*, 51: 101–107, 1990.

23. B.B. Mandelbrot. New methods in statistical economics. *J. Political Economy*, 71:421–440, 1963.

24. P.D. Praetz. Australian share prices and the random walk hypothesis. *Austral. J. Statist.*, 11:123–139, 1969.

25. P.D. Praetz. The distribution of share price changes. *J. Business*, 45:49–55,1972.

26. S.J. Press. A compound events model for security prices. *J. Business*, 40:317–335, 1967.

27. W. Schoutens. *Lévy Processes in Finance: Pricing Financial Derivatives*. Wiley, 2003.

28. E. Seneta. Fitting the Variance-Gamma model to financial data. *Stochastic Methods and Their Applications (C.C. Heyde Festschrift)*, eds. J. Gani and E. Seneta, *J. Appl. Prob.*, 41A:177–187, 2004.

29. E. Seneta and D. Vere-Jones. On the asymptotic behaviour of subcritical branching processes with continuous state-space. *Z. Wahrscheinlichkeutstheorie verw. Geb*, 10:212–225,1968.

30. A. Tjetjep and E. Seneta. Skewed normal variance-mean models for asset pricing and the method of moments. *International Statistical Review*, 74: 109–126, 2006.

Variance-Gamma and Monte Carlo

Michael C. Fu

Robert H. Smith School of Business
Department of Decision & Information Technologies
University of Maryland
College Park, MD 20742, USA
mfu@rhsmith.umd.edu

Summary. The Variance-Gamma (VG) process was introduced by Dilip B. Madan and Eugene Seneta as a model for asset returns in a paper that appeared in 1990, and subsequently used for option pricing in a 1991 paper by Dilip and Frank Milne. This paper serves as a tutorial overview of VG and Monte Carlo, including three methods for sequential simulation of the process, two bridge sampling methods, variance reduction via importance sampling, and estimation of the Greeks.

Key words: Variance-Gamma process; Lévy processes; Monte Carlo simulation; bridge sampling; variance reduction; importance sampling; Greeks; perturbation analysis; gradient estimation.

1 Introduction

1.1 Reflections on Dilip

Dilip and I have been colleagues since 1989, when I joined the faculty of the Business School at the University of Maryland, and just one year after Dilip himself had returned to his (Ph.D.) alma mater after twelve years on the faculty at the University of Sydney, Australia, following the earning of his two Ph.D.s in pure math (1975) and economics (1971). However, because we were in different departments—he in the Finance Department and I in what was then called the Management Science & Statistics Department—we did not really start collaborating until the mid-1990s when I first became interested in financial applications. Since then, Dilip and I have co-chaired five Ph.D. student dissertations (Tong Wang, Rongwen Wu, Xing Jin, Yi Su, Sunhee Kim), and each of us serves regularly on dissertation committees of the other's Ph.D. students. We team-taught a course on computational finance twice (1997, 1999). We have led a research group on computational/mathematical finance since the late 1990s, which became known as the Mathematical Finance Research Interactions Team (RIT) under the interdisciplinary Applied Mathematics and

Scientific Computing program at the University of Maryland. From this group have graduated scores of Ph.D. students (mostly supervised by Dilip; see the photo in the preface) who have almost all gone on to high-octane "quant" positions at financial institutions on Wall Street and in the Washington, D.C. area. Dilip has always been an inspiration to us all and an ideal academic colleague, always energetic and enthusiastic and full of ideas, whether it be at the research meetings with students or during our regular lunch strolls to downtown College Park.

As Dilip has told me numerous times in our discussions, on Wall Street there are really two main numerical models/techniques employed for pricing derivative securities: Monte Carlo simulation or numerical solution of partial differential equations. The former depends on the first fundamental theorem of asset pricing: the existence of a martingale measure so that the price can be expressed as the expectation of an appropriately discounted payoff function. For exotic, hybrid, and complicated path-dependent derivatives, however, simulation is often the only method available. Of course, Dilip's main research interests do not lie in simulation, but he is a regular consumer of the method, and as a result, we collaborated on two papers in the area, on control variates for pricing continuous arithmetic Asian options [12] and on pricing American-style options [11]. Furthermore, many of his recent papers address simulation of processes, for example, [21]. This brief tutorial is intended to be my small tribute to Dilip and also to provide a bridge or segue between the historical perspective on the VG process provided in the opening article by Eugene Seneta [27] and the remarkable properties of the gamma process described in the subsequent piece by Marc Yor [28], who provides a far more advanced and sophisticated view of gamma processes than the rudimentary presentation here.

1.2 The Variance-Gamma Process

The Variance-Gamma (VG) process was introduced to the finance community as a model for asset returns (log-price increments) and options pricing in Madan and Seneta [20], with further significant developments in Madan and Milne [19] and Madan et al. [18]. More recently, it has been applied to American-style options pricing in [15] and [17], the latter using the fast Fourier transform introduced in [6]. For more history on the earlier pioneering years, see the paper in this volume by Eugene Seneta [27], which also discusses distributions similar to VG proposed earlier as a possible alternative to the normal distribution (see also [22], [23, p. 166]).

The VG process is a Lévy process, that is, it is a process of independent and stationary increments (see the appendix for a review of basic definitions). A Lévy process can be represented as the sum of three independent components (cf. [26]): a deterministic drift, a continuous Wiener process, and a pure-jump process. Brownian motion is a special case where the latter is zero, and the Poisson process is a special case on the other end where the first two components are zero. Like the Poisson process, the VG process is pure jump; that is, there is no continuous (Brownian motion and deterministic) component, and thus it can be expressed in terms of its Lévy density, the simplest version with no parameters being

$$k(x) = \frac{1}{|x|} e^{-\sqrt{2}|x|}.$$

For the VG process with the usual (θ, σ, ν) parameterization, the Lévy density is given by

$$k(x) = \frac{1}{\nu|x|} \exp\left(\frac{\theta}{\sigma^2} x - \frac{1}{\sigma} \sqrt{\frac{2}{\nu} + \frac{\theta^2}{\sigma^2}} |x| \right), \qquad (1)$$

where $\nu, \sigma > 0$.

The VG is a special case of the CGMY model of [4, 5], whose Lévy density is given by

$$k(x) = \begin{cases} C \exp(-G|x|)/|x|^{1+Y} & x < 0, \\ C \exp(-M|x|)/|x|^{1+Y} & x > 0, \end{cases}$$

where VG is obtained by taking $Y = 0, C = 1/\nu, G = 2\mu_+/\sigma^2$, and $M = 2\mu_-/\sigma^2$, where $\mu_\pm$ are defined below in (4). Unlike the Poisson process, the VG process may have an infinite number of (infinitesimally small) jumps in any interval, making it a process of infinite activity. Unlike Brownian motion, the VG process has finite variation, so it is in some sense less erratic in its behavior.

The representation of the VG process presented above hides its roots, which come from the following well-known alternative representations:

1. Time-changed (subordinated) Brownian motion, where the subordinator is a gamma process.
2. Difference of two gamma processes.

As described in detail in [27], these are the original ways in which the process was introduced and proposed as a model for asset returns. The CGMY process can also be expressed as time-changed Brownian motion (see Madan and Yor [21]), but it seems that there is no simple representation as the difference of two increasing processes, although in principle such a representation does exist, because all processes with finite variation can be so expressed.

To express the process in terms of the two representations above, let $\{W_t\}$ denote standard Brownian motion (Wiener process), $B_t^{(\mu,\sigma)} \equiv \mu t + \sigma W_t$ denote Brownian motion with constant drift rate μ and volatility σ, $\gamma_t^{(\mu,\nu)}$ the gamma process with drift parameter μ and variance parameter ν, and $\gamma_t^{(\nu)}$ the gamma process with unit drift ($\mu = 1$) and variance parameter ν. Letting $\phi_X(u) = E[e^{iuX}]$ denote the characteristic function (c.f.) of random variable X, the c.f. for the VG process is given simply by

$$\phi_{X_t}(u) = (1 - iu\theta\nu + \sigma^2 u^2 \nu/2)^{-t/\nu},$$

which can be expressed in two forms

$$\phi_{X_t}(u) = [(1 - iu\nu\mu_+)(1 + iu\nu\mu_-)]^{-t/\nu},$$
$$\phi_{X_t}(u) = [1 - i\nu(u\theta + i\sigma^2 u^2/2)]^{-t/\nu},$$

reflecting the two representations above. In particular, using the notation introduced above, the representation of the VG process as time-changed Brownian motion is given by

$$X_t = B^{(\theta,\sigma)}_{\gamma_t^{(\nu)}} = \theta\gamma_t^{(\nu)} + \sigma W_{\gamma_t^{(\nu)}}, \tag{2}$$

whereas the difference-of-gammas representation is given by

$$X_t = \gamma_t^{(\mu_+,\nu_+)} - \gamma_t^{(\mu_-,\nu_-)}, \tag{3}$$

where the two gamma processes are independent (but defined on a common probability space) with parameters

$$\mu_\pm = (\sqrt{\theta^2 + 2\sigma^2/\nu} \pm \theta)/2, \tag{4}$$
$$\nu_\pm = \mu_\pm^2 \nu.$$

In general, there is no unique martingale measure for a Lévy process, due to the jumps, and thus this is the case for the (pure jump) VG process. Assume the asset price dynamics for a Lévy process $\{X_t\}$ (with no dividends and constant risk-free interest rate) are given by

$$S_t = S_0 \exp((r + \omega)t + X_t), \tag{5}$$

where the constant ω is such that the discounted asset price is a martingale; that is, it must satisfy

$$E[e^{-rt}S_t] = S_0,$$

which leads to the condition

$$e^{-\omega} = \phi(-i),$$

where ϕ denotes the characteristic function of the Lévy process. In the case of VG, we have

$$\omega = \ln(1 - \theta\nu - \sigma^2\nu/2)/\nu. \tag{6}$$

2 Monte Carlo Simulation

General books on Monte Carlo (MC) simulation for financial applications include [16], [14] and [23]; see also [2]. Monte Carlo simulation is most fruitful for "high-dimensional" problems, which in finance are prevalent in path-dependent options, such as Asian, lookback, and barrier options, all of which are considered in [1], from which most of our discussion on bridge sampling for simulating VG is taken. The importance sampling discussion following that is based on [29]. Efficiently estimating the Greeks from simulation is included in [10], [16], [14], and [23]. Although extensions such as the CGMY process [4] are not treated explicitly here, comments are occasionally made on how the methods extend to that more general setting.

We begin by presenting sequential sampling and bridge sampling techniques for constructing sample paths of a VG process. Sequential sampling is called incremental path construction in [16], which also includes another technique for the Wiener process based on a spectral decomposition using an orthogonal Hilbert basis.

2.1 Sequential Sampling

There are three main methods to simulate VG (cf. [1, 29]). The first two are based on the two representations presented in the previous section and are "exact" in the sense of having the correct distribution. The third method for simulating VG is to approximate it by a compound Poisson process. The main advantage of the third method is its generality, in that it can be used for any Lévy process, in particular in those settings where a representation as subordinated Brownian motion or as the difference of two other easily simulated processes is not readily available. Representation of the CGMY process as time-changed Brownian motion is treated in [21], which also includes procedures for simulating using that representation; see also [29].

Figure 1 presents the three different algorithms for sequentially generating VG sample paths on $[0, T]$ at time points $0 = t_0 < t_1 < \cdots < t_{N-1} < t_N = T$, where the time spacings $\Delta t_i, i = 1, \ldots, N$ are given as inputs, along with VG parameters. In the compound Poisson process approximation, the positive and negative jumps are separated, and there is a cutoff of ε for the magnitude of the jumps. The Poisson rates are calculated by integrating the Lévy density over the appropriate range, and the jump sizes are then sampled from the corresponding renormalized Lévy density. Smaller jumps (which occur infinitely often) are incorporated into a diffusion process with zero drift and volatility estimated by the second moment of the Lévy density integrated over the range $[-\varepsilon, +\varepsilon]$.

2.2 Bridge Sampling

In [25], bridge sampling for the time-changed Brownian motion representation is introduced, along with stratified sampling and quasi-Monte Carlo to further reduce variance. In [1], bridge sampling for the difference-of-gammas representation is introduced and combined with randomized quasi-Monte Carlo, and bounds (upper and lower) on the discretization error for pricing certain forms of path-dependent options are derived. Here, we present these two bridge sampling approaches, but without the additional variance reduction techniques, to make the exposition easy to follow.

Simulating VG as Gamma Time-Changed Brownian Motion

Input: VG parameters θ, σ, ν; time spacing $\Delta t_1, ..., \Delta t_N$ s.t. $\sum_{i=1}^{N} \Delta t_i = T$.

Initialization: Set $X_0 = 0$.

Loop from $i = 1$ to N:
1. Generate $\Delta G_i \sim \Gamma(\Delta t_i/\nu, \nu), Z_i \sim \mathcal{N}(0,1)$
   independently and independent of past r.v.s.
2. Return $X_{t_i} = X_{t_{i-1}} + \theta \Delta G_i + \sigma \sqrt{\Delta G_i} Z_i$.

Simulating VG as Difference of Gammas

Input: VG parameters θ, σ, ν; time spacing $\Delta t_1, ..., \Delta t_N$ s.t. $\sum_{i=1}^{N} \Delta t_i = T$.

Initialization: Set $X_0 = 0$.

Loop from $i = 1$ to N:
1. Generate $\Delta \gamma_i^- \sim \Gamma(\Delta t_i/\nu, \nu\mu_-), \Delta \gamma_i^+ \sim \Gamma(\Delta t_i/\nu, \nu\mu_+)$
   independently and independent of past r.v.s.
2. Return $X_{t_i} = X_{t_{i-1}} + \Delta \gamma_i^+ - \Delta \gamma_i^-$.

Simulating VG as (Approximate) Compound Poisson Process

Input: VG parameters θ, σ, ν; time spacing $\Delta t_1, ..., \Delta t_N$ s.t. $\sum_{i=1}^{N} \Delta t_i = T$.

Initialization: Set $X_0 = 0$;

$$\sigma_\varepsilon^2 = \int_{-\varepsilon}^{+\varepsilon} x^2 k(x)dx, \quad \lambda_\varepsilon^+ = \int_{+\varepsilon}^{\infty} k(x)dx, \quad \lambda_\varepsilon^- = \int_{-\infty}^{-\varepsilon} k(x)dx,$$
$$k_\varepsilon^+(x) = k(x)\mathbf{1}_{\{x \geq \varepsilon\}}/\lambda_\varepsilon^+, \quad k_\varepsilon^-(x) = k(x)\mathbf{1}_{\{x \leq -\varepsilon\}}/\lambda_\varepsilon^-.$$

Loop from $i = 1$ to N:
1. Generate number of positive and negative jumps in Δt_i (N_i^+ and N_i^-, respectively) and corresponding size of jumps using Lévy density (everything independent of each other and of past generated samples):

$$N_i^+ \sim \text{Poisson}(\lambda_\varepsilon^+ \Delta t_i), \quad N_i^- \sim \text{Poisson}(\lambda_\varepsilon^- \Delta t_i),$$
$$X_{i,j}^+ \sim \{k_\varepsilon^+\}, j = 1, ..., N_i^+, \quad X_{i,j}^- \sim \{k_\varepsilon^-\}, j = 1, ..., N_i^-,$$
$$Z_i \sim \mathcal{N}(0,1).$$

2. Return $X_{t_i} = X_{t_{i-1}} + Z_i \sigma_\varepsilon \sqrt{\Delta t_i} + \sum_{j=1}^{N_i^+} X_{i,j}^+ + \sum_{j=1}^{N_i^-} X_{i,j}^-$.

Fig. 1. Algorithms for sequentially simulating VG process on $[0, T]$.

Instead of sequential sampling, which progresses chronologically forward in time, an alternative method for simulating asset price paths is to use bridge sampling, which samples the end of the path first, and then "fills in" the rest of the path as needed. This can lead to a more efficient simulation. In bridge sampling, fixed times are chosen, and the value of the process at such an arbitrary fixed time is a random variable with known (conditional) distribution. It is particularly effective in combination with quasi-Monte Carlo methods, because the sampling sequence usually means that the first samples are more critical than the latter ones, leading to a lower "effective dimension" than in the usual sequential sampling. It is in this setting that quasi-Monte Carlo methods show the greatest improvement over traditional MC methods, and gamma bridge sampling coupled with quasi-Monte Carlo is treated in [25] and [1].

The main idea of bridge sampling is that the conditional distribution of a stochastic process X_t at time $t \in (T_1, T_2)$, given X_{T_1} and X_{T_2} can be easily obtained; that is, for $T_1 \leq t \leq T_2$, one can apply Bayes' rule to get the necessary conditional distribution:

$$P(X_t|X_{T_1}, X_{T_2}) = \frac{P(X_{T_2}|X_{T_1}, X_t)P(X_t|X_{T_1})}{P(X_{T_2}|X_{T_1})}.$$

The following are the most well-known examples.

Poisson process $\{N_t\}$: Conditional on N_{T_1} and N_{T_2},

$$N_t \sim N_{T_1} + \mathbf{bin}(N_{T_2} - N_{T_1}, (t - T_1)/(T_2 - T_1)),$$

where $\mathbf{bin}(n, p)$ denotes the binomial distribution with parameters n and p (mean np and variance $np(1 - p)$). Note that the conditional distribution has no dependence on the arrival rate.

Brownian motion $\{B_t^{(\mu,\sigma)}\}$: Conditional on B_{T_1} and B_{T_2},

$$B_t \sim \mathcal{N}(\alpha B_{T_1} + (1 - \alpha)B_{T_2}, \alpha(t - T_1)\sigma^2),$$

where $\alpha = (T_2 - t)/(T_2 - T_1)$, and there is dependence on σ, but not on the drift μ.

Gamma process $\{\gamma_t^{(\mu,\nu)}\}$: Conditional on γ_{T_1} and γ_{T_2},

$$\gamma_t \sim \gamma_{T_1} + (\gamma_{T_2} - \gamma_{T_1})Y,$$

where $Y \sim \beta((t - T_1)/\nu, (T_2 - t)/\nu)$, which only depends on the parameter ν, where $\beta(\alpha_1, \alpha_2)$ denotes the beta distribution with mean $\alpha_1/(\alpha_1 + \alpha_2)$, variance $\alpha_1\alpha_2/[(\alpha_1 + \alpha_2)^2(\alpha_1 + \alpha_2 + 1)]$, and density $x^{\alpha_1-1}(1 - x)^{\alpha_2-1}/B(\alpha_1, \alpha_2)$ on $[0, 1]$, where

$$B(x, y) = \int_0^1 t^{x-1}(1 - t)^{y-1}dt = \frac{\Gamma(x)\Gamma(y)}{\Gamma(x + y)}, \qquad \Gamma(x) = \int_0^\infty t^{x-1}e^{-t}dt.$$

Simulating VG via Brownian (Gamma Time-Changed) Bridge

Input: VG parameters θ, σ, ν; number of bridges $N = 2^M (T = t_N)$.

Initialization: Set $X_0 = 0, \gamma_0 = 0$.
  Generate $\gamma_{t_N} \sim \Gamma(t_N/\nu, \nu), X_{t_N} \sim \mathcal{N}(\theta\gamma_{t_N}, \sigma^2\gamma_{t_N})$ independently.

Loop from $k = 1$ to M: $n \leftarrow 2^{M-k}$;
  Loop from $j = 1$ to 2^{k-1}:
  1. $i \leftarrow (2j-1)n$;
  2. Generate $Y_i \sim \beta((t_i - t_{i-n})/\nu, (t_{i+n} - t_i)/\nu)$ independent of past r.v.s;
  3. $\gamma_{t_i} = \gamma_{t_{i-n}} + [\gamma_{t_{i+n}} - \gamma_{t_{i-n}}]Y_i$;
  4. Generate $Z_i \sim \mathcal{N}(0, [\gamma_{t_{i+n}} - \gamma_{t_i}]\sigma^2 Y_i)$ independent of past r.v.s;
  5. Return $X_{t_i} = Y_i X_{t_{i+n}} + (1 - Y_i)X_{t_{i-n}} + Z_i$.

Simulating VG via Difference-of-Gammas Bridge

Input: VG parameters θ, σ, ν; number of bridges $N = 2^M (T = t_N)$.

Initialization: Set $\gamma_0^+ = \gamma_0^- = 0$.
  Generate $\gamma_{t_N}^+ \sim \Gamma(t_N/\nu, \nu\mu_+), \gamma_{t_N}^- \sim \Gamma(t_N/\nu, \nu\mu_-)$ independently.

Loop from $k = 1$ to M: $n \leftarrow 2^{M-k}$;
  Loop from $j = 1$ to 2^{k-1}:
  1. $i \leftarrow (2j-1)n$;
  2. Generate $Y_i^+, Y_i^- \sim \beta((t_i - t_{i-n})/\nu, (t_{i+n} - t_i)/\nu)$ independently;
  3. $\gamma_{t_i}^+ = \gamma_{t_{i-n}}^+ + [\gamma_{t_{i+n}}^+ - \gamma_{t_{i-n}}^+]Y_i^+, \gamma_{t_i}^- = \gamma_{t_{i-n}}^- + [\gamma_{t_{i+n}}^- - \gamma_{t_{i-n}}^+]Y_i^-$;
  4. Return $X_{t_i} = \gamma_{t_i}^+ - \gamma_{t_i}^-$.

Fig. 2. Algorithms for simulating VG process on $[0, T]$ via bridge sampling.

As with the gamma distribution, there is no analytical closed form for the c.d.f.; also $\beta(1, 1) = U(0, 1)$. If either of the parameters is equal to 1, then the inverse transform method can be easily applied. Otherwise, the main methods to generate variates are acceptance–rejection, numerical inversion, or a special algorithm using the following known relationship with the gamma distribution: if $Y_i \sim \Gamma(\alpha_i, \psi)$ independently generated, then $Y_1/(Y_1 + Y_2) \sim \beta(\alpha_1, \alpha_2)$. For more information on efficiently generating samples from the gamma and beta distributions, see [7]. MATLAB and other commercial software generally have built-in functions/subroutines to handle these distributions.

The two algorithms given in Figure 2, adapted from [1] and corresponding to the time-changed Brownian motion representation of (2) and the difference-of-gammas representation of (3), respectively, generate VG sample paths of progressively finer resolution by adding bisecting bridge points, so the number of simulated points on the path should be a power of 2, called the dyadic partition in [1], where it is noted this allows the simulation efficiency to be further improved by using "a fast beta random-variate generator that exploits the symmetry." This partition also makes the methods easier to present in algorithmic form; however, the method is clearly just as easily applicable to general time steps as in the previous sequential versions.

For pricing path-dependent options that depend on the entire continuous sample path—such as continuous Asian, barrier, and lookback options—the advantage of the bridge sampling is that the first samples often capture most of the contribution to the expected payoff that is being estimated. In either case, Richardson extrapolation is one method that can be used to go from the discrete to the continuous case. This is discussed for the difference-of-gammas bridge sampling in [1]; it was also used in [12] in sequential Monte Carlo simulation.

2.3 Variance Reduction

Variance reduction techniques can lead to orders of magnitude of improvement in simulation efficiency, and thus are of practical importance. A simple example where it would be critical is a deep out-of-the-money barrier option, in which the payoff on most sample paths generated by simulation would be zero, so being in the money is essentially a "rare event" in simulation lingo. In this section, we briefly discuss the variance reduction technique of importance sampling, whereby simulation is carried out under a different measure than the one of interest, and then an adjustment is made to the payoff function by way of the Radon–Nikodym derivative (change of measure). In the barrier option example, the resulting measure change would lead to a great increase in the number of generated paths that are in the money. Other useful variance reduction techniques not discussed here that are effective in financial simulations include common random numbers (called "variate recycling" in [16]), conditional Monte Carlo, stratified sampling, and control variates; see [14] for more details. As mentioned earlier, quasi-Monte Carlo combined with gamma bridge sampling for the VG process is described in [25] and [1].

The general form for the Radon–Nikodym derivative of a Lévy process can be found in [26]; see also [28] for gamma processes. For pure-jump Lévy processes, it turns out that sufficient conditions for ensuring equivalence in the measure change are that the corresponding Lévy measures be equivalent plus a constraint relating the drifts and corresponding Lévy measures. For the VG process, under the difference-of-gammas representation, it turns out that the measure change in which the ν parameter is kept the same can be computed based on just the terminal values; that is, the intermediate values on the path are not needed (cf. [29, Propositions 2 and 3 in Chapter 2]). The Radon–Nikodym derivative needed to adjust for going from VG with parameters (θ, σ, ν) to the measure change of VG with parameters (θ', σ', ν) is given by the following exponential twisting:

$$\exp\left(-t \int_{-\infty}^{+\infty} (k(x) - k'(x))dx\right)\phi^+(\tilde{\gamma}_t^+)\phi^-(-\tilde{\gamma}_t^-),$$

$$\phi^\pm(x) = e^{2(\mu'_\mp/(\sigma')^2 - \mu_\mp/\sigma^2|x|)},$$

where $\tilde{\gamma}_t^\pm$ are the independent processes generated in the difference-of-gammas representation of VG with (VG) parameters (θ', σ', ν), and $\mu_\pm$ is given by (4).

2.4 The Greeks

In many cases, it is possible to estimate sensitivities of derivatives prices with respect to various parameters directly using the same sample path that was used to estimate the price itself, that is, without resorting to resimulation. This was first demonstrated in [9] and [3]. In this section, we discuss only infinitesimal perturbation analysis (IPA), or what is called pathwise differentiation in [16] and [14]. We do not treat the likelihood ratio method (also known as the score function method or measure-valued differentiation); see the previous two referenced books [16, 14], or [8] for further discussion of this approach, which is also related to the Malliavin calculus approach proposed by some researchers (see [14]). An extension of IPA based on conditional Monte Carlo, which can handle discontinuous payoff functions, is treated in [10]. None of the previous references treat the estimation of Greeks explicitly in the VG (or general Lévy process) context.

The basis of IPA is quite simple: one differentiates the sample quantities of interest with respect to the parameter of interest. Specifically, one is usually interested in some payoff function h that may depend on the entire path of $\{S_t\}$. For simplicity of illustration here, we assume the payoff depends only on a single point in time, such as a call option with $h(x) = (x - K)^+$, where K is the strike price. Then, the price to be estimated by simulation is given by

$$E[e^{-rT}h(S_T)],$$

for maturity T (European option), whereas the sensitivity to be estimated is given by

$$\frac{dE[e^{-rT}h(S_T)]}{d\chi},$$

where χ is the parameter of interest. The IPA estimator is given by

$$\frac{d(e^{-rT}h(S_T))}{d\chi} = h(S_T)\frac{d(e^{-rT})}{d\chi} + e^{-rT}\left(\frac{\partial h}{\partial \chi} + h'(S_T)\frac{dS_T}{d\chi}\right),$$

so the applicability of the IPA estimator comes down to a question of whether exchanging the operations of expectation (integration) and differentiation (limit) is justified (according to the dominated convergence theorem); that is, whether

$$E\left[\frac{d(e^{-rT}h(S_T))}{d\chi}\right] = \frac{dE[e^{-rT}h(S_T)]}{d\chi},$$

which clearly depends both on the payoff function h and the representation of the VG process $\{X_t\}$, which enters $\{S_t\}$ through (5).

As an example, it can be easily seen from (5) that the current asset price is a scale parameter for a future stock price, so that we have

$$\frac{dS_t}{dS_0} = \frac{S_t}{S_0}.$$

Thus, for example, a call option with the usual payoff $(S_T - K)^+$, which would be estimated in simulation by

$$e^{-rT}(S_T - K)^+,$$

would have its delta estimated by

$$e^{-rT}\frac{S_T}{S_0}\mathbf{1}_{\{S_T>K\}},$$

because $h'(x) = \mathbf{1}_{\{x>K\}}$. However, a digital option with payoff $\mathbf{1}_{\{S_T>K\}}$ would lead to a biased estimator (identically zero). Thus, the IPA estimator for the gamma would be biased. Roughly speaking, if h is almost surely continuous with respect to the parameter of interest, then the IPA estimator will be unbiased. The call payoff function is continuous, with a "kink" at K, which leads to a discontinuity in its first derivative at K, just as for the digital option.

The example above was in some sense the simplest, because S_0 doesn't appear anywhere else in the expression for S_t given by (5) except as a scale factor. Other parameters of interest include time t, the interest rate r, and the VG parameters (θ, σ, ν), and these all make more complicated appearances in S_t, both directly through (5) and indirectly through X_t, where in the latter case the chain rule would be applied. The resulting quantity $dX_t/d\chi$ may depend on the representation of $\{X_t\}$, that is, in how the stochastic process $\{X_t\}$ is constructed, and we saw there are at least three different ways a VG process can be generated. Thus, for the simple call option, we have the following IPA estimators (for any exercise point t).

$$\frac{dS_t}{dr} = tS_t,$$

$$\frac{dS_t}{dt} = S_t\left(r + \omega + \frac{dX_t}{dt}\right),$$

$$\frac{dS_t}{d\theta} = S_t\left(t\frac{d\omega}{d\theta} + \frac{dX_t}{d\theta}\right),$$

$$\frac{dS_t}{d\sigma} = S_t\left(t\frac{d\omega}{d\sigma} + \frac{dX_t}{d\sigma}\right),$$

$$\frac{dS_t}{d\nu} = S_t\left(t\frac{d\omega}{d\nu} + \frac{dX_t}{d\nu}\right),$$

with (from differentiating (6))

$$\frac{d\omega}{d\theta} = -1/(1 - \theta\nu - \sigma^2\nu/2),$$

$$\frac{d\omega}{d\sigma} = -\sigma/(1 - \theta\nu - \sigma^2\nu/2),$$

$$\frac{d\omega}{d\nu} = -[(\theta + \sigma^2/2)/(1 - \theta\nu - \sigma^2\nu/2) + \omega]/\nu,$$

but the $dX_t/d\chi$ term in the IPA estimators for $dS_t/d\chi$ above undetermined without specifying a particular representation of $\{X_t\}$. For example, if we chose the parameter of interest χ to be the θ parameter defining the VG process, then using the time-changed Brownian motion representation given by (2) leads to a very simple

$$\frac{dX_t}{d\theta} = \gamma_t^{(\nu)},$$

whereas for the difference-of-gammas representation given by (3),

$$\frac{dX_t}{d\theta} = \frac{d\gamma_t^{(\mu_+,\nu_+)}}{d\theta} - \frac{d\gamma_t^{(\mu_-,\nu_-)}}{d\theta},$$

which is more complicated to compute. In this case, it is likely that both lead to unbiased IPA estimators, but the resulting estimators would have very different variance properties.

3 Conclusions

In the context of Monte Carlo simulation, there is a large body of work on variance reduction techniques and gradient estimation techniques for the usual Gaussian/diffusion setting in finance, but with the exception of a few recent results such as [25] and [1], the more general Lévy process setting is relatively untouched. After reviewing the three main methods for simulating a VG process and presenting two recently developed bridge sampling approaches, we just scratched the surface here in presenting the importance sampling change of measure and introducing IPA sensitivity estimators for the Greeks in the VG setting. In the latter case, the choice of representation plays a key role in determining the applicability of IPA; see [13] and [10] for more on this theme.

Acknowledgments

This work was supported in part by the National Science Foundation under Grant DMI-0323220, and by the Air Force Office of Scientific Research under Grant FA95500410210. I thank Eugene Seneta, Philip Protter, Andy Hall, Scott Nestler, Qing Xia, and Huiju Zhang for their comments and corrections.

References

1. A.N. Avramidis and P. L'Ecuyer. Efficient Monte Carlo and quasi-Monte Carlo option pricing under the Variance-Gamma model. *Management Science*, 52:1930–1944, 2006.
2. P. Boyle, M. Broadie, and P. Glasserman. Monte Carlo methods for security pricing. *Journal of Economic Dynamics and Control*, 21:1267–1321, 1997.
3. M. Broadie and P. Glasserman. Estimating security price derivatives using simulation. *Management Science*, 42:269–285, 1996.
4. P. Carr, H. Geman, D.B. Madan, and M. Yor. The fine structure of asset returns: An empirical investigation. *Journal of Business*, 75:305–332, 2002.
5. P. Carr, H. Geman, D.B. Madan, and M. Yor. Stochastic volatility for Lévy processes. *Mathematical Finance*, 2:87–106, 2003.
6. P. Carr and D. Madan. Option valuation using the fast Fourier transform. *Journal of Computational Finance*, 2:61–73, 1999.
7. G.S. Fishman. *Monte Carlo Methods: Concepts, Algorithms, and Applications*. Springer, 1996.
8. M.C. Fu. Stochastic gradient estimation. Chapter 19 in *Handbooks in Operations Research and Management Science: Simulation*, eds. S.G. Henderson and B.L. Nelson, Elsevier, 2006.
9. M.C. Fu and J.Q. Hu, Sensitivity analysis for Monte Carlo simulation of option pricing. *Probability in the Engineering and Informational Sciences*, 9:417–446, 1995.
10. M.C. Fu and J.Q. Hu. *Conditional Monte Carlo: Gradient Estimation and Optimization Applications*. Kluwer Academic, 1997.
11. M.C. Fu, S.B. Laprise, D.B. Madan, Y. Su, and R. Wu. Pricing American options: A comparison of Monte Carlo simulation approaches. *Journal of Computational Finance*, 4:39–88, 2001.
12. M.C. Fu, D.B. Madan and T. Wang. Pricing continuous Asian options: A comparison of Monte Carlo and Laplace transform inversion methods. *Journal of Computational Finance*, 2:49–74, 1999.
13. P. Glasserman. *Gradient Estimation Via Perturbation Analysis*. Kluwer Academic, 1991.
14. P. Glasserman. *Monte Carlo Methods in Financial Engineering*. Springer, 2003.
15. A. Hirsa and D.B. Madan. Pricing American options under variance gamma. *Journal of Computational Finance*, 7:63–80, 2004.
16. P. Jäckel. *Monte Carlo Methods in Finance*. Wiley, 2002.
17. S. Laprise. *Stochastic Dynamic Programming: Monte Carlo Simulation and Applications to Finance*. Ph.D. dissertation, Department of Mathematics, University of Maryland, 2002.
18. D.B. Madan, P. Carr, and E.C. Chang. The variance gamma process and option pricing. *European Finance Review*, 2:79–105, 1998.
19. D.B. Madan and F. Milne. Option pricing with V.G. martingale components. *Mathematical Finance*, 1:39–55, 1991.
20. D.B. Madan and E. Seneta. The Variance-Gamma (V.G.) model for share market returns. *Journal of Business*, 63:511–524, 1990.
21. D.B. Madan and M. Yor. Representing the CGMY and Meixner processes as time changed Brownian motions. manuscript, 2006.
22. D.L. McLeish. A robust alternative to the normal distribution. *Canadian Journal of Statistics*, 10:89–102, 1982.

23. D.L. McLeish, *Monte Carlo Simulation and Finance*. Wiley, 2005.
24. P. Protter. *Stochastic Integration and Differential Equations*, 2nd edition. Springer-Verlag, 2005.
25. C. Ribeiro, and N. Webber. Valuing path-dependent options in the Variance-Gamma model by Monte Carlo with a gamma bridge. *Journal of Computational Finance*, 7:81–100, 2004.
26. K. Sato. *Lévy Processes and Infinitely Divisible Distributions*. Cambridge University Press, 1999.
27. E. Seneta. The early years of the Variance-Gamma process. In this volume, 2007.
28. M. Yor. Some remarkable properties of gamma processes. In this volume, 2007.
29. B. Zhang. *A New Lévy Based Short-Rate Model for the Fixed Income Market and Its Estimation with Particle Filter*. Ph.D. dissertation (directed by Dilip B. Madan), Department of Mathematics, University of Maryland, 2006.

Appendix: Review of Basic Definitions

Here we provide the elementary characterizations of the Wiener, Poisson, and gamma processes. (For further properties on the gamma process, see Marc Yor's contribution [28] in this volume.) Each of these is a Lévy process, that is, a process of independent and stationary increments. We consider only homogeneous Lévy processes, in which case the increments are i.i.d. Thus, one way to differentiate among them is to specify the distribution of the increments. For standard Brownian motion $\{W_t\}$, also known as the Wiener process, the increments are normally distributed with zero mean and variance equal to the size of the increment; that is, for any t,

$$W_{t+\delta} - W_t \sim \mathcal{N}(0, \delta),$$

where $\mathcal{N}(\mu, \sigma^2)$ denotes the normal (Gaussian) distribution with mean μ and variance σ^2. Similarly, the gamma process $\{\gamma_t\}$ has gamma distributed increments; that is, for any t, $\gamma_{t+\delta} - \gamma_t \sim \Gamma(\delta, 1)$, where $\Gamma(\alpha, \beta)$ denotes the gamma distribution with mean $\alpha\beta$ and variance $\alpha\beta^2$. Unlike the Wiener process, the gamma process is discontinuous, and like the Poisson process, it is nondecreasing (since the gamma distribution has support on the positive real line). For the two-parameter gamma process $\{\gamma_t^{(\mu,\nu)}\}$, increments are gamma distributed with the mean μ and variance ν both multiplied by the size of the increment; that is, for any t,

$$\gamma_{t+\delta}^{(\mu,\nu)} - \gamma_t^{(\mu,\nu)} \sim \Gamma(\delta\mu^2/\nu, \nu/\mu).$$

The nondecreasing property makes the gamma distribution suitable as a subordinator (time change). Although the difference-of-gammas representation of VG uses the two-parameter gamma distribution, the time-changed representation of VG uses the one-parameter version (taking $\mu = 1$ in the two-parameter version), denoted here $\{\gamma_t^{(\nu)}\}$, where for any t,

$$\gamma_{t+\delta}^{(\nu)} - \gamma_t^{(\nu)} \sim \Gamma(\delta/\nu, \nu),$$

with a mean equal to the size of the increment, making it most suitable for a time change. This is also the version of the gamma process that is used in the following paper by Marc Yor [28] in this volume.

By the Lévy–Khintchine theorem (cf. [24, 26]), a Lévy process $\{X_t\}$, with finite-variation jump component can be specified by its unique characteristic function

$$\phi_{X_t}(u) = E[e^{iuX_t}] = \exp\left(iuat - u^2b^2t/2 + t\int_{-\infty}^{+\infty}(e^{iux}-1)k(x)dx\right),$$

where $k(x)$ is the Lévy density, a is the drift rate, and b is the diffusion coefficient. The three components correspond to a deterministic drift, a continuous Wiener process, and a pure-jump process, where intuitively the Lévy density is a measure on the arrival rate of different jump sizes. Note that because a Lévy process is infinitely divisible, a single time point (e.g., $t = 1$) suffices to characterize the process, in terms of its marginal distribution at any time. In the VG process, the drift rate (a) and diffusion coefficient (b) are both zero, and the process is characterized by three parameters (θ, σ, ν) that appear in the Lévy density according to (1). The symmetric version of the VG process given by $(\theta = 0)$,

$$X_t = \sigma W_{\gamma_t^{(\nu)}},$$

has mean 0, variance $\sigma^2 t$, skewness 0, and kurtosis $3(1+\nu)$, so the parameter ν gives the excess kurtosis over Brownian motion (which has a kurtosis of 3). In the symmetric version, the difference-of-gammas representation involves two (independent) gamma processes with the same distribution. In the asymmetric VG process, the parameter θ controls skewness, with a negative value giving a fatter (heavier) left tail. Brownian motion can be obtained as a limiting case of the VG process, because $\lim_{\nu\to 0}\gamma_t^{(\nu)} = t$.

Some Remarkable Properties of Gamma Processes

Marc Yor

Laboratoire de Probabilités et Modèles Aléatoires
Université Paris VI, 175 rue du Chevaleret - Boîte courrier 188
75013 Paris, France
& Institut Universitaire de France
deaproba@proba.jussieu.fr

Summary. A number of remarkable properties of gamma processes are gathered in this paper, including realisation of their bridges, absolute continuity relationships, realisation of a gamma process as an inverse local time, and the effect of a gamma process as a time change. Some of them are put in perspective with their Brownian counterparts.

Key words: Gamma process; bridges; absolute continuity; inverse local time; time changes.

1 Acknowledging a Debt

1.1 A Good Fairy

Hélyette Geman played the role of a 'good fairy' in two instances of my research activities:

- The first instance was when Hélyette kept asking me, at the end of 1988, about the price of Asian options, that is, finding a closed-form formula for

$$
E\left[\left(\int_0^T ds \ \exp(B_s + \nu s) - K\right)^+\right], \tag{1}
$$

where $(B_s, s \geq 0)$ is a Brownian motion, $\nu \in \mathbb{R}$, $T > 0$ is the maturity of the option, and K its strike. It took us some time to introduce what now seem to be the 'right' tools to deal with this question, namely the use of Lamperti's representation of $\{\exp(B_s + \nu s), s \geq 0\}$ as a time-changed Bessel process, together with the Laplace transform in T. See [21] for a compendium of a number of papers on this subject.

- The second instance was when Hélyette introduced me to Dilip Madan, around January 1996, when Dilip was visiting Hélyette at the University of Dauphine. This was the beginning of multiple collaborations among the three of us, and sometimes other coauthors; these joint works mostly originated from a constant stream of questions by Dilip, formulated while drinking coffees on rue Soufflot, or at the Hotel Senlis nearby, or ....

I have been deeply impressed with the ease with which Dilip, starting from a finance question that arises as he is trying to develop a new model, is able to raise quite challenging related mathematical questions, involving Lévy processes, semimartingales, and so on.

1.2 Propaganda for Gamma Processes

As is well known, Dilip has been an ardent propagandist, over the years, of the gamma processes, that is, the one-parameter family of subordinators $(\gamma_t^{(m)}, t \geq 0)$, with Lévy measure

$$\frac{dx}{x} \exp(-mx), \ x > 0,$$

whose Lévy–Khintchine representation is given by

$$E[\exp(-\lambda \gamma_t^{(m)})] = \left(1 + \frac{\lambda}{m}\right)^{-t} \tag{2}$$

$$= \exp\left(-t \int_0^\infty \frac{dx}{x} e^{-mx}(1 - e^{-\lambda x})\right). \tag{3}$$

Clearly, from Equation (2), we obtain

$$(\gamma_t^{(m)}, t \geq 0) \stackrel{(\text{law})}{=} \left(\frac{1}{m} \gamma_t, t \geq 0\right),$$

with $\gamma_t = \gamma_t^{(1)}$; hence, there is no loss of generality in assuming $m = 1$, although it may be convenient sometimes to avail oneself of the parameter m.

  In fact, Dilip even prefers the Variance-Gamma (VG) processes [11] and [12], which may be presented either as

$$(\gamma_t^{(m)} - \widetilde{\gamma}_t^{(m)}; t \geq 0),$$

or as

$$(\beta_{\gamma_t^{(m)}}, t \geq 0),$$

where $(\beta_u, u \geq 0)$ is a Brownian motion independent of $(\gamma_t^{(m)}, t \geq 0)$ and $(\widetilde{\gamma}_t^{(m)}, t \geq 0)$ is an independent copy of $(\gamma_t^{(m)}, t \geq 0)$.

  This festschrift for Dilip's 60th birthday seems to be a good opportunity to gather a few remarkable properties of the gamma processes, which make them worthy companions of Brownian motion.

2 Brownian Bridges and Gamma Bridges

I begin with a somewhat informal statement. Among Lévy processes, I only know of two processes for which one can present an explicit construction of their bridges in terms of the original Lévy process: Brownian motions (with or without drift) and gamma processes.

2.1 Brownian Bridges

Letting $(B_u, u \geq 0)$ denote a standard Brownian motion, then the process $(B_u - (u/t) B_t, u \leq t)$ is independent of B_t, hence of $(B_v, v \geq t)$. Consequently, as is well known, a Brownian bridge of length t, starting at 0, and ending at y at time t, may be obtained:

$$b_y^{(t)}(u) = \left(B_u - \frac{u}{t} B_t \right) + \frac{u}{t} y, \quad u \leq t.$$

2.2 Gamma Bridges

It follows from the beta-gamma algebra (see, e.g., Dufresne [6] and Chaumont and Yor [3]) that if $(\gamma_t^{(m)}, t \geq 0)$ denotes a gamma process with parameter m, then $(\gamma_u^{(m)}/\gamma_t^{(m)}, u \leq t)$ is independent of $\gamma_t^{(m)}$, hence of $(\gamma_v^{(m)}, v \geq t)$. Moreover, it is obvious that the law of $(\gamma_u^{(m)}/\gamma_t^{(m)}, u \leq t)$ does not depend on m. Hence, a gamma bridge of length t, starting at 0, and ending at $a > 0$, may be obtained by taking $(a\,\gamma_u/\gamma_t, u \leq t)$. The process $D_u^{(t)} \overset{\text{def}}{=} \gamma_u/\gamma_t, u \leq t$, is often called the Dirichlet process with parameter t, and a number of studies have been devoted to the laws of means of Dirichlet processes; that is, the laws of $\int_0^t h(u) dD_u^{(t)}$, for deterministic functions h (see, e.g., [1] and [2]).

2.3 The Filtration of Brownian Bridges

Denote by $(\mathcal{B}_t, t \geq 0)$ the natural filtration of $(B_t, t \geq 0)$, and let

$$\mathcal{G}_t = \sigma\{b_y^{(t)}(u), u \leq t\} \equiv \sigma\{b_o^{(t)}(u), u \leq t\} .$$

It is easily shown that $(\mathcal{G}_t, t \geq 0)$ is an increasing family of σ-fields, which deserves to be called the filtration of Brownian bridges. It has been proven (Jeulin and Yor [8], Yor [20] Chapter 1) that its natural filtration is that of a Brownian motion; more precisely, the process $(B_t - \int_0^t (ds/s) B_s, t \geq 0)$ is a Brownian motion, whose natural filtration is $\{\mathcal{G}_t\}$.

2.4 The Filtration of Gamma Bridges

Inspired by the result just recalled for the filtration of Brownian bridges $\{\mathcal{G}_t\}$, it is natural to ask whether the filtration

$$\mathcal{H}_t = \sigma\left\{\frac{\gamma_u}{\gamma_t}, u \leq t\right\}, \quad t \geq 0,$$

of gamma bridges is also the natural filtration of a gamma process. This question has been solved by Emery and Yor [7].

Theorem 1. *For each $s > 0$, the formula*

$$\int_{u_s(x)}^{\infty} \frac{e^{-z}dz}{z} = \int_x^{\infty} \frac{dy}{y(1+y)^s} \quad (< +\infty, \ for \ x > 0),$$

defines a bijection $u_s : \mathbb{R}^+ \to \mathbb{R}^+$. The sum $\gamma_t^* = \sum_{s \leq t} u_s \left(\Delta\gamma_s/\gamma_{s-}\right), t \geq 0$ is a.s. convergent and defines a gamma process γ^* which generates the filtration $(\mathcal{H}_t, t \geq 0)$.

2.5 Absolute Continuity Relationships

Closely related to the bridges constructions recalled in Sections 2.3 and 2.4 are the following absolute continuity relationships.

Proposition 1. *(i) For every measurable functional F, and every $a > 0$,*

$$E[F(a\gamma_u, u \leq t)] = E\left[F(\gamma_u, u \leq t)\frac{\exp\left(-\left(\frac{1}{a}-1\right)\gamma_t\right)}{a^t}\right].$$

(ii) For every measurable functional F, for every $\nu \in \mathbb{R}$,

$$E[F(B_u + u\nu; u \leq t)] = E\left[F(B_u, u \leq t)\exp\left(\nu B_t - \frac{\nu^2 t}{2}\right)\right].$$

Proof. We only give it for the gamma process, as the proof for Brownian motion is well known.

$$\begin{aligned}
E[F(a\gamma_u, u \leq t)] &= E\left[F\left(a\frac{\gamma_u}{\gamma_t}\gamma_t; u \leq t\right)\right] \\
&= \frac{1}{\Gamma(t)}\int_0^{\infty} dx\, x^{t-1}e^{-x}E\left[F\left(a\frac{\gamma_u}{\gamma_t}x; u \leq t\right)\right] \\
&= \frac{1}{\Gamma(t)}\int_0^{\infty} \frac{dy\, y^{t-1}}{a^t}e^{-y/a}E\left[F\left(\frac{\gamma_u}{\gamma_t}y; u \leq t\right)\right] \\
&= \frac{1}{\Gamma(t)}\int_0^{\infty} dy\, y^{t-1}e^{-y}\left(\frac{e^{-y(1/a-1)}}{a^t}\right)E\left[F\left(\frac{\gamma_u}{\gamma_t}y; u \leq t\right)\right] \\
&= E\left[F(\gamma_u; u \leq t)\frac{e^{-\gamma_t(1/a-1)}}{a^t}\right].
\end{aligned}$$

$\square$

2.6 Brownian Motion and the Gamma Process as Special Harnesses

Let $X = (X_t, t \geq 0)$ be a Lévy process such that $E(|X_t|) < \infty$ for every $t \geq 0$. It is known that the following conditional expectation result holds,

$$E\left[\frac{X_c - X_b}{c - b} \bigg| \mathcal{F}_{a,d}^X\right] = \frac{X_d - X_a}{d - a}, \quad a < b < c < d, \tag{4}$$

where $\mathcal{F}_{a,d}^X$ denotes the σ-field generated by $(X_u, u \leq a)$ and $(X_v, v \geq d)$. A process that satisfies the property (4) is called a harness, following Hammersley's terminology (see e.g., [3] and [13]). The Brownian motion $B = (B_t)$ and the gamma process $\gamma = (\gamma_t)$ are special harnesses in the following sense.

(i) For $X = B$, $\dfrac{X_c - X_b}{c - b} - \dfrac{X_d - X_a}{d - a}$ is independent of $\mathcal{F}_{a,d}^X$,

(ii) For $X = \gamma$, $\dfrac{X_c - X_b}{X_d - X_a}$ is independent of $\mathcal{F}_{a,d}^X$.

It is quite plausible that these properties characterize, respectively, Brownian motion and the gamma process, say, for example, among Lévy processes. Indeed, in the same spirit, we may recall D. Williams ([19] and unpublished manuscript), who asserts that the only continuous harnesses are (essentially) Brownian motion with drift.

3 Space–Time Harmonic Functions for Brownian Motion and the Gamma Process

If $X = (X_t, t \geq 0)$ is a real-valued stochastic process, we say that $f : \mathbb{R} \times \mathbb{R}^+$ is a space–time harmonic function if

$$\{f(X_t, t), t \geq 0\}$$

is a martingale (with respect to the filtration of X). For now, let X denote either Brownian motion or the gamma process (or the VG process). In this section, we describe the following.

1. Polynomials $P(x, t)$ in both variables x and t, which are space–time harmonic functions of X
2. All positive space–time harmonic functions for X, a Brownian motion, or a gamma process

3.1 Polynomial Space–Time Harmonic Functions

We start, for Brownian motion (B_t) and the gamma process (γ_t), with the exponential martingales

$$\exp\left(\lambda\beta_t - \frac{\lambda^2 t}{2}\right) \quad \text{and} \quad (1+\lambda)^t \exp(-\lambda\gamma_t),$$

which have appeared naturally in the expressions of the Radon–Nikodym densities in Proposition 1. From the classical series expansions

$$\exp\left(\lambda x - \frac{\lambda^2 t}{2}\right) = \sum_{n=0}^{\infty} \frac{\lambda^n}{n!} H_n(x,t),$$

$$(1+\lambda)^u \exp(-\lambda g) = \sum_{n=0}^{\infty} \frac{\lambda^n}{n!} \widetilde{C}_n(u,g),$$

where

$$H_n(x,t) = t^{n/2} h_n\left(\frac{x}{\sqrt{t}}\right),$$

with (h_n) the classical Hermite polynomials, and where $\widetilde{C}_n(u,g)$ are the so-called monic Charlier polynomials (see, e.g., Schoutens [15, p. 66]), we obtain the following martingales,

$$(H_n(B_t,t), t \geq 0) \quad \text{and} \quad (\widetilde{C}_n(u,\gamma_u), u \geq 0).$$

It is noteworthy that the latter gamma martingales admit counterparts for the standard Poisson process $(N_t, t \geq 0)$, with the roles of time and space interchanged. Thus, $((1+\lambda)^{N_t}\exp(-\lambda t), \ t \geq 0)$ and $(\widetilde{C}_n(N_t,t), t \geq 0)$ are Poisson martingales. That this results from the corresponding ones for the gamma process may be (partly) understood from the fact that the sequence $T_n \equiv \inf\{t : N_t = n\}, n = 1, 2, \ldots$, is the trace of a gamma process $(\gamma_u, u \geq 0)$ on the integers; that is, the sequences $(T_n)_{n\geq1}$ and $(\gamma_n)_{n\geq1}$ are identical in law.

3.2 On Positive Space–Time Harmonic Functions

We now show the following.

Theorem 2. *(i) Every $\mathbb{R}^+$-valued space–time harmonic function of $(B_t, t \geq 0)$ such that $f(0,0) = 1$, may be written as*

$$f(x,t) = \int_{-\infty}^{\infty} d\mu(\lambda) \exp\left(\lambda x - \frac{\lambda^2 t}{2}\right), \tag{5}$$

for some probability measure $d\mu(\lambda)$ on $\mathbb{R}$.
(ii) Every $\mathbb{R}^+$-valued space–time harmonic function of $(\gamma_t, t \geq 0)$ may be written as

$$f(g,t) = \int_0^{\infty} d\nu(a) \frac{1}{a^t} \exp\left(-\left(\frac{1}{a} - 1\right)g\right), \tag{6}$$

for some probability measure $d\nu(a)$ on $\mathbb{R}^+$.

Proof. We consider the new probabilities induced by the Radon–Nikodym densities $f(X_t, t)$, with respect to the law of Brownian motion (respectively, gamma process), which we denote P^0. Thus, we have

$$E[F(X_s, s \leq t)] = E^{(0)}[F(X_s, s \leq t)f(X_t, t)]. \tag{7}$$

To prove the desired result, we show that under P, $(X_s, s \geq 0)$ may be written as

(i) $X_s = s\Lambda + B_s$, $s \geq 0$,

where Λ is a real-valued r.v. independent of the Brownian motion $(B_t, t \geq 0)$;

(ii) $X_s = A\gamma_s$, $s \geq 0$,

where A is a $\mathbb{R}^+$-valued r.v. independent of the gamma process $(\gamma_s, s \geq 0)$.

Then, denoting μ (respectively, ν) the law of Λ (respectively, A), we deduce from Proposition 1 in the Brownian case that

$$E[F(X_s, s \leq t)] = \int d\mu(\lambda) E\left[F(B_s, s \leq t) \exp\left(\lambda B_t - \frac{\lambda^2 t}{2}\right)\right]. \tag{8}$$

It now remains to use Fubini's theorem on the RHS, and to compare the expression thus obtained with the RHS of (7). A similar argument holds in the gamma case.

It now remains to prove (i) and (ii). This follows easily enough from the facts that

- In the Brownian case, the two-parameter process

$$\left(\frac{X_u}{u} - \frac{X_v}{v} ; 0 < u < v \leq t\right)$$

is distributed under P as under P^0, and is independent of X_t,
- In the gamma case, the two-parameter process

$$\left(\frac{X_u}{X_v} ; 0 < u < v \leq t\right)$$

is distributed under P as under P^0, and is independent of X_t. $\square$

4 The Gamma Process as Inverse Local Time of a Diffusion

Roughly, it is a consequence of Krein's theory of strings that any subordinator without drift, and whose Lévy measure $\nu(dx)$ is of the form

$$\nu(dx) = dx \left(\int \mu(dy) \exp(-yx)\right), \tag{9}$$

for $\mu(dy)$ a suitable Radon measure on $\mathbb{R}^+$, may be obtained as the inverse local time of a diffusion (see, e.g., Kotani–Watanabe [10] and Knight [9] for precise statements). The Lévy measure $(dx/x)\exp(-mx)$ of the gamma process with parameter m is a particular instance of (9), with

$$\mu(dy) = dy\, \mathbf{1}_{\{y \geq m\}}.$$

Hence, it is natural to ask for the description of a diffusion whose inverse local time at zero is the gamma process $\gamma^{(m)}$. This question was solved by Donati-Martin and Yor [4]; see also [5] for further examples.

Theorem 3. *Let $m > 0$. The diffusion on $[0, \infty)$, with 0 instantaneously reflecting, and infinitesimal generator*

$$\mathcal{L}^{m\downarrow} = \frac{1}{2}\frac{d^2}{dx^2} + \left(\frac{1}{2x} + \sqrt{2m}\frac{K_0'}{K_0}\left(\sqrt{2mx}\right)\right)\frac{d}{dx},$$

where K_0 denotes the Bessel–McDonald function with index 0, admits $(\gamma_t^{(m)}, t \geq 0)$ as inverse local time at 0.

5 The Gamma Process Time Changes Many 'Erratic' Processes into Processes with Bounded Variation

5.1 Why Does the VG Process Have Bounded Variation?

The following statement gives a partial answer to a more general question of D. Madan.

Proposition 2. *Consider a process $(X_u, u \geq 0)$ such that there exist two constants $\mathcal{C}$ and $\alpha > 0$ for which*

$$E[|X_u - X_v|] \leq \mathcal{C}|u - v|^\alpha \quad \text{for all } u, v \geq 0. \tag{10}$$

Then the process $(Y_t = X_{\gamma_t}, t \geq 0)$, where $(\gamma_t, t \geq 0)$ is a gamma process independent of $(X_u, u \geq 0)$, has bounded variation on any finite interval $[0, T]$. More precisely,

$$E\left[\int_0^T |dY_s|\right] \leq \mathcal{C}\Gamma(\alpha)T < \infty.$$

Proof. It follows from our hypothesis (10) that for any pair (s, t), with $s < t$,

$$E[|Y_t - Y_s|] \leq \mathcal{C}E[(\gamma_t - \gamma_s)^\alpha].$$

Now, the gamma process satisfies

$$E[(\gamma_t - \gamma_s)^\alpha] = E[(\gamma_{t-s})^\alpha] = \frac{\Gamma(\alpha + t - s)}{\Gamma(t - s)}.$$

Thus,

$$E[(\gamma_{s+h} - \gamma_s)^\alpha] = \frac{\Gamma(\alpha+h)h}{\Gamma(1+h)} \underset{(h\to 0)}{\sim} \Gamma(\alpha)h.$$

As a consequence of these estimates, if $(\tau_n)_{n\in\mathbb{N}}$ is a refining sequence of subdivisions of $[0, T]$, then the increasing sequence

$$E\left[\sum_{\tau_n} \left|Y_{t_{k+1}} - Y_{t_k}\right|\right], \quad n \geq 1,$$

is majorized by $C\Gamma(\alpha)T$, which implies the desired result.   □

In fact, the property that was crucial in the proof of Proposition 2 is that for every $\alpha > 0$, the gamma process is of α-bounded variation on compact sets (in time).

The statement of Proposition 2 deserves a number of further comments.

(i) The condition (10) is not sufficient to yield the existence of a continuous version of (X_t), unless $\alpha > 1$, in which case Kolmogorov's continuity criterion applies. However, if (X_t) is a centred Gaussian process, then for any $p > 1$, there exist universal constants $0 < c_p < C_p < \infty$ such that

$$c_p(E[|X_u - X_v|^p])^{1/p} \leq E[|X_u - X_v|] \leq C_p E[|X_u - X_v|^p]^{1/p}.$$

Thus, if $\alpha > 0$ is given such that (10) is satisfied, then there exists $p > 1$ sufficiently large such that $p\alpha > 1$; hence,

$$E[|X_u - X_v|^p] \leq K_p|u - v|^{p\alpha},$$

and (X_u) admits a continuous version, which is locally Hölder with exponent β, for any $\beta < \alpha$.

(ii) In the same vein as in (i), note that if $(B_u^H, u \geq 0)$ is a fractional Brownian motion with Hurst index H, that is,

$$E[(B_u^H - B_v^H)^2] = C(u - v)^{2H},$$

then Proposition 2 applies, and $(B_{\gamma_t}^H, t \geq 0)$ has bounded variation.

(iii) In the particular case when $H = 1/2$, then $\{B^{1/2} \equiv (B_u, u \geq 0)\}$ is a Brownian motion, and there is the identity in law

$$(\gamma_t - \gamma_t'; t \geq 0) \overset{(\text{law})}{=} (\sqrt{2}B_{\gamma_t'}; t \geq 0),$$

where γ and γ' are two independent copies, and B is independent of γ'.

(iv) Another interesting question is whether for any $H \in (0, 1)$, there exist two independent copies (γ_t^H) and $(\tilde{\gamma}_t^H)$ such that

$$(B_{\gamma_t}^H, t \geq 0) \overset{(\text{law})}{=} (\gamma_t^H - \tilde{\gamma}_t^H, t \geq 0) \quad ?$$

5.2 Some Occurrences of Gamma Time Changes with Brownian Motion and Markov Processes

Here we consider $(X_t, t \geq 0)$ a real-valued diffusion such that 0 is regular for itself; we denote $(L_t,\ t \geq 0)$ a choice of the local time at 0, and $\tau_\ell = \inf\{t \geq 0 : L_t > \ell\}$ the inverse local time. We assume that $L_\infty = \infty$ a.s. It is well known that if $A_t = \int_0^t ds\, f(X_s)$, for $f \geq 0$, Borel, then $(A_{\tau_\ell},\ \ell \geq 0)$ is a subordinator. We note here, following Salminen et al. [14], that if $(\psi(\theta),\ \theta \geq 0)$ denotes the Bernstein function associated with $(\tau_\ell,\ \ell \geq 0)$, then for $\mathbf{e}_\theta$ an exponential variable with parameter θ independent of $(X_t, t \geq 0)$, the following hold,

$$(i)\ \ L_{\mathbf{e}_\theta} \overset{(law)}{=} \mathbf{e}_{\psi(\theta)};$$
$$(ii)\ \ P(A_{g_{\mathbf{e}_\theta}} \in du | L_{\mathbf{e}_\theta} = \ell) = P(A_{\tau_\ell} \in du;\ \exp(-\theta\tau_\ell + \psi(\theta)\ell)),$$

where $g_t = \sup\{s < t : X_s = 0\}$. As a consequence, if $(\gamma_t,\ t \geq 0)$ denotes a gamma process independent of $(X_u,\ u \geq 0)$, then $(A_{g_{\gamma_t}},\ t \geq 0)$ is a subordinator time changed by an independent gamma process.

6 Conclusion

Although I have tried hard to present the main remarkable properties of gamma processes, I still feel that – due perhaps to the richness of the subject – I have not quite succeeded, and would like to refer the reader to the papers by N. Tsilevich and A. Vershik ([16–18]) which may complete the picture somewhat. In one sentence, these papers interpret gamma processes as '0-stable' processes, which may be obtained as the limit, when $\alpha \to 0$, of tilted α-stable subordinators; links with Poisson-Dirichlet measures are also established.

References

1. D.M. Cifarelli and E. Melilli. Some new results for Dirichlet priors. *Ann. Stat.*, 28:1390–1413, 2000.
2. D.M. Cifarelli and E. Regazzini. Distribution functions of means of a Dirichlet process. *Ann. Stat.*, 18:429–442, 1990.
3. L. Chaumont and M. Yor. *Exercises in Probability.* Cambridge University Press, 2003.
4. C. Donati-Martin and M. Yor. Some explicit Krein representations of certain subordinators, including the gamma process. *Publ. RIMS*, Kyoto, 42(4):879–895, 2006.
5. C. Donati-Martin and M. Yor. Further examples of explicit Krein representations of certain subordinators. *Publ. RIMS*, Kyoto, 43(4), 2007.
6. D. Dufresne. Algebraic properties of beta and gamma distributions, and applications. *Adv. in App. Math.*, 20(3):285–299, 1998.

7. M. Emery and M. Yor. A parallel between Brownian bridges and gamma bridges. *Publ. RIMS*, Kyoto Univ., 40(3):669–688, 2004.

8. T. Jeulin and M. Yor. Filtration des ponts Browniens et équations différentielles stochastiques linéaires. *Séminaire de Probabilités XXIV*, Springer LNM 1426, 227–265, 1990.

9. F. Knight. Characterization of the Lévy measure of inverse local times of gap diffusion. *Seminar on Stochastic Processes*, eds. E. Cinlar, K. Chung, and R. Getoor, Birkhäuser, 53–78, 1981.

10. S. Kotani and S. Watanabe. Krein's spectral theory of strings and generalized diffusion process. *Functional Analysis in Markov Processes* (Katata/Kyoto 1981), Springer LNM 923, 235–259, 1982.

11. D.B. Madan, P. Carr, and E.C. Chang. The variance gamma process and option pricing. *European Finance Review*, 2:79–105, 1998.

12. D. Madan and E. Seneta. The variance gamma (V.G.) model for share market returns. *J. of Business*, 63(4):511–524, 1990.

13. R. Mansuy and M.Yor. Harnesses, Lévy bridges and Monsieur Jourdain. *Stoch. Proc. and Applications*, 115(2):329–338, February 2005.

14. P. Salminen, P. Vallois, and M. Yor. On the excursion theory for linear diffusions. *Jap. J. Maths*, in honor of K. Itô, 2(1)97–127, 2007.

15. W. Schoutens. *Stochastic Processes and Orthogonal Polynomials*. LNS 146, Springer, 2000.

16. N. Tsilevich and A. Vershik. Quasi-invariance of the gamma process and multiplicative properties of the Poisson-Dirichlet measures. *C. R. Acad. Sci.*, Paris, Sér. I, 329:163–168, 1999.

17. N. Tsilevich, A. Vershik, and M. Yor. An infinite-dimensional analogue of the Lebesgue measure and distinguished properties of the gamma process. *J. Funct. Anal.*, 185:274–296, 2001.

18. A. Vershik and M. Yor. Multiplicativité du processus gamma et étude asymptotique des lois stables d'indice α, lorsque α tend vers 0. Prépublication 289, Laboratoire de Probabilités, Paris, 1995.

19. D. Williams. Some basic theorems on harnesses. *Stochastic Analysis (A Tribute to the Memory of Rollo Davidson)*, eds. D. Kendall and H. Harding, Wiley, 1973.

20. M. Yor. *Some Aspects of Brownian Motion, Part I: Some Special Functionals*. ETH Zürich, Birkhäuser, 1992.

21. M. Yor. *On Exponential Functionals of Brownian Motion and Related Processes*. Springer, 2001.

A Note About Selberg's Integrals in Relation with the Beta-Gamma Algebra

Marc Yor

Laboratoire de Probabilités et Modèles Aléatoires
Université Paris VI, 175 rue du Chevaleret - Boîte courrier 188
75013 Paris, France
& Institut Universitaire de France
deaproba@proba.jussieu.fr

Summary. To prove their formulae for the moments of the characteristic polynomial of the generic matrix of $U(N)$, Keating and Snaith [8] (see also Keating [7]) use Selberg's integrals as a 'black box.' In this note, we point out some identities in law which are equivalent to the expressions of Selberg's integrals and which involve beta, gamma, and normal variables. However, this is a mere probabilistic translation of Selberg's results, and does not provide an independent proof of them. An outcome of some of these translations is that certain logarithms of (Vandermonde) random discriminants are self-decomposable, which hinges on the self-decomposability of the logarithms of the beta (a, b) $(2a + b \geq 1)$ and gamma $(a > 0)$ variables. Such self-decomposability properties have been of interest in some joint papers with D. Madan.

Key words: Selberg's integrals; beta-gamma algebra; self-decomposable distributions; Vandermonde determinant; characteristic polynomial.

1 Introduction and Statement of Results

Recall that for $a, b > 0$, a beta random variable $\beta_{a,b}$ with parameters a and b is distributed as

$$P(\beta_{a,b} \in dx) = \frac{dx \, x^{a-1}(1-x)^{b-1}}{B(a,b)}, \quad 0 < x < 1, \tag{1}$$

whereas a gamma random variable γ_a with parameter $a > 0$ is distributed as

$$P(\gamma_a \in dt) = \frac{dt}{\Gamma(a)} \, t^{a-1} \exp(-t), \quad t > 0. \tag{2}$$

The beta-gamma algebra mentioned in the title of this paper may be summarized as follows. If γ_a and γ_b are two independent gamma variables, then there is the identity in law:

$$(\gamma_a, \gamma_b) \overset{(\text{law})}{=} (\beta_{a,b}, (1 - \beta_{a,b}))\gamma_{a+b}, \tag{3}$$

where on the RHS, the variables $\beta_{a,b}$ and γ_{a+b} are independent. For various extensions of this fact, see Dufresne [4].

In the following, we also need to use the Mellin transforms of the laws of $\beta_{a,b}$ and γ_a, which are given by, for any $m \geq 0$:

$$E[(\gamma_a)^m] = \frac{\Gamma(a+m)}{\Gamma(a)},$$

$$E[(\beta_{a,b})^m] = \frac{\Gamma(a+m)}{\Gamma(a)} \Big/ \frac{\Gamma(a+b+m)}{\Gamma(a+b)}.$$

The second identity follows from the first one, together with (3).

The celebrated Selberg integrals (see, e.g., [1, Theorem 8.1.1]) may be expressed partly in probabilistic terms as follows. For any integer $n \geq 2$ and any $\gamma > 0$,

$$E[(\Delta(\beta_{a,b}^{(i)}; 1 \leq i \leq n))^{2\gamma}] \prod_{j=1}^{n} \left(\frac{\Gamma(a+b+(n+j-2)\gamma)\Gamma(1+\gamma)}{\Gamma(a+b)} \right)$$

$$= \prod_{j=1}^{n} \left(\frac{\Gamma(a+(j-1)\gamma)}{\Gamma(a)} \frac{\Gamma(b+(j-1)\gamma)}{\Gamma(b)} \right) \Gamma(1+j\gamma), \tag{4}$$

where $\Delta(x^{(i)}; 1 \leq i \leq n) = \prod_{i<j}(x^{(i)} - x^{(j)})$ is the value of the Vandermonde determinant associated with the $x^{(i)}$s.

In order to present below an identity in law involving (and characterizing the law of) $(\Delta(\beta_{a,b}^{(i)}; 1 \leq i \leq n))^2$, we need the following.

Lemma 1. *Let* **e** *be a standard exponential variable (with expectation 1), so that* $\mathbf{e} \overset{(law)}{=} \gamma_1$. *Then, for any* $\mu \in (0,1)$,

$$\mathbf{e}^{1/\mu} \overset{(law)}{=} \frac{\mathbf{e}}{T_\mu}, \tag{5}$$

where T_μ *denotes the one-sided stable variable with index* μ, *whose law is characterized by*

$$E[\exp(-\lambda T_\mu)] = \exp(-\lambda^\mu), \quad \lambda \geq 0.$$

Moreover, in (5), T_μ *is assumed to be independent of* **e**.

A proof and extensions of this lemma are presented in Chaumont and Yor [3, Exercise 4.19]. The identity in law (5) goes back at least to Shanbhag and Sreehari [10].

Comment about notation for identities in law. Throughout this note, it should be understood that when an identity in law such as

$$\Phi(X, Y) \overset{(law)}{=} \Psi(Z, T)$$

is presented, with Φ and Ψ two functions, then the r.v.'s X and Y on one hand, and Z and T, are assumed to be independent. Moreover, these variables are often multidimensional; then their components are assumed to be independent.

We now present three identities in law, which involve, respectively,

$$\Delta(\beta_{a,b}^{(i)}; 1 \le i \le n), \quad \Delta(\gamma_a^{(i)}; 1 \le i \le n), \quad \Delta(\mathcal{N}^{(i)}; 1 \le i \le n),$$

where $\mathcal{N}^{(i)}$, $i = 1, \ldots, n$, are independent standard Gaussian variables.

Proposition 1. *The following identity in law holds,*

$$(\Delta(\beta_{a,b}^{(i)}; 1 \le i \le n))^2 \prod_{j=1}^{n} (\gamma_{a+b}^{(j)})^{n+j-2}$$

$$\stackrel{(law)}{=} \prod_{j=2}^{n} \left\{ (\beta_{a,b}^{(j)}(1 - \beta_{a,b}^{(j)}))^{j-1} (\gamma_{a+b}^{(j)})^{2(j-1)} \frac{1}{T_{1/j}} \right\}. \tag{6}$$

Proposition 2. *The following identity in law holds,*

$$(\Delta(\gamma_a^{(i)}; 1 \le i \le n))^2 \stackrel{(law)}{=} \prod_{j=2}^{n} (\gamma_a^{(j)})^{j-1} \frac{1}{T_{1/j}}. \tag{7}$$

Proposition 3. *The following identity in law holds,*

$$(\Delta(\mathcal{N}^{(i)}; 1 \le i \le n))^2 \stackrel{(law)}{=} \prod_{j=2}^{n} \left(\frac{1}{T_{1/j}} \right). \tag{8}$$

Consequently,

$$E[(\Delta(\mathcal{N}^{(i)}; 1 \le i \le n))^2] = \prod_{j=2}^{n} (j!). \tag{9}$$

The two following identities in law are immediate consequences of (7) and (8).

Corollary 1. *For any $a, b > 0$,*

$$(\Delta(\gamma_a^{(i)}; 1 \le i \le n))^2 \stackrel{(law)}{=} \left(\prod_{j=2}^{n} (\beta_{a,b}^{(j)})^{j-1} \right) (\Delta(\gamma_{a+b}^{(i)}; 1 \le i \le n))^2, \tag{10}$$

$$(\Delta(\gamma_a^{(i)}; 1 \le i \le n))^2 \stackrel{(law)}{=} \left(\prod_{j=2}^{n} (\gamma_a^{(j)})^{j-1} \right) (\Delta(\mathcal{N}^{(i)}; 1 \le i \le n))^2. \tag{11}$$

One may note the parenthood between (10) and the identity in law,

$$\gamma_a \stackrel{(law)}{=} \beta_{a,b} \gamma_{a+b},$$

which is the main ingredient of the beta-gamma algebra, already presented in a slightly different manner in (3).

2 Proofs of the Three Propositions

Proof (of Proposition 1). In order to prove the identity in law (6), we first show the following,

$$(\Delta(\beta_{a,b}^{(i)}; 1 \leq i \leq n))^2 \prod_{j=1}^{n} (\gamma_{a+b}^{(j)})^{n+j-2} \mathbf{e}^{(j)}$$

$$\stackrel{(\text{law})}{=} \prod_{j=1}^{n} \left\{ (\beta_{a,b}^{(j)}(1 - \beta_{a,b}^{(j)}))^{j-1} (\gamma_{a+b}^{(j)})^{2(j-1)} (\mathbf{e}^{(j)})^j \right\}. \tag{12}$$

Indeed, assuming that (12) holds, we then use the identity (5) to write

$$(\mathbf{e}^{(j)})^j \stackrel{(\text{law})}{=} \frac{\mathbf{e}^{(j)}}{T_{1/j}},$$

and we then proceed to the simplification on both sides of (12) of all the variables $\mathbf{e}^{(j)}$. Thus, it now remains to prove (12), or the equivalent identity in law (i.e., (12)) with its RHS written as

$$\prod_{j=1}^{n} \{ (\gamma_a^{(j)} \gamma_b^{(j)})^{j-1} (\mathbf{e}^{(j)})^j \}. \tag{13}$$

Now, in order to prove the identity in law between the LHS of (12) and (13), it suffices to show the equality of the Mellin transforms of both sides, which boils down to (4). □

We now indicate that Proposition 2 may be deduced from Proposition 1, thanks to an application of the law of large numbers, whereas Proposition 3 follows from Proposition 2, thanks to an application of the central limit theorem.

Proof (of Proposition 2). On the LHS of (6), we multiply every $\beta_{a,b}^{(i)}$ by b, and divide every $\gamma_{a+b}^{(j)}$ by b (or $a+b$). Then, letting $b \to \infty$, we obtain that the LHS of (6) converges in law towards the LHS of (7). The operation we have just effected on the LHS of (6) translates, on the RHS of (6), into the replacement of $\beta_{a,b}^{(j)}(1 - \beta_{a,b}^{(j)})$ by $b\beta_{a,b}^{(j)}(1 - \beta_{a,b}^{(j)})$ and of $\gamma_{a+b}^{(j)}$ by $(\gamma_{a+b}^{(j)})/b$. Letting $b \to \infty$, this RHS of (6), so modified, converges in law towards the RHS of (7). □

Proof (of Proposition 3). We start from (7), where, on the LHS, we replace $\gamma_a^{(i)}$ by $(\gamma_a^{(i)} - a)/\sqrt{a}$, whereas on the right-hand side, we may replace $\gamma_a^{(i)}$ by $(\gamma_a^{(i)})/a$. Then, letting $a \to \infty$, and applying the central limit theorem on the LHS, and the law of large numbers on the RHS, we obtain (8). The identity in law (9) follows from (8) and (5). (I only added it here because it is easy to memorize, and it has been given a combinatorial proof in [5].) □

3 Comments and Further References

Some of the results reported above are closely related with those of Lu and Richards [9], as already mentioned in the Comments about Exercise 4.19 in Chaumont and Yor [3].

If we compare the identities in law (6), (7), and (8), we find that in the first case of (6), the term $(\Delta(\beta_{a,b}^{(i)}; 1 \le i \le n))^2$ needs to be multiplied by a certain independent r.v., say $X_{a,b}^{(n)}$, to obtain an identity in law with the RHS of (6), which we denote $Y_{a,b}^{(n)}$, whereas for the two other cases (7) and (8), the square of the Vandermonde determinant stands alone on the RHS of these identities in law. Thus, coming back to (6), it seems reasonable to ask whether one might find $Z_{a,b}^{(n)}$ such that

$$Y_{a,b}^{(n)} \overset{\text{(law)}}{=} X_{a,b}^{(n)} Z_{a,b}^{(n)},$$

in which case we could conclude with the identity in law between $(\Delta(\beta_{a,b}^{(i)}; i \le n))^2$ and $Z_{a,b}^{(n)}$. I have not been able to solve this question so far, but it is clearly closely related to the self-decomposability of the variables $\log(\gamma_a)$ and $\log(1/\beta_{a,b})$, which we discuss below, before attempting to simplify the identity in law (6).

As a consequence of Propositions 2 and 3, it may be pointed out that the variables

$$\log|\Delta(\gamma_a^{(i)}; 1 \le i \le n)|, \quad \log|\Delta(\mathcal{N}^{(i)}; 1 \le i \le n)|, \tag{14}$$

are infinitely divisible, and even self-decomposable, which follows from the well-known results.

Proposition 4. *(a) For any $a > 0$, the characteristic function of $(-\log \gamma_a)$ admits the Lévy–Khintchine representation:*

$$E[\exp iv(-\log \gamma_a)] = \exp\left\{ iv(-\psi(a)) + \int_0^\infty \frac{ds\, e^{-as}}{s(1-e^{-s})} (e^{ivs} - 1 - ivs) \right\},$$

$v \in \mathbb{R}$, where $\psi(a) = \Gamma'(a)/\Gamma(a)$.
(b) For any $a, b > 0$, the Laplace transform of $\log(1/\beta_{a,b})$ admits the Lévy–Khintchine representation:

$$E[\exp(-\lambda \log(1/\beta_{a,b}))] = \exp\left\{ -\int_0^\infty ds\, \frac{e^{-as}(1-e^{-bs})}{s(1-e^{-s})} (1 - e^{-\lambda s}) \right\}.$$

Proof (of Proposition 4).
(a) is a well-known result; see, for example, Carmona et al. [2] and Gordon [6].
(b) follows from (a), thanks to the identity

$$-\log(\gamma_a) \overset{\text{(law)}}{=} \log(1/\beta_{a,b}) - \log(\gamma_{a+b}). \qquad \square$$

Corollary 2. *(a) For any $a > 0$, $\log \gamma_a$ is self-decomposable.*
(b) For any $a, b > 0$, $\log(1/\beta_{a,b})$ is self-decomposable if and only if $2a + b \geq 1$.

Proof. In both cases, the self-decomposability property boils down to the property that, if $(h(s))/s$ denotes the Lévy density of the variable in question, then h is decreasing on $\mathbb{R}_+$. Now, with obvious notation, we have

$$h_a(s) = \frac{e^{-as}}{1 - e^{-s}} \equiv k_a(e^{-s}),$$

$$h_{a,b}(s) = \frac{e^{-as}(1 - e^{-bs})}{1 - e^{-s}} \equiv k_{a,b}(e^{-s}).$$

Thus, the result follows from the elementary facts that k_a is increasing, whereas $k_{a,b}$ is increasing if and only if $2a + b \geq 1$. $\qquad\square$

Comment and reference. The results for Corollary 2 may be found, for example, in Steutel and van Harn's book [11], respectively, in Example 9.18, p. 322 and Example 12.21, p. 422.

To prove the result concerning (14), we also need the self-decomposability property of $\log(1/T_\mu)$ (see, e.g., (5)).

Proposition 5. *Let $0 < \mu < 1$. The Lévy–Khintchine representation of $\log(T_\mu)$ is given by*

$$E\left(\left(\frac{1}{T_\mu}\right)^k\right) \equiv E[e^{-k \log T_\mu}]$$

$$= \exp\left\{\frac{k}{\mu}(1 - \mu)\psi(1) + \int_0^\infty \frac{du}{u}(\varphi(\mu u) - \varphi(u))(e^{-ku} - 1 + ku)\right\},$$

where $\varphi(u) = 1/(e^u - 1)$, and $\psi(1) = \Gamma'(1)$ is Euler's constant. Consequently, $\log(T_\mu)$ is self-decomposable.

Proof. The Lévy–Khintchine representation is easily deduced from that of $(\log \mathbf{e})$, thanks to (5). $\qquad\square$

Comment and reference. As for Corollary 2, the results of Proposition 5 may be found in [11, p. 321, Example 9.17].

Corollary 3. $\log |\Delta(\gamma_a^{(i)}; 1 \leq i \leq n)|$ *and* $\log |\Delta(\mathcal{N}^{(i)}; 1 \leq i \leq n)|$ *are self-decomposable.*

4 On the Joint Distribution of the Vandermonde Determinant and the Characteristic Polynomial Associated to a Uniform Unitary Matrix

My starting point here is formula (110) in [8]. If $\underline{\theta}_N = (\theta_1, \ldots, \theta_N)$ denotes the random vector consisting of N independent, uniform variables on $[0, 2\pi]$, then

$$E[|\Delta(e^{i\underline{\theta}_N})|^{2\beta}|Z_N(e^{i\underline{\theta}_N})|^s] = \left\{\prod_{j=0}^{N-1}\frac{\Gamma(1+j\beta)\Gamma(1+s+j\beta)}{(\Gamma(1+\frac{s}{2}+j\beta))^2}\right\}\frac{\Gamma(1+N\beta)}{(\Gamma(1+\beta))^N},$$

(15)

where we assume that $\beta, s \geq 0$ (for simplicity). On the LHS of (15), I have adopted the shorthand notation $\Delta(\underline{x}_N)$ and $Z_N(\underline{x}_N)$ for, respectively, the Vandermonde determinant

$$\prod_{i<j\leq N}(x_i - x_j),$$

and the 'characteristic polynomial'

$$\prod_{j\leq N}(1 - x_j)$$

associated with the vector $\underline{x}_N = (x_j)_{1\leq j\leq N}$.

In this section, I provide some probabilistic interpretations – in the form of identities in law – of the general formula (15). To give some flavor of these interpretations, let us consider the case $s = 0$ in (15); in this case, (15) reduces to

$$\Gamma(1 + N\beta) = E[|\Delta(e^{i\underline{\theta}_N})|^{2\beta}](\Gamma(1+\beta))^N,$$

(16)

which, by injectivity of the Mellin transform, may be written as

$$\mathbf{e}^N \overset{(\text{law})}{=} |\Delta(e^{i\underline{\theta}_N})|^2\mathbf{e}_1...\mathbf{e}_N,$$

(17)

where $\mathbf{e}$ on the LHS, and $\mathbf{e}_1,\ldots,\mathbf{e}_N$ on the RHS denote standard exponential variables (with mean 1), with $\Delta(e^{i\underline{\theta}_N})$ and $(\mathbf{e}_1,\ldots,\mathbf{e}_N)$ independent on the RHS of (17). Also, note that on the RHS of (17), it is the ordinary product of the variables $\mathbf{e}_1,\ldots,\mathbf{e}_N$ which occurs. In the sequel, without mentioning it anymore, we always understand that, in identities in law such as (17) algebraic quantities $f(X_1,\ldots,X_N)$ concern independent random variables.

Taking logarithms on both sides of (17), we obtain

$$\log\mathbf{e} \overset{(\text{law})}{=} \frac{1}{N}\left(\sum_{i=1}^{N}\log\mathbf{e}_i\right) + \frac{2}{N}\log|\Delta(e^{i\underline{\theta}_N})|.$$

(18)

The identity (18) suggests introducing the notion of an *arithmetically decomposable random variable* (*AD* for short), that is, a r.v. X such that for any integer N,

$$X \overset{(\text{law})}{=} \frac{1}{N}\left(\sum_{i=1}^{N}X_i\right) + Y_N,$$

where on the RHS the X_i's are independent copies of X, and Y_N is independent from the X_i's. Note the parenthood between an *AD* random variable and the notion of a *self-decomposable random variable*, which satisfies

$$X \stackrel{\text{(law)}}{=} cX + X_{(c)},$$

for every $c \in (0,1)$, with $X_{(c)}$ independent of X. Thus, from (18), log $\mathbf{e}$ is AD and, in Section 4.1, we discuss further the notion of AD variables, and exploit formula (15) to provide other examples.

In random matrix theory, it is not so much random vectors such as $\underline{\theta}_N$, with uniform components, which are of interest, but rather the probabilities

$$P_N^{(\beta)}(d\underline{\theta}_N) \equiv \left\{ \frac{|\Delta(e^{i\underline{\theta}_N})|^{2\beta}}{C_\beta} \right\} \cdot P_N^{(o)}(d\underline{\theta}_N),$$

where $C_\beta = (\Gamma(1 + N\beta))/((\Gamma(1 + \beta))^N)$ and under $P_N^{(o)}$ the θ_is are independent and uniform. In fact, it is the particular cases $\beta = 1, 2, 4$, which are relevant (see [8]). It is, of course, immediate to rewrite formula (15) so as to give the joint Mellin transform of $|\Delta(e^{i\underline{\theta}_N})|^2$ and $|Z_N(e^{i\underline{\theta}_N})|$ under $P_N^{(\beta)}$. As an example, we write

$$E_N^{(\beta)}(|Z_N(e^{i\underline{\theta}_N})|^s) = \prod_{j=0}^{N-1} \frac{\Gamma(1 + j\beta)\Gamma(1 + s + j\beta)}{(\Gamma(1 + \frac{s}{2} + j\beta))^2},$$

which, when written in the equivalent form

$$\prod_{j=0}^{N-1} \frac{\Gamma(1 + j\beta + s)}{\Gamma(1 + j\beta)} = E_N^{(\beta)}(|Z_N(e^{i\underline{\theta}_N})|^s) \prod_{j=0}^{N-1} \left(\frac{\Gamma(1 + \frac{s}{2} + j\beta)}{\Gamma(1 + j\beta)} \right)^2,$$

may be translated as

$$\prod_{j=0}^{N-1} \gamma_{(1+j\beta)} \stackrel{\text{(law)}}{=} |Z_N^{(\beta)}(e^{i\underline{\theta}_N})| \sqrt{\prod_{j=0}^{N-1} \gamma_{(1+j\beta)} \gamma'_{(1+j\beta)}}, \qquad (19)$$

where $Z_N^{(\beta)}(e^{i\underline{\theta}_N})$ denotes $Z_N(e^{i\underline{\theta}_N})$ under $P_N^{(\beta)}$. Clearly, the identities in law (18) and (19) are two instances of consequences of the general formula (15), which we exploit more systematically later on. In the case $\beta = 1$, formula (19) has been the starting point of the discussion in [12] of the understanding of the factor $((G(1 + \lambda))^2)/(G(1 + 2\lambda))$ in the expression of the Keating–Snaith conjecture for the asymptotics of the moments of $|\zeta(1/2 + it)|$; see [12] for details.

The remainder of this section consists of a discussion of the arithmetic decomposability of $\log(\gamma_a)$, for any $a \geq 1$, in the light of, for example, formula (16). This is done below in Section 4.1. In Section 4.2, an identity in law equivalent to the general formula (15) is presented and briefly discussed.

Notation. Unfortunately, I use the Greek letters beta and gamma for two altogether different items: the beta and gamma random variables; and, on the other hand, the parameters in the usual random matrix models. Nevertheless, this should not lead to any confusion.

4.1 The Arithmetic Decomposability of $\log(\gamma_a)$, $a \geq 1$, and the Distribution of $|\Delta^{(\beta)}(e^{i\underline{\theta}_N})|$

The aim of this section is to understand better the identities in law (17) and (18), where exponential random variables are featured; moreover, some adequate extensions to gamma variables are proposed.

We first discuss the arithmetic decomposability of $\log(\gamma_a)$, $a \geq 1$. It has been shown in [12], as a consequence of the beta-gamma algebra, that

$$\gamma_a \overset{\text{(law)}}{=} (\gamma_a^{(1)} \dots \gamma_a^{(N)})^{1/N} \, N \left(\prod_{k=0}^{N-1} \beta_{((a+k)/N, a-((a+k)/N))} \right)^{1/N}, \qquad (20)$$

which clearly implies the AD property of $\log(\gamma_a)$. Comparing formulae (20) and (17), we obtain

$$|\Delta(e^{i\underline{\theta}_N})|^2 \overset{\text{(law)}}{=} (N^N) \prod_{k=1}^{N} \beta_{(k/N, 1-(k/N))}.$$

In order to obtain an adequate extension of this formula, we consider formula (16), in which we change β in $(\beta + \gamma)$; thus, we obtain

$$\frac{\Gamma(1 + N\beta + N\gamma)}{\Gamma(1 + N\beta)} = E[|\Delta^{(\beta)}(e^{i\underline{\theta}_N})|^{2\gamma}] \left(\frac{\Gamma(1 + \beta + \gamma)}{\Gamma(1 + \beta)} \right)^N,$$

which translates as

$$(\gamma_{1+N\beta})^N \overset{\text{(law)}}{=} |\Delta^{(\beta)}(e^{i\underline{\theta}_N})|^2 \, \gamma_{1+\beta}^{(1)} \dots \gamma_{1+\beta}^{(N)}. \qquad (21)$$

From the beta-gamma algebra, we can write

$$\gamma_{1+\beta}^{(i)} \overset{\text{(law)}}{=} \beta_{1+\beta, (N-1)\beta}^{(i)} \, \gamma_{1+N\beta}^{(i)}.$$

We first plug these in (21), and then compare (21), thus modified, with (20). We then obtain, with $a = 1 + N\beta$:

$$N^N \prod_{k=0}^{N-1} \beta_{(((a+k)/N), a-((a+k)/N))} \overset{\text{(law)}}{=} |\Delta^{(\beta)}(e^{i\underline{\theta}_N})|^2 \left\{ \prod_{i=1}^{N} \beta_{1+\beta, (N-1)\beta}^{(i)} \right\}.$$

Note in particular how this formula simplifies for $\beta = 0$:

$$|\Delta^{(0)}(e^{i\underline{\theta}_N})|^2 \overset{\text{(law)}}{=} N^N \prod_{j=1}^{N} \beta_{(j/N, 1-(j/N))},$$

as discussed earlier.

4.2 An Interpretation of Formula (15)

As we did in (16) and (17), bringing the denominator in (15) on the LHS, it is not difficult to see that this identity may be interpreted as the following (complicated!) identity in law between two 2-dimensional random variables.

$$\left(\prod_{j=0}^{N-1} (\mathbf{e}_j \mathbf{e}'_j)^j \mathbf{e}^N, \prod_{j=0}^{N-1} \mathbf{e}'_j \right)$$

$$\overset{(\text{law})}{=} \left\{ \prod_{j=0}^{N-1} (\mathbf{e}_j \mathbf{e}'_j)^j |\Delta(e^{i\theta_N})|^2 \prod_{j=0}^{N-1} \mathbf{e}''_j; |Z_N(e^{i\theta_N})| \prod_{j=0}^{N-1} \sqrt{\mathbf{e}_j \mathbf{e}'_j} \right\}. \quad (22)$$

We note in particular that the identity in law between the two first components of (22) reduces to the identity (17), whereas the identity in law between the two second components of (22) is the particular case $\beta = 0$ of (19).

Finally, we leave to the interested reader the task of interpreting the law of $\{|\Delta(e^{i\theta_N})|^2, |Z_N(e^{i\theta_N})|\}$ under the probability

$$P_N^{(\beta,s)} = \frac{|\Delta(e^{i\theta_N})|^{2\beta} |Z_N(e^{i\theta_N})|^s}{C_N(\beta,s)} \cdot P.$$

References

1. G.E. Andrews, R.A. Askey, and R. Roy. Special functions. *Encyclopedia of Mathematics and its Applications*, 71. Cambridge Univ. Press, 1999.
2. P. Carmona, F. Petit, and M. Yor. On the distribution and asymptotic results for exponential functionals of Lévy processes. *Exponential Functionals and Principal Values Related to Brownian Motion. A Collection of Research Papers*. Biblioteca de la Revista Matematica Ibero-Americana, ed. M. Yor, 73–126, 1997.
3. L. Chaumont and M. Yor. *Exercises in Probability: A Guided Tour from Measure Theory to Random Processes, via Conditioning*. Cambridge Univ. Press, 2003.
4. D. Dufresne. Algebraic properties of beta and gamma distributions, and applications. *Adv. in App. Math.*, 20(3):385–399, 1998.
5. R. Ehrenborg. A bijective answer to a question of Zvonkin. *Ann. Comb.*, 4:195–197, 2000.
6. L. Gordon. A stochastic approach to the gamma function. *Amer. Math. Monthly*, 101:858–865, 1994.
7. J. Keating. *L*-functions and the characteristic polynomials of random matrices. *Recent Perspectives in Random Matrix Theory and Number Theory*, eds. F. Mezzadri, N. Snaith, London Mathematical Society LN 322, 2004.
8. J. Keating and N. Snaith. Random matrix theory and $\zeta(\frac{1}{2} + it)$. *Comm. Math. Physics*, 214:57–89, 2000.
9. I.L. Lu and R. Richards. Random discriminants. *Ann. Stat*, 21:1982–2000, 1993.
10. D.N. Shanbhag and M. Sreehari. On certain self-decomposable distributions. *Zeit für Wahr.*, 38:217–222, 1977.
11. F.W. Steutel and K. van Harn. *Infinite Divisibility of Probability Distributions on the Real Line*. M. Dekker, 2003.
12. M. Yor. Reading Notes from [7]. Preprint 2006.

Itô Formulas for Fractional Brownian Motion

Robert J. Elliott[1] and John van der Hoek[2]

[1] Haskayne School of Business
Scurfield Hall
University of Calgary
2500 University Drive NW
Calgary, Alberta T2N 1N4, Canada
relliott@ucalgary.ca

[2] Discipline of Applied Mathematics
University of Adelaide
Adelaide, South Australia 5005, Australia
john.vanderhoek@adelaide.edu.au

Summary. This article reviews the theory of fractional Brownian motion (fBm) in the white noise framework, and we present a new approach to the proof of Itô-type formulas for the stochastic calculus of fractional Brownian motion.

Key words: Itô formulas; fractional Brownian motion; white noise analysis.

1 Introduction

There are various approaches to this topic. We adopt the white noise analysis (WNA) approach, whose origins go back to the work of Hida [8]. Literature closest to this approach includes Holden, et al. [11], Hu and Øksendal [12], Huang and Yan [13], Hida et al. [9], Kuo [17], and papers of Lee [19–21]. The Malliavin-style approach is represented by Nualart [23; 24]. The WNA approach was used by Elliott and van der Hoek [6].

We present an introduction to white noise analysis in Section 2, a construction of fractional Brownian motion in Sections 3 through 5, and a new approach to the study of Itô-type formulas in Section 6. Only the outlines of proofs have been provided.

2 White Noise Analysis

We first want to construct a probability space $(\Omega, \mathcal{F}, P)$ with $\Omega = S'(R)$ such that for all $f \in S(R)$, we have

$$E\left[e^{\langle f, \cdot \rangle}\right] = e^{1/2\|f\|^2}, \tag{1}$$

where $S(R)$ is the usual space of rapidly decreasing C^∞ functions on R and $S'(R)$ its dual space of tempered distributions. It is important to note that $S(R)$ is what is called a nuclear space. The pairing $\langle f, g \rangle$ represents a dual pairing, with $f \in S(R)$ and $g \in S'(R)$. The basic theory of these objects can be found in Gel'fand and Vilenkin [7] and in Reed and Simon [26]. We cannot replace $S'(R)$ by a separable Hilbert space H, say. Let $\{e_n\}$ be a complete orthonormal system for H. Then we would have

$$E\left[e^{\langle e_n, \cdot \rangle}\right] = e^{1/2} \neq 1,$$

but the left-hand side converges to 1 as $n \to \infty$, giving a contradiction. This should suggest that the construction of such a probability space is nontrivial. The existence of such a probability is due to the Böchner–Minlos theorem. For the basic ideas, see Kuo [17] and Holden et al. [11].

We have the Gel'fand triple:

$$S(R) \subset L^2(R) \subset S'(R).$$

Here $\|f\|^2 = \int_R |f(t)|^2 dt$ and $\langle f, \rangle : S'(R) \to R$ with $\langle f, \omega \rangle \in R$ for all $\omega \in \Omega$. It is called a Gaussian space, because the random variable $\langle f, \rangle$ is $\mathcal{N}(0, \|f\|^2)$ for any $f \in S(R)$. Indeed, later $\langle f, \rangle$ is defined for all $f \in L^2(R)$.

We can also consider $\langle f, \rangle : S'(R) \to R^m$ when $f \in S(R)^m$ and $\|f\|^2 = \sum_{i=1}^m \|f_i\|^2$, where $f = (f_1, f_2, \dots, f_m)$. We concentrate in this paper on the case $m = 1$. For any real t, we have

$$E\left[e^{\langle tf, \cdot \rangle}\right] = e^{1/2t^2\|f\|^2},$$

or

$$\sum_{n=0}^{\infty} \frac{t^n}{n!} E\left[\langle f, \cdot \rangle^n\right] = \sum_{k=0}^{\infty} \frac{t^{2k}}{2^k k!} \|f\|^{2k}, \tag{2}$$

and we can obtain various identities by comparing powers in t on the left- and right-hand sides of Equation (2). So

$$E\left[\langle f, \cdot \rangle\right] = 0,$$
$$E\left[\langle f, \cdot \rangle^2\right] = \|f\|^2,$$
$$E\left[\langle f, \cdot \rangle^{2k}\right] = (2k-1)!!\|f\|^{2k},$$

where $(2k-1)!! = 1.3.5...(2k-1)$. Likewise, we can show

$$E\left[\langle f_1, \cdot \rangle \langle f_2, \cdot \rangle\right] = \langle f_1, f_2 \rangle,$$
$$E\left[\langle f_1, \cdot \rangle \langle f_2, \cdot \rangle \langle f_3, \cdot \rangle \langle f_4, \cdot \rangle\right] = \langle f_1, f_2 \rangle \langle f_3, f_4 \rangle + \langle f_1, f_3 \rangle \langle f_2, f_4 \rangle + \langle f_2, f_3 \rangle \langle f_1, f_4 \rangle,$$

by taking $f = t_1 f_1 + t_2 f_2 + t_3 f_3 + t_4 f_4$ in Equation (1). Clearly,

$$E\left[(\langle f, \cdot \rangle - \langle g, \cdot \rangle)^2\right] = \|f - g\|^2, \tag{3}$$

for $f, g \in S(R)$. We can now extend the definition of $\langle f, \cdot \rangle$ from $f \in S(R)$ to $f \in L^2(R)$, for if $f \in L^2(R)$, there exists a sequence $\{f_n\} \subset S(R)$ such that $\|f_n - f\| \to 0$ as $n \to \infty$. Then by (3), $\langle f_n, \cdot \rangle$ converges in mean-square. We define (using this meaning of limit)

$$\langle f, \cdot \rangle = \lim_{n \to \infty} \langle f_n, \cdot \rangle,$$

and, of course, by standard theorems, $\langle f, \cdot \rangle$ is also Gaussian for $f \in L^2(R)$, with

$$E\left[\langle f, \cdot \rangle\right] = 0,$$
$$E\left[\langle f, \cdot \rangle^2\right] = \|f\|^2.$$

We write $\mathcal{L}^2$ for the space of square integrable random variables on Ω. Thus $\langle f, \cdot \rangle \in \mathcal{L}^2$ for each $f \in L^2(R)$. These results are classical and are discussed by various authors, for example, in Janson [15], Neveu [22], and Kuo [16].

2.1 Brownian Motion

We can now give the WNA definition of Brownian motion. Let $\mathbf{1}(0, t](x) = 1$ if $x \in (0, t]$ and $\mathbf{1}(0, t](x) = 0$ if $x \notin (0, t]$. Then Brownian motion B can be defined by

$$B(t)(\omega) = \langle \mathbf{1}(0, t], \omega \rangle.$$

Technically speaking, we can consider a continuous (in t) version of this definition. We then have the standard properties:

$$B(0)(\omega) = 0,$$
$$E\left[B(t)B(s)\right] = E\left[\langle \mathbf{1}(0, t], \cdot \rangle \langle \mathbf{1}(0, s], \cdot \rangle\right] = \min\{t, s\},$$
$$B(t) - B(s) = \langle \mathbf{1}(s, t], \cdot \rangle \sim \mathcal{N}(0, t - s) \quad \text{for } t > s.$$

Remark. If we wish to construct m-dimensional Brownian motion, we proceed in a similar way, but with $S(R)$ replaced by $S(R)^m$ and $S'(R)$ by $S'(R)^m$, and we require a probability measure space $(\Omega, \mathcal{F}, P)$ satisfying Equation (1), where $\langle f, \omega \rangle = \langle f_1, \omega_1 \rangle + \cdots + \langle f_m, \omega_m \rangle$, and $\|f\|^2 = \|f_1\|^2 + \cdots + \|f_m\|^2$. Then

$$B(t)(\omega) = (B_1(t)(\omega), \ldots, B_m(t)(\omega)) = (\langle \mathbf{1}(0, t], \omega_1 \rangle, \ldots, \langle \mathbf{1}(0, t], \omega_m \rangle).$$

See Holden et al. [11] for more discussion.

2.2 Hermite Polynomials and Functions

We define Hermite polynomials $\{h_n\}$ on $\mathbb{R}$ by

$$h_n(x) = (-1)^n e^{x^2/2} \frac{d^n}{dx^n} e^{-(x^2/2)},$$

for $n = 0, 1, 2, \ldots$. The first few are

$$h_0(x) = 1,$$
$$h_1(x) = x,$$
$$h_2(x) = x^2 1,$$
$$h_3(x) = x^3 - 3x,$$

and so on. In fact, we can establish the following useful properties:

$$h_{n+1}(x) = xh_n(x) - nh_{n-1}(x),$$

and we have the generating function

$$e^{xt-(1/2)t^2} = \sum_{n=0}^{\infty} \frac{t^n}{n!} h_n(x). \tag{4}$$

Expanding the left-hand side of (4) in powers of t

$$h_n(x) = \sum_{l=0}^{[n/2]} (-1)^l \frac{n!}{2^l l!(n-2l)!} x^{n-2l} = \sum_{l=0}^{[n/2]} (-1)^l \binom{n}{2l} (2l-1)!! \, x^{n-2l},$$

where $[y]$ is the integer part of y, and starting with

$$e^{xt} = e^{(1/2)t^2} \sum_{n=0}^{\infty} \frac{t^n}{n!} h_n(x),$$

we obtain

$$x^n = \sum_{l=0}^{[n/2]} \binom{n}{2l} (2l-1)!! \, h_{n-2l}(x).$$

Also

$$h_n(x) = \int_R (x \pm iy)^n \frac{1}{\sqrt{2\pi}} e^{-(1/2)y^2} dy,$$

and

$$x^n = \int_R h_n(x+y) \frac{1}{\sqrt{2\pi}} e^{-(1/2)y^2} dy.$$

We also use

$$h_n(x)h_m(x) = \sum_{p=0}^{m \wedge n} p! \binom{n}{p} \binom{m}{p} h_{n+m-2p}(x) \tag{5}$$

and many other nice identities. An important orthogonality identity is the following,

$$\int_R h_n(x)h_m(x) \, e^{-(x^2/2)} dx = \sqrt{2\pi} \, n! \, \delta_{n,m}.$$

The **Hermite functions** $\{\xi_n | n = 1, 2, 3, \ldots\}$ are defined by

$$\xi_n(x) = \pi^{-(1/4)} [(n-1)!]^{-(1/2)} h_{n-1}(\sqrt{2} \, x)e^{-(x^2/2)},$$

and constitute a complete orthonormal sequence for $L^2(R)$. These properties are discussed in various places, for example, in Lebedev [18]. The completeness is established in Hille and Phillips [10].

2.3 Chaos Expansions in $\mathcal{L}^2$

We also talk about the Wiener–Itô chaos expansion. Any $\Phi \in \mathcal{L}^2$ has the unique representation:

$$\Phi(\omega) = \sum_{\alpha \in \mathcal{I}} c_\alpha H_\alpha(\omega),$$

where $\mathcal{I}$ is the collection of all multi-indices $\alpha = (\alpha_1, \alpha_2, \ldots)$, where $\alpha_i \in \{0, 1, 2, 3, \ldots\}$ and only a finite number of these indices are nonzero. We write for $|\alpha| = \alpha_1 + \alpha_2 + \cdots < \infty$,

$$H_\alpha(\omega) = \prod_{i=1}^{\infty} h_{\alpha_i}(\langle \xi_i, \omega \rangle).$$

For each α, $H_\alpha \in \mathcal{L}^2$ and for $\alpha, \beta \in 1$,

$$E[H_\alpha H_\beta] = \alpha! \, \delta_{\alpha,\beta},$$

where

$$\alpha! = \prod_{i=1}^{\infty} \alpha_i!,$$

$$\delta_{\alpha,\beta} = \prod_{i=1}^{\infty} \delta_{\alpha_i,\beta_i}.$$

In fact,

$$E\left[H_\alpha H_\beta\right] = \prod_{i=1}^{\infty} E\left[h_{\alpha_i}(\langle \xi_i, \cdot \rangle)h_{\beta_i}(\langle \xi_i, \cdot \rangle)\right]$$

$$= \prod_{i=1}^{\infty} \int_R h_{\alpha_i}(z)h_{\beta_i}(z)\frac{1}{\sqrt{2\pi}}\,e^{-(z^2/2)}dz$$

$$= \prod_{i=1}^{\infty} \alpha_i!\,\delta_{\alpha_i,\beta_i} = \alpha!\,\delta_{\alpha,\beta}.$$

We also have

$$c_\alpha = \frac{1}{\alpha!}E\left[\Phi H_\alpha\right],$$

$$E\left[\Phi^2\right] = \sum_\alpha \alpha!\,c_\alpha^2 < \infty.$$

Before concluding this section, we also mention that an alternative chaos expansion can be given in terms of multiple integrals with respect to Brownian motion. We do not use this form of the chaos expansion in this paper, but refer readers to the beautiful paper of Itô [14] or to Holden et al. [11].

2.4 Hida Test Functions and Distributions

If $f \in L^2(R)$, then $f = \sum_n c_n\xi_n$, where $c_n = \int_R f(x)\xi_n(x)dx$ and $\sum_n c_n^2 = \|f\|^2 < \infty$. It can be shown (see Reed and Simon [26]) that $f \in S(R)$ iff $f = \sum_n c_n\xi_n$ and for all integers $p \geq 0$, $|f|_p^2 \equiv \sum_n n^{2p}c_n^2 < \infty$. In a similar way, we can express $f \in S'(R)$ in the form $f = \sum_n c_n\xi_n$, where $|f|_{-q}^2 \equiv \sum_n n^{-2q}c_n^2 < \infty$ for some $q \geq 0$. To illustrate this latter representation, if $f \in S'(R)$, then $c_n = \langle \xi_n, f \rangle$. For the Dirac-delta function δ, we write

$$\delta = \sum_n \xi_n(0)\,\xi_n.$$

By the estimate

$$\sup_x |\,\xi_n(x)| \leq Cn^{-(1/4)}, \tag{6}$$

then

$$\sum_n n^{-2q}\xi_n(0)^2 \leq C^2 \sum_n n^{-2q}n^{-(1/2)} < \infty,$$

for $q > \frac{1}{4}$. If $f = \sum_n c_n\xi_n \in S(R)$ and $g = \sum_n d_n\xi_n \in S'(R)$, then

$$\langle f, g \rangle = \sum_n c_n d_n \tag{7}$$

is always a finite sum. These ideas are now generalized to the Hida test functions (S) and distributions $(S)^*$. Just as

$$S(R) \subset L^2(R) \subset S'(R),$$

so we have

$$(S) \subset \mathcal{L}^2 \subset (S)^*.$$

In fact, $\Phi \in (S)$ iff $\Phi = \sum_\alpha c_\alpha H_\alpha$ and

$$\|\Phi\|_p^2 = \sum_{\alpha \in \mathcal{I}} \alpha! \, c_\alpha^2 (2N)^{p\alpha} < \infty,$$

for all $p \in \{0, 1, 2, \ldots\}$, where

$$(2N)^{p\alpha} = \prod_{i=1}^{\infty} (2i)^{p\alpha_i}.$$

Likewise, we can represent $\Psi \in (S)^*$ as $\Psi = \sum_\alpha d_\alpha H_\alpha$, and for some integer $q \geq 0$,

$$\|\Psi\|_{-q}^2 = \sum_{\alpha \in \mathcal{I}} \alpha! \, c_\alpha^2 (2N)^{-q\alpha} < \infty,$$

and then

$$\langle\langle \Phi, \Psi \rangle\rangle = \sum_\alpha \alpha! \, c_\alpha d_\alpha.$$

We show that $(S)^*$ is a very large class of objects that is closed under many different operations. One of these is the Wick product. If $\Phi = \sum c_\alpha H_\alpha$ and $\Psi = \sum d_\alpha H_\alpha$ are elements of $(S)^*$, then their Wick product $\Phi \diamond \Psi \in (S)^*$ is defined by

$$\Phi \diamond \Psi = \sum_{\alpha, \beta} c_\alpha d_\beta H_{\alpha+\beta}.$$

A useful operator is the S-transform. If $\Phi \in (S)^*$, then the S-transform $\mathcal{S}\Phi$ is defined for $h \in S(R)$ by

$$(\mathcal{S}\Phi)(h) = \langle\langle \Phi, \mathcal{E}_h \rangle\rangle = \sum_\alpha c_\alpha \langle \xi, h \rangle^\alpha,$$

where

$$\mathcal{E}_h(\omega) = \exp^\diamond(\langle h, \omega \rangle) = \exp(\langle h, \omega \rangle - \frac{1}{2}\|h\|^2) \in (S),$$

$$\langle \xi, h \rangle^\alpha = \prod_{i=1}^{\infty} \langle \xi_i, h \rangle^{\alpha_i}.$$

The characterization of T such that $T = \mathcal{S}\Psi$ for some $\Psi \in (S)^*$ is given in Chapter 8 of Kuo [17]. We can define the Wick product of $\Psi, \Phi \in (S)^*$ as the unique Hida distribution $T \in (S)^*$ such that $\mathcal{S}T = \mathcal{S}\Psi\mathcal{S}\Phi$. In WNA, the S-transform plays a role analogous to a Fourier transform. Other transforms are also used (see Kuo [17] and Holden et al. [11]).

We now give some examples.

(1) **Brownian motion**

$$B(t)(\omega) = \sum_{i=1}^{\infty} \langle 1(0,t], \xi_i \rangle \langle \xi_i, \omega \rangle = \sum_{i=1}^{\infty} \left[\int_0^t \xi_i(u)du \right] H_{\varepsilon_i}(\omega),$$

and

$$\sum_{i=1}^{\infty} \left[\int_0^t \xi_i(u)du \right]^2 = t < \infty.$$

(2) **Brownian white noise**

$$W(t)(\omega) = \sum_{i=1}^{\infty} \xi_i(t) H_{\varepsilon_i}(\omega),$$

which is the "derivative" of B with respect to t, and we have (using (6))

$$\|W(t)\|_{-q}^2 = \sum_{i=1}^{\infty} (\xi_i(t))^2 (2i)^{-q} \leq C^2 \sum_{i=1}^{\infty} i^{-(1/2)} (2i)^{-q} < \infty,$$

for $q > \frac{1}{2}$, and so $W(t) \in (S)^*$ for all $t \geq 0$. Of course $W(t) \notin \mathcal{L}^2$.

In both examples we used $H_{\varepsilon_i}(\omega) = \langle \xi_i, \omega \rangle$, where $\varepsilon_i = (0,0,\ldots,0,1,0,\ldots)$, with the 1 in the ith place.

(3) **A Wick product**

$$B(t) \diamond W(t) = \sum_{i,j=1}^{\infty} \left[\xi_i(t) \int_0^t \xi_j(u)du \right] H_{\varepsilon_i + \varepsilon_j}$$

$$= B(t)W(t) - \sum_{i=1}^{\infty} \xi_i(t) \int_0^t \xi_i(u)du, \qquad (8)$$

because

$$H_{\varepsilon_i + \varepsilon_j} = \begin{cases} H_{\varepsilon_i} H_{\varepsilon_j} & \text{if } i \neq j, \\ H_{\varepsilon_i}^2 - 1 & \text{if } i = j. \end{cases}$$

So by (8),

$$\int_0^T B(t) \diamond W(t) \, dt = \sum_{i,j=1}^{\infty} \left[\int_0^T \xi_i(t) \int_0^t \xi_j(u) du \, dt \right] H_{\varepsilon_i + \varepsilon_j}$$

$$= \frac{1}{2} \sum_{i,j=1}^{\infty} \left[\int_0^T \xi_i(u) du \int_0^T \xi_j(u) du \right] H_{\varepsilon_i + \varepsilon_j}$$

$$= \frac{1}{2} \left[\sum_{i=1}^{\infty} \left(\int_0^T \xi_i(u) du \right) H_{\varepsilon_i} \right]^2 - \frac{1}{2} \sum_{i=1}^{\infty} \left[\int_0^T \xi_i(u) du \right]^2$$

$$= \frac{1}{2} B(T)^2 - \frac{1}{2} T.$$

So (as we show),

$$\int_0^T B(t) dB(t) = \int_0^T B(t) \diamond W(t) \, dt = \frac{1}{2} B(T)^2 - \frac{1}{2} T.$$

(4) Wick exponential

We define $X^{\diamond n} = X \diamond X \diamond \cdots \diamond X$ (n times) and

$$\exp^{\diamond}(X) = \sum_{n=0}^{\infty} \frac{1}{n!} X^{\diamond n}.$$

Then we have

$$\exp^{\diamond}(\langle f, \omega \rangle) = \exp\left(\langle f, \omega \rangle - \frac{1}{2} \|f\|^2 \right).$$

2.5 Wiener Integrals

When $f \in L^2(R)$, we can define

$$\int_R f(t) dB(t)(\omega) = \langle f, \omega \rangle.$$

We now exploit this observation to define fractional Brownian motion (fBm). We also note that the chaos expansion of $\int_R f(t) dB(t)$ is

$$\int_R f(t) dB(t) = \sum_{i=1}^{\infty} \left[\int_R f(u) \xi_i(u) du \right] H_{\varepsilon_i}.$$

We make a comment about the important role played by the Wick product. For $f, g \in L^2(R)$, we define

$$I_2(f \hat{\otimes} g) = \int_{R^2} f(s) g(t) dB(s) dB(t) = 2 \int_R g(t) \left[\int_{-\infty}^t f(s) dB(s) \right] dB(t).$$

The left-hand side is not equal to

$$\left(\int_R f(s)dB(s)\right)\left(\int_R g(t)dB(t)\right),$$

but rather to

$$\left(\int_R f(s)dB(s)\right)\diamond\left(\int_R g(t)dB(t)\right).$$

Likewise, it can be shown for any multi-index α with $|\alpha| = n$, we have the formula for the multiple Wiener integral

$$\int_{R^n}\xi^{\hat\otimes\alpha}dB^{\otimes n} = H_\alpha,$$

where we have used tensor product and symmetric tensor product notation. If we consider symmetric functions $f_n = \sum_{|\alpha|=n} c_\alpha \xi^{\hat\otimes\alpha}$, then

$$\Phi = \sum_\alpha c_\alpha H_\alpha = \sum_{n=0}^\infty \sum_{|\alpha|=n} c_\alpha H_\alpha = \sum_{n=0}^\infty I_n(f_n)$$

gives the alternate chaos expansion of Φ in terms of multiple Wiener integrals. This is another classical starting point for WNA when $\Phi \in \mathcal{L}^2$. The mapping $\Phi \to (f_0, f_1, f_2, \ldots)$ establishes

$$\mathcal{L}^2 \simeq \bigoplus_{n=0}^\infty \hat{L}^2(R^n).$$

2.6 Convergence of Distributions

Many of the results that we discuss below can be proved by approximation and taking limits in (S) or $(S)^*$.

We can let $(S)_p = \{\Phi \in \mathcal{L}^2 \mid ||\Phi||_p < \infty\}$ and then $(S) = \bigcap_{p>0}(S)_p$, and $\Phi_n \to \Phi$ in (S) as $n \to \infty$ if for all $p > 0$, $\{\Phi, \Phi_n \ n = 1, 2, 3, \ldots\} \subseteq (S_p)$ and $||\Phi_n - \Phi||_p \to 0$ as $n \to \infty$.

In a similar way, we define $(S)^*_{-q} = \{\Phi \in (S)^* \mid ||\Phi||_{-q} < \infty\}$ and then $(S)^* = \bigcup_{q\geq0}(S)^*_{-q}$, and $\Phi_n \to \Phi$ in $(S)^*$ as $n \to \infty$ if for some $q \geq 0$, $\{\Phi, \Phi_n \ n = 1, 2, 3, \ldots\} \subseteq (S)^*_{-q}$ and $||\Phi_n - \Phi||_{-q} \to 0$ as $n \to \infty$.

Convergence is often demonstrated using S-transforms (see Kuo [17]). The following two results from this reference are used.

Let $\Phi \in (S)^*$. Then the S-transform $F = S\Phi$ satisfies:

(a) For any $\xi, \eta \in S_c(R)$ (the complexification of $S(R)$), $F(z\xi+\eta)$ is an entire function of complex variable z,
(b) There exist nonnegative K, a, and p such that for all $\xi \in S_c(R)$,

$$|F(\xi)| \leq K\exp\left(a|\xi|_p^2\right).$$

Conversely, if F is defined on $S_c(R)$ and satisfies (a) and (b) above, then there is a unique $\Phi \in (S)^*$ with $F = \mathcal{S}\Phi$. This is one of the characterization theorems for $(S)^*$.

Now for $n = 1, 2, 3, \ldots$, let $\Phi_n \in (S)^*$ and $F_n = \mathcal{S}\Phi_n$. Then $\{\Phi_n\}$ converges in $(S)^*$ if and only if the following conditions are satisfied.

(a) $\lim_{n \to \infty} F_n(\xi)$ exists for each $\xi \in S(R)$,
(b) There exist nonnegative K, a, and p independent of n such that for all $n = 1, 2, 3, \ldots$, and $\xi \in S(R)$,

$$|F_n(\xi)| \leq K \exp(a|\xi|_p^2).$$

3 Fractional Brownian Motion

There are several constructions of fractional Brownian motion. See Samorodnitsky and Taqqu [27] (Chapter 7), for several approaches. In [6], we introduced an alternative approach for all Hurst indices $0 < H < 1$, and this is the approach that is reviewed here.

A fBm with Hurst parameter H $(0 < H < 1)$ is a zero mean Gaussian process $B^H = \{B^H(t) \mid t \in R\}$ with covariance

$$E\left[B^H(t)B^H(s)\right] = \frac{1}{2}\left\{|t|^{2H} + |s|^{2H} - |t - s|^{2H}\right\}.$$

We take $B^H(0) = 0$. For $H = \frac{1}{2}$, $B^{1/2}$ is standard Brownian motion.

We define the operator M_H on $S(R)$ by

$$(M_H f)(x) = -\frac{d}{dx} C_H \int_R (t - x)| t - x|^{H-(3/2)} f(t) dt,$$

for each $0 < H < 1$. We should remark that M_{1-H} is the inverse operator to M_H (up to a constant multiple); see [6]. This operator also makes sense when $f = \mathbf{1}(0, t]$, and then we write $M_H(0, t)$ for the resultant function which lies in $L^2(R)$. In fact,

$$M_H(0, t)(x) = C_H\left[(t - x)| t - x|^{H-(3/2)} + x| x|^{H-(3/2)}\right].$$

The constant C_H is chosen such that

$$\int_R [M_H(0, t)(x)]^2 dx = t^{2H}, \tag{9}$$

because it is easy to show that the right-hand side of (9) is a multiple of t^{2H}. We can then define B^H by

$$B^H(t)(\omega) = \langle M_H(0, t), \omega \rangle. \tag{10}$$

Technically, we take the continuous version of the process defined in (10). It is a technical exercise to show that this definition does verify the conditions for the definition of fBm with Hurst index H.

We can define multiple fBms in various ways, either in the way we have just described based on a number of independent standard Brownian motions, or we can define various fBms with different Hurst parameters based on a common standard Brownian motion, or we can adopt a combination of both approaches. For this paper we focus on a single fBm process with Hurst parameter $0 < H < 1$.

A second-order stationary process $X = \{X(t),\ t \geq 0\}$ has Hurst index defined by

$$H = \inf\{h : \limsup_{t \to \infty} t^{-2h}\ \mathrm{var}\,(X(t) - X(0)) < \infty\}\,.$$

B^H has Hurst index H, but so does $B^H + B^K$ when $H > K$. This indicates that using fBm to model long-range dependence is not a trivial issue. We do not digress into this area, except to say that our machinery can deal with multiple fBm models. However, as we said before, we focus on the stochastic calculus with a simple fBm with a fixed Hurst parameter $0 < H < 1$.

The chaos expansion for B^H is

$$B^H(t)(\omega) = \sum_{i=1}^{\infty} \left[\int_0^t (M_H \xi_i)\,(u)du \right] H_{\varepsilon_i}(\omega),$$

and leads to a definition of fractional white noise $W^H : R \to (S)^*$ by

$$W^H(t)(\omega) = \sum_{i=1}^{\infty} [(M_H \xi_i)\,(t)]\,H_{\varepsilon_i}(\omega).$$

In fact, because

$$\sup_t |M_H \xi_n(t)| \leq C\ n^{(2/3)-(H/2)},$$

so

$$\|W^H(t)\|_{-q}^2 = \sum_{i=1}^{\infty} |M_H \xi_i(t)|^2\ (2i)^{-q} \leq C^2 2^{-q} \sum_{i=1}^{\infty} i^{(2/3)-(H/2)-q} < \infty,$$

for $q > (5/3) - (H/2)$.

With these definitions, we can now generalize the Wiener integral. For $f \in S(R)$ we can define

$$\int_R f(t)dB^H(t)(\omega) = \langle M_H f, \omega \rangle,$$

for $\omega \in \Omega$. This definition can readily be extended to functions f for which $\int_R [(M_H f)(x)]^2\,dx < \infty$.

4 Integrals with Respect to fBm

If $Z : R \to (S)^*$ is such that $\langle\langle Z(t), F \rangle\rangle \in L^1(R)$ for all $F \in (S)$, then $\int_R Z(t)dt$ is defined to be that unique element of $(S)^*$ such that

$$\left\langle\left\langle \int_R Z(t)dt, F \right\rangle\right\rangle = \int_R \langle\langle Z(t), F \rangle\rangle dt \quad \text{for all } F \in (S).$$

We then say that $Z(t)$ is integrable in $(S)^*$.

If $Y : R \to (S)^*$ is such that $Y(t) \diamond W^H(t)$ is integrable in $(S)^*$, then define

$$\int_R Y(t)dB^H(t) \triangleq \int_R Y(t) \diamond W^H(t)dt.$$

We note that (with limits in $(S)^*$),

$$\int_0^t Y(s)dB^H(s) = \lim_{n\to\infty} \sum_{i=0}^{n-1} Y(t_i) \diamond \left(B^H(t_{i+1}) - B^H(t_i)\right),$$

using suitable partitions. Furthermore,

$$E\left[Y(t_i) \diamond \left(B^H(t_{i+1}) - B^H(t_i)\right)\right] = 0,$$

because

$$E\left[H_\alpha \diamond \langle M_H(t_i, t_{i+1}), \cdot\rangle\right] = E\left[\sum_{i=1}^{\infty} \left(\int_{t_i}^{t_{i+1}} (M_H\xi_i)(u)du\right) H_\alpha \diamond H_{\varepsilon_i}\right]$$

$$= \sum_{i=1}^{\infty} \left(\int_{t_i}^{t_{i+1}} (M_H\xi_i)(u)du\right) E\left[H_\alpha \diamond H_{\varepsilon_i}\right]$$

$$= \sum_{i=1}^{\infty} \left(\int_{t_i}^{t_{i+1}} (M_H\xi_i)(u)du\right) E\left[H_{\alpha+\varepsilon_i}\right]$$

$$= \sum_{i=1}^{\infty} \left(\int_{t_i}^{t_{i+1}} (M_H\xi_i)(u)du\right) E\left[H_{\alpha+\varepsilon_i}H_0\right] = 0,$$

and so having displayed the crucial steps,

$$E\left[\int_0^t Y(s)dB^H(s)\right] = 0.$$

From now on, we write this Hitsuda–Skorohod integral as

$$\int_R Y(t) \diamond dB^H(t),$$

to distinguish it from some other stochastic integrals. When $H = \frac{1}{2}$, this integral is more general than the Itô integral, because we did not require that

Y be adapted. However, in this case when Y is adapted to Brownian motion, the Hitsuda–Skorohod integral is the same as the classical Itô integral.

Path integrals are based on approximations

$$\sum_{i=0}^{n-1} Y(t_i) \left(B^H(t_{i+1}) - B^H(t_i) \right),$$

and a satisfactory theory exists for $H > \frac{1}{2}$. We refer readers to Chapter 5 of the forthcoming book by Biagini et al. [4] for details and references. However, in this case

$$E \left[\int_R Y(t) dB^H(t) \right] \neq 0.$$

In fact, if $Y(t) = B^H(t)$, then

$$E \left[B^H(t_i) \left(B^H(t_{i+1}) - B^H(t_i) \right) \right] \neq 0,$$

by a direct computation. For $H > \frac{1}{2}$, using these path integrals

$$\int_0^t B^H(s) dB^H(s) = \frac{1}{2} \left[B^H(t) \right]^2,$$

and so

$$E \left[\int_0^t B^H(s) dB^H(s) \right] = \frac{1}{2} t^{2H} \neq 0.$$

When $H = \frac{1}{2}$, the path integral and the Itô integral agree for suitably adapted integrands.

To show how this all works we provide the following example,

$$\int_0^t B^H(s) \diamond dB^H(s) = \frac{1}{2} \left[B^H(t) \right]^2 - \frac{1}{2} t^{2H}.$$

To show this, recall the chaos expansions for B^H and W^H:

$$B^H(s)(\omega) = \sum_{i=1}^{\infty} \left(\int_0^s M_H \xi_i(u) du \right) H_{\varepsilon_i}(\omega),$$

$$W^H(s)(\omega) = \sum_{j=1}^{\infty} M_H \xi_j(s) H_{\varepsilon_j}(\omega).$$

Then

$$\int_0^t B^H(s) \diamond dB^H(s) = \sum_{i,j}^{\infty} \int_0^t \left(\int_0^s M_H \xi_i(u) du \right) M_H \xi_j(s) ds H_{\varepsilon_i + \varepsilon_j}(\omega)$$

$$= \frac{1}{2} \sum_{i,j}^{\infty} \int_0^t M_H \xi_i(u) du \int_0^t M_H \xi_j(s) ds H_{\varepsilon_i + \varepsilon_j}(\omega)$$

$$= \frac{1}{2} \sum_{i,j}^{\infty} \int_0^t M_H \xi_i(u) du \int_0^t M_H \xi_j(s) ds \left[H_{\varepsilon_i}(\omega) H_{\varepsilon_j}(\omega) - \delta_{i,j} \right]$$

$$= \frac{1}{2} \left[B^H(t) \right]^2 - \frac{1}{2} \sum_{i=1}^{\infty} \left(\int_0^t M_H \xi_i(u) du \right)^2$$

$$= \frac{1}{2} \left[B^H(t) \right]^2 - \frac{1}{2} t^{2H}.$$

The same proof works with $H = \frac{1}{2}$.   □

Under weak conditions, $\int_R Y(t) \diamond dB^H(t) \in (S)^*$, and under stronger conditions, it is in $\mathcal{L}^2$.

5 Hida Derivatives

A good reference for this topic is Kuo [17]. For $\Phi \in (S)^*$ and $y \in S'(R)$, define

$$(D_y \Phi)(\omega) = \lim_{\varepsilon \to 0} \frac{\Phi(\omega + \varepsilon y) - \Phi(\omega)}{\varepsilon}.$$

This can be computed in terms of the chaos expansion of Φ:

$$\Phi(\omega) = \sum_{\alpha} c_\alpha H_\alpha(\omega),$$

for which we have

$$(D_y \Phi)(\omega) = \sum_{i=1}^{\infty} \sum_{\alpha} c_\alpha \alpha_i H_{\alpha - \varepsilon_i}(\omega) \langle \xi_i, y \rangle.$$

For $y = \delta_t$, we then write $D_y = \partial_t$:

$$(\partial_t \Phi)(\omega) = \sum_{i=1}^{\infty} \sum_{\alpha} c_\alpha \alpha_i H_{\alpha - \varepsilon_i}(\omega) \xi_i(t).$$

We always apply the Hida derivative to the chaos expansion. Kuo [17] points out the following example where the left-hand sides are formally different, but the right-hand sides are equal,

$$\langle \mathbf{1}(0, t], \omega \rangle = \sum_{i=1}^{\infty} \left(\int_0^t \xi_i(u) du \right) H_{\varepsilon_i}(\omega)$$

and

$$\langle \mathbf{1}(0,t),\omega\rangle = \sum_{i=1}^{\infty} \left(\int_0^t \xi_i(u)du \right) H_{\varepsilon_i}(\omega).$$

Using the left-hand sides,

$$\partial_t \langle \mathbf{1}(0,t],\omega\rangle = 1,$$
$$\partial_t \langle \mathbf{1}(0,t),\omega\rangle = 0,$$

but using the right-hand sides,

$$\partial_t B(t)(\omega) = \sum_{i=1}^{\infty} \left(\int_0^t \xi_i(u)du \right) \xi_i(t).$$

A basic identity is the following,

$$\Phi(\omega)\left[\Psi(\omega) \diamond \langle h,\omega\rangle\right] = (\Phi(\omega)\Psi(\omega)) \diamond \langle h,\omega\rangle + \Psi(\omega)D_h\Phi(\omega), \qquad (11)$$

which is proved first by taking $\Phi = H_\alpha$, $\Psi = H_\beta$, and $h = \xi_i$ and then using linearity and limits in $(S)^*$. We also use

$$H_\alpha(\omega)H_\beta(\omega) = \sum_{\gamma \leq \alpha \wedge \beta} \gamma! \binom{\alpha}{\gamma}\binom{\beta}{\gamma} H_{\alpha+\beta-2\gamma}(\omega),$$

which generalizes (5).

We now define the generalized Hida derivatives. It is easy to check that

$$\langle M_H \phi, \psi \rangle = \langle \phi, M_H \psi \rangle,$$

for all $\phi, \psi \in S(R)$. By analogy to what was done above, we define

$$\left(D_y^H \Phi\right)(\omega) = \lim_{\varepsilon \to 0} \frac{\Phi(\omega + \varepsilon M_H y) - \Phi(\omega)}{\varepsilon}.$$

Define for each $\phi \in S(R)$,

$$\langle M_H y, \phi \rangle \triangleq \langle y, M_H \phi \rangle = \langle y, \sum_{i=1}^{\infty} \langle M_H \phi, \xi_i \rangle \xi_i \rangle$$

$$= \sum_{i=1}^{\infty} \langle M_H \phi, \xi_i \rangle \langle y, \xi_i \rangle = \sum_{i=1}^{\infty} \langle \phi, M_H \xi_i \rangle \langle y, \xi_i \rangle.$$

This all makes sense when y has compact support. We write ∂_t^H for D_y^H when $y = \delta_t$, and if $\Phi = \sum_\alpha c_\alpha H_\alpha$, then

$$\left(\partial_t^H \Phi\right)(\omega) = \sum_{i=1}^{\infty} \sum_{\alpha} c_\alpha \alpha_i H_{\alpha-\varepsilon_i}(\omega) M_H \xi_i(t).$$

Here are some examples:

(1) We have (for $H > \frac{1}{2}$)

$$\partial_t^H B^H(u) = \begin{cases} H \cdot \left(t^{2H-1} - (t-u)^{2H-1}\right) & \text{if } t \geq u, \\ H \cdot \left(t^{2H-1} + (u-t)^{2H-1}\right) & \text{if } t \leq u. \end{cases} \tag{12}$$

This follows from two observations,

$$\partial_t^H B^H(u) = M_H^2(0, u)(t),$$

$$\int_0^s M_H^2(0, t)(x)dx = \int_R M_H(0, t)(x) M_H(0, s)(x)dx$$
$$= \frac{1}{2}\left[|t|^{2H} + |s|^{2H} - |s-t|^{2H}\right],$$

and from differentiating both sides of this identity with respect to s.

(2) $(M_H \delta_t)(x) \equiv \gamma_t^H(x) = C_H \cdot \left(H - \frac{1}{2}\right) |t-x|^{H-(2/3)} \in S'(R)$ for all $0 < H < 1$. For $H > \frac{1}{2}$, $\gamma_t^H \in L_{loc}^1(R) \subset S'(R)$, and for $H < \frac{1}{2}$, it is the distributional derivative of such a function.

(3) Let $X(t) = \int_0^t v(s)dB^H(s)$ for some deterministic function v. Then

$$\partial_t^H X(t) = H(2H-1)\int_0^t v(z)(t-z)^{2H-2}dz,$$

for $H > \frac{1}{2}$. In fact, for such H,

$$(M_H^2 f)(x) = H(2H-2)\int_R |x-z|^{2H-2}f(z)dz, \tag{13}$$

which follows from

$$\int_R (M_H f)(x)(M_H g)(x)dx = H(2H-2)\int_{-\infty}^{\infty}\int_{-\infty}^{\infty} |x-z|^{2H-2}f(x)g(z)dxdz.$$

So

$$\partial_t^H X(t) = \langle M_H(\mathbf{1}(0,t)v), M_H \delta_t \rangle = \langle \mathbf{1}(0,t)v, M_H^2 \delta_t \rangle$$
$$= \int_0^t v(z)(M_H^2 \delta_t)(z)dz$$
$$= H(2H-1)\int_0^t v(z)(t-z)^{2H-2}dz.$$

On the other hand, when $H < \frac{1}{2}$, we first write

$$X(t) = v(t)B^H(t) - \int_0^t v'(s)B^H(s)ds,$$

so

$$\partial_t^H X(t) = Hv(t)t^{2H-1} - \int_0^t v'(s)\partial_t^H B^H(s)ds,$$

and by (12),

$$\partial_t^H X(t) = Hv(t)t^{2H-1} - H\int_0^t v'(s)\left[t^{2H-1} - (t-s)^{2H-1}\right]ds.$$

We formally define the adjoint of ∂_t^H, written $(\partial_t^H)^*$, by

$$\langle\langle(\partial_t^H)^*\Psi, \Phi\rangle\rangle = \langle\langle\Psi, \partial_t^H \Phi\rangle\rangle.$$

By (7) we can find a formula for the $(\partial_t^H)^*\Psi$,

$$(\partial_t^H)^*\Psi = \sum_{\alpha,i} c_\alpha M_H \xi_i(t)H_{\alpha+\varepsilon_i} = \Psi \diamond W^H(t).$$

We obtain the generalization

$$\int_R Y(t) \diamond dB^H(t) = \int_R Y(t) \diamond W^H(t)dt = \int_R (\partial_t^H)^* Y(t)dt,$$

which is known as a representation of the Hitsuda–Skorohod integral when $H = \frac{1}{2}$; see Kuo [17].

(4) If $S(t) = S(0)\exp\left[B^H(t) - \frac{1}{2}t^{2H}\right]$, then $\partial_t^H S(t) = H \cdot t^{2H-1}S(t)$, which follows from an obvious chain rule.

(5) $\partial_t^H W^H(r) = H(2H-1)|t-r|^{2H-2} \in S'(R)$ for $t \neq r$.

(6) By the identity (11) with $h = \delta_t$, we can deduce the following two identities.

$$\Psi(\omega)\left[\Phi(\omega) \diamond W_t^H(\omega)\right] = \left[\Psi(\omega)\Phi(\omega)\right] \diamond W_t^H(\omega) + \Psi(\omega)\partial_t^H \Phi(\omega),$$

$$\Psi \int_R \Phi(t) \diamond dB_t^H = \int_R \left[\Psi\Phi(t)\right] \diamond dB_t^H + \int_R \Phi(t)\partial_t^H \Psi dt, \qquad (14)$$

so

$$\int_R Y(t)dB^H(t) = \int_R Y(t) \diamond dB^H(t) + \int_R \partial_t^H Y(t)dt, \qquad (15)$$

where Equation (14) leads to (15) as follows. If $\Psi = Y(t_i)$ and $\Phi = \mathbf{1}(t_i, t_{i+1}]$ in (14), then

$$Y(t_i)\left[B^H(t_{i+1}) - B^H(t_i)\right] = \int_{t_i}^{t_{i+1}} Y(t_i) \diamond dB^H(t) + \int_{t_i}^{t_{i+1}} \partial_t^H Y(t_i)dt.$$

One then does some approximations and limits in $(S)^*$. Equation (15) can be regarded as an alternative formulation of the path integral, which may exist as an object in $(S)^*$ for all $0 < H < 1$. Sottinen and Valkeila [28] and Øksendal [25] give alternative equivalent formulations of (14), but we think the one presented here is more transparent.

We note that our generalized Hida derivative is the composition $M_H^2 \circ D_t^{(H)}$, where $D_t^{(H)}$ is a Malliavin derivative (see Biagini and Øksendal [3]), and this can be compressed by writing $\Gamma = M_H^2$ in (13), an operator used by several authors for $H > \frac{1}{2}$.

6 Itô-Type Formulas

We now present some new ideas on how to prove Itô-type formulas based on the identity (14). If

$$X(t) = X(0) + \int_0^t a(u)du + \int_0^t b(u) \diamond dB^H(u),$$

and $F \equiv F(t,x)$ is sufficiently smooth, then we wish to obtain an expression for $F(t, X(t))$. For simplicity of presentation, we set $a(u) \equiv 0$ and $F \equiv F(x)$. To adapt this to the more general case is not difficult.

The idea is as follows. Let $0 = t_0 < t_1 < \cdots < t_n = t$. Then

$$
\begin{aligned}
F(X(t)) - F(X(0)) &= \sum_{j=0}^{n-1} F(X(t_{j+1})) - F(X(t_j)) \\
&= \sum_{j=0}^{n-1} \int_0^1 \frac{d}{d\theta} F(X(t_j) + \theta(X(t_{j+1}) - X(t_j)))d\theta \\
&= \sum_{j=0}^{n-1} \Phi_j \left[X(t_{j+1}) - X(t_j) \right] \\
&= \sum_{j=0}^{n-1} \Phi_j \int_{t_j}^{t_{j+1}} b(u) \diamond dB^H(u) \\
&= \sum_{j=0}^{n-1} \int_{t_j}^{t_{j+1}} [\Phi_j b(u)] \diamond dB^H(u) + \sum_{j=0}^{n-1} \int_{t_j}^{t_{j+1}} b(u) \partial_u^H \Phi_j du \\
&= T_1 + T_2,
\end{aligned}
$$

where

$$\Phi_j = \int_0^1 F'(X(t_j) + \theta(X(t_{j+1}) - X(t_j)))d\theta.$$

Now,

$$T_1 \to \int_0^t [b(u)F'(X(u))] \diamond dB^H(u),$$

where the convergence is in $(S)^*$ as the mesh$\{t_j\} \to 0$. We now consider T_2.

$$\partial_u^H \Phi_j = \int_0^1 F''(X(t_j) + \theta(X(t_{j+1})-X(t_j)))[(1-\theta)\partial_u^H X(t_j) + \theta\partial_u^H X(t_{j+1})] \, d\theta.$$

Now,

$$X(t) = X(0) + \int_0^t b(u) \diamond dB^H(u),$$

and for $H > \frac{1}{2}$ and b deterministic,

$$\partial_u^H X(t) = H(2H-1) \int_0^t b(r)|u-r|^{2H-2}dr,$$

and so

$$\partial_u^H \Phi_j = H(2H-1) \int_0^1 F''(X(t_j) + \theta(X(t_{j+1}) - X(t_j))).$$

$$\left[(1-\theta)\int_0^{t_j} b(r)|u-r|^{2H-2}dr + \theta\int_0^{t_{j+1}} b(r)|u-r|^{2H-2}dr\right] d\theta$$

$$\to H(2H-1)b(u)F''(X(u)) \left[\int_0^u b(r)|u-r|^{2H-2}dr\right],$$

so

$$F(X(t)) = F(X(0)) + \int_0^t [b(u)F'(X(u))] \diamond dB^H(u)$$

$$+ H(2H-1) \int_0^t b(u)F''(X(u)) \left[\int_0^u b(r)|u-r|^{2H-2}dr\right] du. \quad (16)$$

This agrees with the results of Sottinen and Valkeila [28] and Øksendal [25]. Sottinen and Valkeila also write the Itô-type formula when b is not deterministic, but by different methods. Writing $X(t)$ as a chaos expansion, we are able to compute $\partial_t^H X(t)$ and compare the resulting Itô-type formula with the one obtained by Sottinen and Valkeila; see also the paper by Bender [2].

We could also note that

$$F(X(t)) = F(X(0)) + \int_0^t [b(u)F'(X(u))] \diamond dB^H(u)$$

$$+ \int_0^t b(u)F''(X(u))\partial_u^H X(u) \, du,$$

and if $H > \frac{1}{2}$, then

$$X(t) = \int_0^t b(r) \diamond dB^H(r) = \int_0^t b(r) \diamond W^H(r) \, dr$$

implies

$$\partial_t^H X(t) = \int_0^t \partial_t^H b(r) \diamond W^H(r) \, dr + \int_0^t b(r) \diamond \partial_t^H W^H(r) \, dr$$

$$= \int_0^t \partial_t^H b(r) \diamond dB^H(r) + H(2H-1) \int_0^t b(r)(t-r)^{2H-2} dr.$$

This leads to a result that agrees with the results of Alos and Nualart [1, (Theorem 8)].

For $H < \frac{1}{2}$, and b deterministic, we could proceed as follows. From

$$X(t) = b(t)B^H(t) - \int_0^t b'(u)B^H(u)du,$$

we have

$$\partial_t^H X(t) = b(t)\partial_t^H B^H(t) - \int_0^t b'(u)\partial_t^H B^H(u)du$$

$$= Hb(t)t^{2H-1} - H \int_0^t b'(u) \left[t^{2H-1} - (t-u)^{2H-1}\right] du,$$

so

$$F(X(t)) = F(X(0)) + \int_0^t [F'(X(u))b(u)] \diamond dB^H(u)$$

$$+ \int_0^t b(u)F''(X(u))\partial_u^H X(u)du$$

$$= F(X(0)) + \int_0^t [F'(X(u))b(u)] \diamond dB^H(u)$$

$$+ H \int_0^t b(u)^2 F''(X(u))u^{2H-1} du$$

$$- H \int_0^t b(u)F''(X(u)) \int_0^u b'(r) \left[u^{2H-1} - (u-r)^{2H-1}\right] drdu,$$

$$(17)$$

which is the usual Itô formula when $H = \frac{1}{2}$, where we used Equation (12) for $t \geq u$. Note that the case $H = \frac{1}{2}$ follows from taking limit at $H \to \frac{1}{2}-$ in (17) but not from taking limits as $H \to \frac{1}{2}+$ in (16). This makes intuitive sense, because Brownian motion can be obtained (in some sense) from B^H by fractional integration if $H < \frac{1}{2}$, but by fractional differentiation if $H > \frac{1}{2}$.

The case when b is not deterministic is not treated here but will be discussed along with other details in a future paper.

Various Itô-type formulas are presented in Biagini et al. [4], where an extensive bibliography is provided, and we mention also a paper of Decreusefond [5], which used the Malliavin calculus approach to obtain Itô-type formulas for $0 < H < 1$.

References

1. E. Alos and D. Nualart. Stochastic integration with respect to the fractional Brownian motion, *Stochastics*, 75:129–152, 2003.
2. C. Bender. An Itô formula for generalized functionals of fractional Brownian motion with arbitrary Hurst parameter. Working Paper, University of Konstanz, 2002.
3. F. Biagini and B. Øksendal. Forward integrals and an Itô formula for fractional Brownian motion. Preprint, University of Oslo, 2005.
4. F. Biagini, Y. Hu, B. Øksendal, and T. Zhang. *Fractional Brownian Motion and Applications*. Springer Verlag, forthcoming.
5. L. Decreusefond. Stochastic integration with respect to Volterra processes. *Ann. Inst. H. Poincar Probab. Statist.*, 41:123–149, 2005.
6. R.J. Elliott and J. van der Hoek. A general fractional white noise theory and applications to finance. *Mathematical Finance*, 13:301–330, 2003.
7. I.M. Gel'fand and N.Y. Vilenkin. *Generalized Functions*, vol.4, Academic Press, 1964.
8. T. Hida, *Brownian Motion*. Springer-Verlag, 1983.
9. T. Hida, H.-H. Kuo, J. Pothoff, and L. Streit. *White Noise - An Infinite Dimensional Analysis*. Kluwer Academic, 1993.
10. E. Hille and R.S. Phillips. *Functional Analysis and Semigroups*. American Mathematical Society, 1957.
11. H. Holden, B. Øksendal, J. Ubøe, and T. Zhang. *Stochastic Partial Differential Equations*. Birkhäuser, 1996.
12. Y. Hu and B. Øksendal. Fractional white noise calculus and applications to finance. *Infinite Dimensional Analysis, Quantum Probability and Related Topics*, 6:1–32, 2003.
13. Z.-Y. Huang and J.-A. Yan. *Introduction to Infinite Dimensional Stochastic Analysis*. Kluwer Academic, 2000.
14. K. Itô. Multiple Wiener integral. *J. Math. Soc. Japan*, 3:157–169, 1951.
15. S. Janson. *Gaussian Hilbert Spaces*. Cambridge University Press, 1997.
16. H.-H. Kuo. *Gaussian Measures in Banach Spaces*. LNM, 463. Springer-Verlag, 1975.
17. H.-H. Kuo. *White Noise Distribution Theory*. CRC Press, 1996.
18. N.N. Lebedev. *Special Functions and their Applications*. Prentice-Hall, 1965.
19. Y.J. Lee. Generalized functions on infinite dimensional spaces and its applications to white noise calculus. *J. Funct. Anal.*, 82:429–464, 1989.
20. Y.J. Lee. A characterization of generalized functions on infinite dimensional spaces and Bargman-Segal analytic functions. *Gaussian Random Fields*, eds. K. Itô and T. Hida, World Scientific, 272–284, 1991.

21. Y.J. Lee. Analytic version of test functionals - Fourier transform and a characterization of measures in white noise calculus. *J. Funct. Anal.*, 100:359–380, 1991.

22. J. Neveu. *Processus Aléatoires Gaussiens.* Montréal, Presses de l'Université de Montréal, 1968.

23. D. Nualart. *The Malliavin Calculus and Related Topics.* Springer-Verlag, 1995.

24. D. Nualart. Stochastic integration with respect to fractional Brownian motion and applications. *Stochastic Models* (Mexico City, 2002), Contemp. Math. 336, American Math. Soc., Providence, RI, 3–39, 2003.

25. B. Øksendal. Fractional Brownian motion in finance. *Stochastic Economic Dynamics*, eds. B.S. Jensen and T. Palokangas, Cambridge Univ. Press (to appear).

26. M. Reed and B. Simon. *Functional Analysis*, Vol. 1, Academic Press, 1980.

27. G. Samorodnitsky and M.S. Taqqu. *Stable Non-Gaussian Random Processes.* Chapman and Hall, 1994.

28. A. Sottinen and B. Valkeila. On arbitrage and replication in the fractional Black-Scholes pricing model. Working Paper, University of Helsinki, 2003.

Part II

Asset and Option Pricing

A Tutorial on Zero Volatility and Option Adjusted Spreads

Robert Jarrow

Johnson Graduate School of Management
Cornell University
Ithaca, NY 14853, USA
raj15@cornell.edu

Summary. This paper provides a brief tutorial on the notions of a zero volatility (ZV) spread and an option adjusted spread (OAS), as applied to fixed income securities. Using the standard definitions, it is shown that the zero volatility spread measures the percentage of a security's spread due to any embedded options and any mispricings. The mispricings could be due to either market or model error. In contrast, the OAS only measures the percentage of the security's spread due to mispricings. Refinements and alternative measures of a bond's embedded optionality and mispricings are also provided.

Key words: Option adjusted spreads; zero volatility spreads; HJM model; arbitrage opportunities.

1 Introduction

Zero volatility (ZV) and option adjusted spreads (OAS) apply to bonds with embedded options. The embedded options could be call provisions, prepayment provisions, or even credit risk (viewed as the option to default). ZV spreads and OAS were first used in the residential mortgage-backed securities market to adjust for prepayment risk.

The purpose of this paper is to provide a brief tutorial on the notions of a ZV spread and OAS, using as a frame of reference the HJM [4] arbitrage-free term structure models. The HJM model, as applied to bonds, provides an objective method for valuing embedded options and determining market mispricings. Using these objective measures, we can more easily define and characterize both ZV spreads and OAS. We show that ZV spreads measure the excess spread on a bond due to both the embedded options and any mispricings. The mispricings could be due to either model or market errors. A market error represents an arbitrage opportunity. A model error represents a misspecified model perhaps due to the selection of an incorrect stochastic process, omitted risk premia, omitted risks (e.g., liquidity risk), or market imperfections (e.g., transaction costs). In contrast, we show that OAS is a measure of the residual spread in a bond due to only mispricings (after the removal of all embedded options and interest rate risk); see [2; 3], and [8] for background material. Refinements and alternative measures of a bond's embedded optionality and mispricings are provided.

An outline for this paper is as follows. Section 2 presents the theory, Section 3 discusses ZV spreads, and Section 4 studies OAS. Section 5 provides a numerical example to illustrate the previous concepts. Section 6 concludes.

2 The Theory

Assume a typical Heath–Jarrow–Morton (HJM [4]) economy. Given is a filtered probability space $(\Omega, F_\tau, P, (F_t)_{t \in [0,\tau]})$ satisfying the usual conditions (see [7]), where Ω is the state space with generic element ω, F_τ is a set of events, P the statistical probability measure, and $(F_t)_{t \in [0,\tau]}$ is the filtration. The state ω can be thought of as a possible interest rate scenario. Traded are a collection of default-free zero coupon bonds and various other (as needed) fixed income securities with embedded options. The spot rate of interest is denoted by $r_s(\omega)$, time 0 forward rates of maturity T by $f(0,T)$, and time 0 default-free zero coupon bond prices of maturity T by $p(0,T)$.

We assume that the market for interest rate risk is arbitrage-free and complete; hence, this implies the existence and uniqueness of an equivalent martingale probability Q, with expectation denoted by $E[\cdot]$, that can be used for valuation. This is the standard model used for valuing interest rate risk. For simplicity, we only investigate securities that reflect interest rate risk. The analysis, however, readily extends for the inclusion of credit risk; see [1] for a good review of this extension.

Let us consider a traded fixed income security, called a bond, that has an embedded option(s). The bond is, otherwise, default-free. For concreteness, we can think of the embedded option as a prepayment provision (the bond is short this option), although as evidenced below, the analysis applies in complete generality.

Let the bond have a maturity T, with discrete cash flows given by $(C_t)_{t=1,2,\dots,T}$ if no embedded options are exercised. For these bonds, we impose the condition that $C_t \geq 0$ for all t and $C_t > 0$ for some t. Let $(c_t(\omega))_{t=1,2,\dots,T}$ be the bond's cash flows in state ω given embedded options are possibly exercised where $c_t(\omega) \geq 0$ with probability one for all t and for some t there is a set of positive probability where $c_t(\omega) > 0$. This excludes interest rate swaps from consideration. (In fact, we argue below that the notion of an OAS is not useful for comparing fixed income securities that violate this nonnegative cash flow condition due to nonuniqueness of computed OAS.) The cash flows with embedded options $(c_t(\omega))_{t=1,2,\dots,T}$ will differ from the cash flows without $(C_t)_{t=1,2,\dots,T}$. For example, for a bond with a prepayment provision, the two cash flows are the same except if prepayment occurs, and afterwards.

On the prepayment date (assuming it occurs on a coupon payment date), the cash flow with embedded options includes the coupon and remaining principal (the prepayment). There are no further cash flows if prepayment occurs. In contrast, on the prepayment date, the cash flow without embedded options is just the coupon payment. The remaining cash flows without the embedded option consist of the coupons on the coupon payment dates and the principal on the maturity date.

Given this structure, the time 0 price of the bond with embedded options, denoted V_0, is given by

$$V_0 = E\left[\sum_{t=1}^{T} c_t(\omega)e^{-\int_0^t r_s(\omega)ds}\right] > 0. \tag{1}$$

This represents the present value of the discounted cash flows from holding the bond. The expectation is taken under the martingale measure Q, which implies that the formula adjusts for interest rate risk premia. An intuitive way to understand this is to note that instead of increasing the discount rate in computing the present value, the probabilities are adjusted to reflect interest rate risk. Note that the price of the bond is positive due to the nonnegativity conditions imposed on the bond's cash flows. This is called the fair or arbitrage-free price of the bond.

It is convenient to consider an otherwise identical bond that has no embedded options. This bond also has a maturity T, and its cash flows (across all states) are given by $(C_t)_{t=1,2,...,T}$. Its time 0 value, denoted B_0, is given by

$$B_0 = E\left[\sum_{t=1}^{T} C_t e^{-\int_0^t r_s(\omega)ds}\right] > 0. \tag{2}$$

Again, this value is strictly positive. Using the properties of the martingale measure, some algebra yields the following equivalent expressions for B_0,

$$\sum_{t=1}^{T} C_t E\left[e^{-\int_0^t r_s(\omega)ds}\right] = \sum_{t=1}^{T} C_t p(0,t) = \sum_{t=1}^{T} C_t e^{-\int_0^t f(0,s)ds}. \tag{3}$$

These alternative expressions prove useful below.

Given expressions (1)–(3), the dollar value of the bond's embedded options can be computed as

$$V_0 - B_0.$$

If the bond holder is long an embedded option, then $V_0 - B_0 \geq 0$. Conversely, as in the case of a prepayment provision, if the bondholder is short an embedded option, then $V_0 - B_0 \leq 0$.

In the above structure, both V_0 and B_0 represent the fair or arbitrage-free value of the bonds given a particular cash flow pattern. These are the bond prices in markets with no mispricings. To complete the setup for the discussion of ZV spreads and OAS, we need to introduce some notation for the market price of the bonds, denoted V_0^{mkt} and B_0^{mkt}, respectively. These prices could differ from the fair values.

From the interest rate derivatives literature, standard measures of mispricings are readily available including the dollar mispricing, represented by

$$V_0^{mkt} - V_0 = V_0^{mkt} - E\left[\sum_{t=1}^{T} c_t(\omega)e^{-\int_0^t r_s(\omega)ds}\right],$$

and the percentage dollar mispricing, represented by

$$\frac{V_0^{mkt} - V_0}{V_0^{mkt}} = \frac{V_0^{mkt} - E\left[\sum_{t=1}^{T} c_t(\omega)e^{-\int_0^t r_s(\omega)ds}\right]}{V_0^{mkt}}.$$

The mispricings characterized by these measures could be due to either model or market errors. A market error represents an arbitrage opportunity. A model error represents a misspecified model perhaps due to the selection of an incorrect stochastic process, omitted risk premia, omitted risks (e.g., liquidity risk), or market imperfections (e.g., transaction costs). To obtain a spread mispricing measure, we next discuss zero volatility and option-adjusted spreads.

3 ZV Spread

This section defines a ZV spread (sometimes called a static spread); see [3, p. 340]. Here, we show that the ZV spread captures both mispricings and the value of any embedded options.

Definition 1 (ZV spread). *The* static or ZV spread $zv(mkt)$ *is the solution to the following expression,*

$$V_0^{mkt} = \sum_{t=1}^{T} C_t e^{-\int_0^t [f(0,s)+zv(mkt)]ds}$$

$$= \sum_{t=1}^{T} e^{-zv(mkt)\cdot t} C_t p(0,t). \tag{4}$$

$zv(mkt)$ is that spread such that the discounted cash flows to the bond C_t (excluding any embedded options) equal the market price. Note that because the market price (the left side) includes both the value of the embedded options and any possible mispricings, $zv(mkt)$ is a joint measure of the excess spread in the bond due to both the embedded options and mispricings.

The first question to ask with respect to ZV spreads concerns their existence and uniqueness. Using Sturm's Theorem [5, p. 298], it can be easily shown that there is always one relevant real root to this equation (the proof is in the appendix). This is due to the fact that both the bond's cash flows and market price are nonnegative.

To obtain a spread measure of the bond's mispricing alone, using ZV spreads, we need to define another quantity, $zv(fair)$, as the solution to the following expression,

$$V_0 = \sum_{t=1}^{T} C_t e^{-\int_0^t [f(0,s)+zv(fair)]ds}.$$

Note that the computation of $zv(fair)$ uses the bond's fair price V_0, and not the market price on the left side of the expression. (This $zv(fair)$ exists and is unique by the same argument used previously.) This computation, therefore, requires the use of a model for the evolution of the term structure of interest rates.

Then, a spread mispricing measure is defined by

$$spread\ mispricing = zv(market) - zv(fair).$$

If this quantity is positive, then the bond is earning too large a spread and it is undervalued. If it is negative, the converse is true. As a spread, this mispricing measure can be compared across bonds with different maturities, coupons, or embedded options. This spread measure of mispricing is an alternative to an OAS, discussed in the next section.

4 OAS

This section defines OAS and explains its meaning. We show that OAS is a spread measure of mispricing. OAS is defined in the literature in the following fashion (cf. [2, p. 2; 3, p. 359], and [8, p. 253]).

Definition 2 (OAS). OAS, φ, is the solution to the following expression,

$$V_0^{mkt} = E\left[\sum_{t=1}^{T} c_t(\omega) e^{-\int_0^t [r_s(\omega)+\varphi]ds}\right]$$

$$= \sum_{t=1}^{T} e^{-\varphi t} E\left[c_t(\omega) e^{-\int_0^t r_s(\omega)ds}\right]. \tag{5}$$

OAS is that spread φ that equates the expected value of the bond's cash flows (including any embedded options) to the market price. The probabilities are computed under the martingale measure so that they are adjusted to reflect any relevant risk premia. As before, it can be shown that there is always one relevant real root to this equation (the proof is in the appendix). This is due to the fact that both the bond's cash flows and market price are nonnegative.

As defined, OAS is a measure of mispricings. This can be easily proven by referring back to expression (1). If the bond is properly priced (i.e., $V_0^{mkt} = V_0$) then we get

$$V_0^{mkt} = \sum_{t=1}^{T} E\left[c_t(\omega)e^{-\int_0^t r_s(\omega)ds}\right]. \tag{6}$$

This is similar in appearance to expression (5) in the definition of an OAS. In expression (6), a zero spread equates the expected discounted cash flows to the market price. Hence, any nonzero spread added to expression (6) represents a mispricing. As before, the mispricing reflects both market error and model error. Market errors represent arbitrage opportunities, whereas model errors represent a misspecified model.

The notion of an OAS cannot be extended to swaps or futures. The reason is that if the cash flows $c_t(\omega)$ can be nonpositive, then multiple relevant real roots are possible to expression (6). For this more general situation, OAS is no longer a useful measure of the security's mispricing. Indeed, given multiple roots, which one to use for relative comparisons is indeterminate. This problem is similar to the well-known difficulties with using the internal rate of return for comparing different (capital budgeting) investment projects.

5 An Example

This section presents a numerical example to illustrate the previous points with respect to ZV spreads and OAS. The example represents a discrete time version of an HJM model. For a complete discussion of these discrete time models see [6]. Because the example is in discrete time, we use discrete instead on continuously compounded interest rates. Otherwise, the notation follows the theory section above.

5.1 The Economy

Consider a three-period economy with time periods 0,1,2. Trading are zero-coupon bonds that mature at times 1 and 2. The bond prices are given in the following tree.

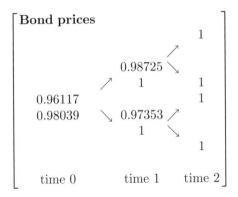

The first column represents time 0. The bond price at the top of the first vector at time 0 represents the price of the two-period zero coupon bond, 0.96117. The bond price at the bottom of the first vector at time 0 represents the one-period zero coupon bond's price, 0.98039. At time 1 these bond prices move randomly to the two possible states of the economy (up and down). The prices for both bonds in these two states are given in the column labeled time 1. In the up state, the price of the two-period bond is 0.98725 and the price of the one-period bond is 1. After time 1 only one bond remains, the two-period bond. Its price is unity for all possible movements of the tree.

The time 0 forward rates (discrete) implied by the bond prices are $f(0,0) = 0.02$ and $f(0,1) = 0.02$.

These bond prices imply the following spot interest rate tree.

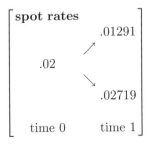

It is easy to show that this term structure evolution is arbitrage-free and complete. The easiest way is to identify the unique martingale probabilities at time 0. They are equal to 0.5. One can check that the time 0 price of the two-period bond equals its discounted expected value:

$$0.96117 = .5(0.98725 + 0.97353)/1.02,$$

making the tree arbitrage-free.

5.2 The Bonds and the Prepayment Option

Next, consider a coupon bond with coupon rate 0.015, principal 100, maturity time 2. Its value plus cash flow tree is:

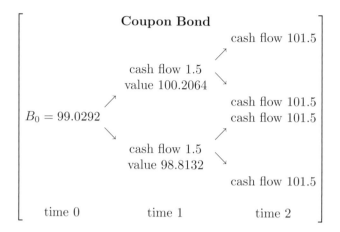

Coupon Bond

cash flow 101.5

cash flow 1.5
value 100.2064

cash flow 101.5
cash flow 101.5

$B_0 = 99.0292$

cash flow 1.5
value 98.8132

cash flow 101.5

time 0   time 1   time 2

Consider a prepayment option with maturity time 1 and strike price $K = 100$. Its value tree is:

option

$\max(100.2064 - K, 0) = 0.2064$

$C_0 = 0.101105$

$\max(98.8132 - K, 0) = 0$

time 0   time 1

where $C_0 = .5(\max(100.2064 - K, 0) + \max(98.8132 - K, 0))/1.02 = 0.101105$.

Next, consider a bond with an embedded prepayment option. Its value is $V_0 = B_0 - C_0 = 98.928074$. One subtracts the prepayment option, because it is exercised by the issuer of the bond. The cash flow to the bond with the prepayment option is:

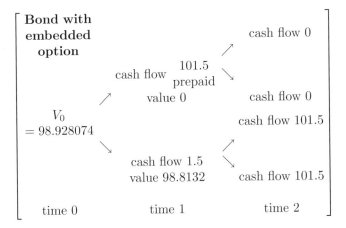

This cash flow tree is needed for the computation of OAS below.

5.3 The ZV Spread Computation

Using the arbitrage-free price for the bond 98.928074, the $zv(fair)$ is obtained by solving the following equation,

$$98.928074 = \frac{1}{(1 + zv(fair))}(1.5)(.98039) + \frac{1}{(1 + zv(fair))^2}(101.5)(.96117).$$

There are two solutions to this expression: $zv(fair) = 0.0005155, -1.9856503$. There is only one root that keeps the discount factor nonnegative. This is the relevant root (0.0005155).

Supposing the market value of the bond is 99, we can compute the $zv(mkt)$ by solving a similar equation,

$$99.00 = \frac{1}{(1 + zv(mkt))}(1.5)(.98039) + \frac{1}{(1 + zv(mkt))^2}(101.5)(.96117).$$

There are two solutions to this expression: $zv(mkt) = 0.0001493, -1.9852949$. There is only one root that keeps the discount factor nonnegative. This is the relevant root (0.0001493).

The mispricing measure based on ZV spreads is $zv(market) - zv(fair)$ $= 0.0001493 - 0.0005155 = -.0003662$. The bond is overvalued by -3.7 basis points.

5.4 The OAS Computation

Using the arbitrage-free price for the bond, 98.928074, the OAS is obtained by solving the following equation,

$$98.928074 = \frac{.5}{(1+\varphi)}\left[\frac{1.5}{1+.02} + \frac{101.5}{1+.02}\right]$$

$$+ \frac{.5}{(1+\varphi)^2}\left[\frac{101.5}{(1+.02)(1+.02719)}\right]\frac{1}{(1+\varphi)^2}.$$

There are two solutions to this expression: $\varphi = 0$ and -1.48963. There is only one root that keeps the discount factor nonnegative. This is the proper root ($\varphi = 0$). This example shows that OAS reflects the fact that there are no mispricings in this economy.

Next, let us suppose that the bond trades for 99.00 dollars; that is, it is overvalued. OAS is then the solution to:

$$99.00 = \frac{.5}{(1+\varphi)}\left[\frac{1.5}{1+.02} + \frac{101.5}{1+.02}\right]$$

$$+ \frac{.5}{(1+\varphi)^2}\left[\frac{101.5}{(1+.02)(1+.02719)}\right]\frac{1}{(1+\varphi)^2}.$$

The solutions are $-.0004878$ and -1.48951. Ignoring the second root again, the OAS of -4.9 basis points shows the bond is overvalued.

Note that the mispricing measure based on ZV spreads versus OAS differs, although both show an overvaluation.

6 Conclusion

This paper provides a tutorial on two commonly used measures: a zero volatility spread and an option adjusted spread. The Heath–Jarrow–Morton model is used as a frame of reference for understanding these two quantities. We show that the ZV spread is a measure of the excess spread due to both a bond's embedded options and any mispricings, whereas the OAS provides only a valid measure of a bond's mispricings.

If mispricings exist, they are due to either market error or model error. If it is market error, then it represents an arbitrage opportunity. If it is model error, then it represents either an incorrect stochastic process, omitted risk premia, omitted risks (e.g., liquidity risk), or market imperfections (e.g., transaction costs). To distinguish between these two alternatives, one can backtest the model with the arbitrage trading strategies implied by the mispricing (necessarily including all market imperfections). If it is market error, then the trading strategy will earn abnormal returns. If not, then it is model error.

References

1. T. Bielecki and M. Rutkowski. *Credit Risk: Modeling, Valuation, and Hedging.* Springer, 2002.
2. J. Cilia. Product summary: Advanced CMO analytics for bank examiners: Practical applications using Bloomberg. Federal Reserve Bank of Chicago, May 1995.
3. F. Fabozzi. *Bond Markets, Analysis and Strategies,* 4th edition. Prentice-Hall, 2000.
4. D. Heath, R. Jarrow, and A. Morton. Bond pricing and the term structure of interest rates: A new methodology for contingent claims valuation. *Econometrica,* 60:77–105, 1992.
5. N. Jacobson. *Basic Algebra I.* W. H. Freeman, 1974.
6. R. Jarrow. *Modeling Fixed Income Securities and Interest Rate Options,* 2nd edition. Stanford University Press, 2002.
7. P. Protter. *Stochastic Integration and Differential Equations,* 2nd edition. Springer, 2005.
8. B. Tuckman. *Fixed Income Securities.* John Wiley and Sons, 1996.

Appendix: Real Root Solution to Equations (4) and (5)

Both Equations (4) and (5) take the form

$$\sum_{t=1}^{T} e^{-\beta \cdot t} \alpha_t - \alpha_0 = 0, \tag{7}$$

where $\alpha_t \geq 0$ for all t and $\alpha_0 > 0$. Let us write $x = e^{-\beta}$. Then the relevant range for x is $(0, \infty)$. We seek a real root of the polynomial equation

$$f(x) = \sum_{t=1}^{T} x^t \alpha_t - \alpha_0$$

on $[0, M]$ for M chosen large enough so that $f(M) > 0$. Using Sturm's theorem (see [5, p. 298]), we note that because $f(0) = -\alpha_0 < 0$, $f(M) > 0$, and all derivatives of $f(x)$ up to degree T are nonnegative for all x, we get that the number of variations in the sign of $\{f(0), f'(0), \ldots, f^T(0)\}$ is 1, where $f^T(0)$ denotes the Tth derivative, and the number of variations in the sign of $\{f(M), f'(M), \ldots, f^T(M)\}$ is 0. Sturm's theorem then tells us the number of distinct real roots of $f(x)$ in $(0, M)$ is 1.

Inasmuch as this is true for all $x > M$, we see that there is one distinct real root of $f(x)$ for $x \in (0, \infty)$.

Finally, because $x = e^{-\beta}$, this gives us one real root β of (7).

Asset Price Bubbles in Complete Markets

Robert A. Jarrow,[1] Philip Protter,[2] and Kazuhiro Shimbo[2]

[1] Johnson Graduate School of Management
Cornell University
Ithaca, NY 14850, USA
raj15@cornell.edu

[2] School of Operations Research and Industrial Engineering
College of Engineering
Cornell University
Ithaca, NY, 14850, USA
pep4@cornell.edu, ks266@cornell.edu

Summary. This paper reviews and extends the mathematical finance literature on bubbles in complete markets. We provide a new characterization theorem for bubbles under the standard no-arbitrage framework, showing that bubbles can be of three types. Type 1 bubbles are uniformly integrable martingales, and these can exist with an infinite lifetime. Type 2 bubbles are nonuniformly integrable martingales, and these can exist for a finite, but unbounded, lifetime. Last, Type 3 bubbles are strict local martingales, and these can exist for a finite lifetime only. When one adds a no-dominance assumption (from Merton [24]), only Type 1 bubbles remain. In addition, under Merton's no-dominance hypothesis, put–call parity holds and there are no bubbles in standard call and put options. Our analysis implies that if one believes asset price bubbles exist and are an important economic phenomena, then asset markets must be incomplete.

Key words: Bubbles; no free lunch with vanishing risk (NFLVR); complete markets; local martingale; put–call parity; derivative pricing.

1 Introduction

Although asset price bubbles, their existence and characterization, have enthralled the imagination of economists for many years, only recently has this topic been studied using the tools of mathematical finance; see in particular Loewenstein and Willard [22], Cox and Hobson [7], Jarrow and Madan [20], Gilles [15], Gilles and Leroy [16], and Huang and Werner [17]. The purpose of this paper is to review and to extend this mathematical finance literature in order to increase our understanding of asset price bubbles. In this paper, we restrict our attention to arbitrage-free economies that satisfy both the no-free-lunch-with-vanishing-risk (NFLVR) and complete markets hypotheses, in order that both the first and second fundamental theorems of asset pricing apply. Equivalently, there exists a unique equivalent local martingale measure. We exclude the study of incomplete markets. (We study incomplete market asset price bubbles in a companion paper, see Jarrow et al. [21].) We also exclude the study of charges, because charges require a stronger notion of no-arbitrage (see Jarrow and Madan [20], Gilles [15], and Gilles and Leroy [16]).

We make two contributions to the bubbles literature. First, we provide a new characterization theorem for asset price bubbles. Second, we study the effect of additionally imposing Merton's [24] no-dominance assumption on the existence of bubbles in an economy. Our new results in this regard are:

(i) Bubbles can be of three types: an asset price process that is (1) a uniformly integrable martingale, (2) a martingale that is not a uniformly integrable martingale, or (3) a strict local martingale that is not a martingale. Bubbles of Type 1 can be viewed as the asset price process containing a component analogous to fiat money (see Example 2). Type 2 bubbles are generated by the fact that all trading strategies must terminate in finite time, and Type 3 bubbles are caused by the standard admissibility condition used to exclude doubling strategies.

(ii) Bubbles cannot be started—"born"—in a complete market. (In contrast, they can be born in incomplete markets.) They either exist at the start or not, and if they do exist, they may disappear as the economy evolves.

(iii) Bubbles in standard European call and put options can only be of Type 3, because standard options have finite maturities. Under NFLVR, any assets and contingent claims can have bubbles and put–call parity does not hold in general.

(iv) Under NFLVR and no-dominance, in complete markets, there can be no Type 2 or Type 3 asset price bubbles. Consequently, standard options have no bubbles and put–call parity holds.

The economic conclusions from this paper are threefold. First, bubbles of Type 1 are uninteresting from an economic perspective because they represent a permanent but stochastic wedge between an asset's fundamental value and its market price, generated by a perceived residual value at time infinity.

Second, Type 2 bubbles are the result of trading strategies being of finite time duration, although possibly unbounded. To try to profit from a bubble of Type 2 or Type 3, one would short the asset in anticipation of the bubble bursting. Because a Type 2 bubble can exist, with positive probability, beyond any trading strategy, these bubbles can persist as they do not violate the NFLVR assumption. Type 3 bubbles occur in assets with finite maturities. For these asset price bubbles, unprotected shorting is not feasible, because due to the admissibility condition, if the short's value gets low enough, the trading strategy must be terminated with positive probability before the bubble bursts. This admissibility condition removes downward selling pressure on the asset's price, and hence enables these bubbles to exist.

Third, modulo Type 1 bubbles, under both the NFLVR and no-dominance hypotheses, there can be no asset pricing bubbles in complete markets. This implies that if one believes asset pricing bubbles exist and are an important economic phenomenon, and if one accepts Merton's "no-dominance" assumption, then asset markets must be incomplete.

An outline of this paper is as follows. Section 2 presents our model structure and defines an asset price bubble. Section 3 characterizes the properties of asset price bubbles. Section 4 provides the economic intuition underlying the mathematics, and Section 5 extends the analysis to contingent claims bubbles. Finally, Section 6 concludes.

2 Model Description

This section presents the details of our economic model.

2.1 No Free Lunch with Vanishing Risk (NFLVR)

Traded in our economy is a risky asset and a money market account. For simplicity, and without loss of generality, we assume that the spot interest rate is 0 in our economy, so that the money market account has constant unit value. Let τ be a maturity (life) of the risky asset. Let $(D_t)_{0 \leq t < \tau}$ be a càdlàg semimartingale representing the cumulative dividend process of the risky asset, with X_τ its terminal payoff or liquidation value at time τ. We assume that X_τ, $D_t \geq 0$ for each $t \in (0, \tau)$. The market price of the risky asset is given by a nonnegative càdlàg semimartingale $S = (S_t)_{t \geq 0}$ defined on a filtered complete probability space $(\Omega, \mathcal{F}, \mathbb{F}, P)$ where $\mathbb{F} = (\mathcal{F}_t)_{t \geq 0}$. We assume that the filtration $\mathbb{F}$ satisfies the usual hypotheses (cf. [25]). Note that for t such that $\triangle D_t > 0$, S_t denotes a price ex-dividend, because S is càdlàg. Let $W = (W_t)_{t \geq 0}$ be a wealth process from owning the asset, given by

$$W_t = S_t + \int_0^{t \wedge \tau} dD_u + X_\tau \mathbf{1}_{\{\tau \leq t\}}.$$

A key notion in our economy is an equivalent local martingale measure.

Definition 1 (Equivalent Local Martingale Measure). *Let Q be a probability measure equivalent to P such that the wealth process W is a Q-local martingale. We call Q an* Equivalent Local Martingale Measure (ELMM). *We denote the set of ELMMs by $\mathcal{M}_{loc}^e(W)$.*

A trading strategy is defined to be a pair of processes $(\pi, \eta) = (\pi_t, \eta_t)_{t \geq 0}$ representing the number of units of the risky asset and money market account held at time t with $\pi \in L(W)$ (see [25] for the definition of the space of integrable processes $L(W)$). The wealth process of the trading strategy (π, η)

is given by $V^{\pi,\eta} = (V_t^{\pi,\eta})_{t\geq0}$, where $V_t^{\pi,\eta} = \pi_t S_t + \eta_t$. Assume temporarily that π is a semimartingale. Then a self-financing trading strategy with $V_0^{\pi,\eta} = 0$ is a trading strategy (π, η) such that the associated wealth process $V^{\pi,\eta}$ is given by

$$
\begin{aligned}
V_t^{\pi,\eta} &= \int_0^t \pi_u dW_u = \int_0^t \pi_u dS_u + \int_0^{t\wedge\tau} \pi_u dD_u + \pi_\tau X_\tau \mathbf{1}_{\{\tau\leq t\}} \\
&= \left(\pi_t S_t - \int_0^t S_{u-} d\pi_u - [\pi^c, S^c]_t \right) + \int_0^{t\wedge\tau} \pi_u dD_u + \pi_\tau X_\tau \mathbf{1}_{\{\tau\leq t\}} \\
&= \pi_t S_t + \eta_t,
\end{aligned}
\tag{1}
$$

where we have used integration by parts, and where

$$
\eta_t = \int_0^{t\wedge\tau} \pi_u dD_u + \pi_\tau X_\tau \mathbf{1}_{\{\tau\leq t\}} - \int_0^t S_{u-} d\pi_u - [\pi^c, S^c]_t.
\tag{2}
$$

If we now discard the temporary assumption that π is a semimartingale, we simply define a self-financing trading strategy (π, η) to be a pair of processes, with π predictable and η optional and such that:

$$
V_t^{\pi,\eta} = \pi_t S_t + \eta_t = \int_0^t \pi_u dW_u.
$$

As noted, a self-financing trading strategy starts with zero units of the money market account $\eta_0 = 0$, and it reflects proceeds from purchases/sales of a risky asset that accumulate holdings in the money market account as the cash flows from the risky asset are deposited. In particular, Equation (2) shows that η is uniquely determined by π if a trading strategy is self-financing. Therefore without loss of generality, we represent (π, η) by π.

To avoid doubling strategies, we further restrict the class of self-financing trading strategies.

Definition 2 (Admissibility). *Let $V^{\pi,\eta}$ be the wealth process given by (1). We say that the trading strategy π is a-admissible if it is self-financing and $V_t^{\pi,\eta} \geq -a$ a.s. for all $t \geq 0$. We say a trading strategy is admissible if it is self-financing and $V_t^{\pi,\eta} \geq -a$ for all $t \geq 0$ for some $a \in \mathbb{R}^+$.*

The notion of admissibility corresponds to a lower bound on the wealth process, an implicit inability to borrow if one's debt becomes too large. (For example, see Loewenstein and Willard [22, Equation (5), p. 23].) There are several alternative definitions of admissibility that could be employed and these are discussed in Section 4.3. However, all of our results are robust to these alternative formulations.

We want to explore the existence of bubbles in arbitrage-free markets, hence, we need to define the NFLVR hypothesis. Let

$$
\begin{aligned}
\mathcal{K} &= \{W_\infty^\pi = (\pi \cdot W)_\infty : \pi \text{ is admissible}\}, \\
\mathcal{C} &= (\mathcal{K} - L_+^0) \cap L_\infty.
\end{aligned}
$$

Definition 3 (NFLVR). *We say that a semimartingale S satisfies the* no free lunch with vanishing risk (NFLVR) *condition with respect to admissible integrands, if*

$$\bar{C} \cap L^\infty_+ = \{0\},$$

where $\bar{C}$ denotes the closure of C in the sup-norm topology of L^∞.

Given NFLVR, we impose the following assumption.

Assumption 1 *The market satisfies NFLVR hypothesis.*

By the first fundamental theorem of asset pricing [9], this implies that the market admits an equivalent σ-martingale measure. By Proposition 3.3 and Corollary 3.5 [1, pp. 307, 309], a σ-martingale bounded from below is a local martingale. (For the definition and properties of σ-martingales, see [25; 14; 9; 19, Section III.6e].) Thus we have the following theorem.

Theorem 1 (First Fundamental Theorem). *A market satisfies the NFLVR condition if and only if there exists an ELMM.*

Theorem 1 holds even if the price process is not locally bounded, due to the assumption that W is nonnegative. (In [9], the driving semimartingale (price process) takes values in $\mathbb{R}^d$ and is not locally bounded from below.)

We are interested in studying the existence and characterization of bubbles in complete markets. A market is complete if for all $X_\infty \in L^2(\Omega, \mathcal{F}_\infty, P)$, there exists a self-financing trading strategy (π, η) and $c \in \mathbb{R}$ such that

$$X_\infty = c + \int_0^\infty \pi_u dW_u.$$

For the subsequent analysis, we also assume that the market is complete, hence by the second fundamental theorem of asset pricing (cf. [18]), the ELMM is unique.

Assumption 2 *Given the market satisfies NFLVR, the ELMM is unique.*

This assumption is key to a number of the subsequent results. For the remainder of the paper we assume that both Assumptions 1 and 2 hold, that is, that the markets are arbitrage-free and complete.

2.2 No-Dominance

In addition to Assumption 1, we also study the imposition of Merton's [24] *no-dominance assumption.* To state this assumption in our setting, assume that there are two assets or contingent claims characterized by the pair of cash flows $(\{D^1_t\}_{t\geq 0}, X^1_\tau)$, $(\{D^2_t\}_{t\geq 0}, X^2_\tau)$. Let V^1_t, V^2_t denote their market prices at time t.

Assumption 3 (No-Dominance) *For any stopping time* $\tau_0 \leq \tau$ *a.s., if*

$$D^2_{\tau_0+u} - D^2_{\tau_0} \geq D^1_{\tau_0+u} - D^1_{\tau_0} \quad and \quad X^2_\tau \mathbf{1}_{\{\tau>\tau_0\}} \geq X^1_\tau \mathbf{1}_{\{\tau>\tau_0\}} \quad for\ u > 0, \quad (3)$$

then $V^2_{\tau_0} \geq V^1_{\tau_0}$. *Furthermore, if for some stopping time* τ_0,

$$E\left[\mathbf{1}_{(\{D^2_\infty - D^2_{\tau_0} > D^1_\infty - D^1_{\tau_0}\} \cup \{X^2_\tau \mathbf{1}_{\{\tau>\tau_0\}} > X^1_\tau \mathbf{1}_{\{\tau>\tau_0\}}\})} \middle| \mathcal{F}_{\tau_0}\right] > 0$$

with positive probability, then $V^2_{\tau_0} > V^1_{\tau_0}$.

Note that (3) implies that $X^2_\tau \mathbf{1}_{\{\tau>\tau_0\}} \geq X^1_\tau \mathbf{1}_{\{\tau>\tau_0\}}$ for any stopping time τ_0 such that $\tau_0 \leq \tau$.

   This assumption rephrases Assumption 1 of [24] in modern mathematical terms, we believe for the first time. In essence, it codifies the intuitively obvious idea that, all things being equal, financial agents prefer more to less. Assumption 3 is violated only if there is an agent who is willing to buy a dominated security at the higher price.

   Assumption 3 is related to Assumption 1, but they are not equivalent.

Lemma 1. *Assumption 3 implies Assumption 1. However, the converse is not true.*

Proof. Assume that W allows for a free lunch with vanishing risk. There is $f \in L^\infty_+(P)\backslash\{0\}$ and sequence $\{f_n\}^\infty_{n=0} = \{(H^n \cdot W)_\infty\}^\infty_{n=0}$ where H^n is a sequence of admissible integrands and $\{g_n\}$ satisfying $g_n \leq f_n$ such that

$$\lim_n \|f - g_n\|_\infty = 0.$$

In particular, the negative part $\{(f_n)^-\}$ tends to zero uniformly. (See [10, p. 131]). Applying Assumption 3 to two terminal payoffs f and 0, we have $0 = V_0(f) > 0$, a contradiction. Therefore Assumption 3 implies Assumption 1. For the converse, see Example 1.   □

The domain of Assumption 3 contains a domain of Assumption 3, $\bar{C} \cap L^\infty_+(P)$. This explains why Assumption 3 implies Assumption 1.

   The following is an example consistent with Assumption 1 but excluded by Assumption 3.

Example 1. Consider two assets maturing at τ with payoffs X_τ and Y_τ, respectively. Suppose that $X_\tau \geq Y_\tau$ a.s. Then,

$$X^*_t = E_Q[X_\tau|\mathcal{F}_t]\mathbf{1}_{\{t<\tau\}} \geq E_Q[Y_\tau|\mathcal{F}_t]\mathbf{1}_{\{t<\tau\}} = Y^*_t.$$

Let β be a nonnegative local martingale such that $\beta_\tau = 0$ and $\beta_t > X^*_t - Y^*_t$ for some $t \in (0, \tau)$. (The existence of such a process follows, for example, from Example 3.) Suppose further that the prices of asset $X_t = X^*_t$ and $Y_t = \beta_t + Y^*_t$. Then, Assumption 3 is violated because $Y_t > X_t$.

To see that this is not an NFLVR, consider a strategy that would attempt to take advantage of this mispricing. One would want to sell Y and to buy X, say at time t. Then, if held until maturity, this would generate a cash flow equal to $\beta_t - (X_t^* - Y_t^*) > 0$ at time t and $X_\tau - Y_\tau \geq 0$ at time τ. However, for any u with $t < u \leq \tau$, the market value of this trading strategy is $-Y_u + X_u = -\beta_u + (X_u^* - Y_u^*)$. Because $-\beta$ is negative and unbounded, this strategy is inadmissible and not a FLVR. We discuss issues related to admissibility further in Section 4.3.

One situation Assumption 3 is meant to exclude is often called a suicide strategy (see Harrison and Pliska [18] for the notion of a suicide strategy). An alternative approach for dealing with suicide strategies is to restrict the analysis to the set of maximal assets. An outcome $(\pi \cdot S)_\infty$ of an admissible strategy π is called *maximal* if for any admissible strategy π' such that $(\pi' \cdot S)_\infty \geq (\pi \cdot S)_\infty$, then $\pi' = \pi$.

2.3 Bubbles

This section provides the definition of an asset pricing bubble in our economy. To do this, we must first define the asset's fundamental price.

The Fundamental Price

We define the *fundamental price* as the expected value of the asset's future payoffs with respect to the ELMM $Q \in \mathcal{M}_{loc}^e(W)$. (Recall that we assume that the market is complete, Assumption 2.)

Definition 4 (Fundamental Price). *The* fundamental price S_t^* *of an asset with market price S_t is defined by*

$$S_t^* = E_Q \left[\int_t^\tau dD_u + X_\tau \mathbf{1}_{\{\tau < \infty\}} | \mathcal{F}_t \right] \mathbf{1}_{\{t < \tau\}}. \tag{4}$$

Note that the fundamental price is just the conditional expected value of the asset's cash flows, under the valuation measure Q. (Note that because the random variable is positive, the conditional expectation is always defined; however, in Lemma 2 that follows, we show that it is actually in L^1 and thus is classically defined.) Also, note that if the asset has a payoff at $\tau = \infty$, then this payoff $X_\tau \mathbf{1}_{\{\tau = \infty\}}$ does not contribute to the fundamental price S_t^*. We do this because an agent cannot consume the payoff $X_\tau \mathbf{1}_{\{\tau = \infty\}}$ by employing an admissible trading strategy. Indeed, although unbounded in time, for a given $\omega \in \Omega$, all such admissible trading strategies must terminate in finite time.

Lemma 2. *The fundamental price in (4) is well defined. Furthermore, $(S_t)_{t \geq 0}$ converges to $S_\infty \in L^1(Q)$ a.s. and $(S_t^*)_{t \geq 0}$ converges to 0 a.s.*

Proof. Fix $Q \in \mathcal{M}^e_{loc}(W)$. To show that S^*_t is well defined, it suffices to show that $\int_0^\tau dD_u + X_\tau \mathbf{1}_{\{\tau < \infty\}} \in L_1(Q)$ because for all t,

$$0 \leq \int_t^\tau dD_u + X_\tau \mathbf{1}_{\{t < \tau\}} \leq \int_0^\tau dD_u + X_\tau \mathbf{1}_{\{t < \tau\}}.$$

By hypothesis, W is a nonnegative supermartingale. By the martingale convergence theorem (see [11, VI.6, p. 72]), there exists $W_\infty \in L^1(Q)$ such that W converges to W_∞ a.s. To show the convergence of S, observe that

$$W_\infty = \lim_{t \to \infty} W_t = \lim_{t \to \infty} \left(S_t + \int_0^{t \wedge \tau} dD_u + X_\tau \mathbf{1}_{\{\tau \leq t\}} \right)$$

$$= \lim_{t \to \infty} S_t + \int_0^\tau dD_u + X_\tau \mathbf{1}_{\{\tau < \infty\}} \quad \text{a.s.}$$

It follows that there exist $S_\infty \in L^1(Q)$ and $\int_0^\tau dD_u + X_\tau \mathbf{1}_{\{\tau < \infty\}} \in L^1(Q)$ because $S \geq 0$. Therefore S^*_t is well defined for all $t \geq 0$. Observe that

$$E_Q \left[\int_t^\tau dD_u + X_\tau \mathbf{1}_{\{\tau < \infty\}} | \mathcal{F}_t \right]$$
$$= - \int_0^t dD_u + E_Q \left[\left(\int_0^\tau dD_u + X_\tau \mathbf{1}_{\{\tau < \infty\}} \right) | \mathcal{F}_t \right] \tag{5}$$

and

$$E_Q \left[\left(\int_0^\tau dD_u + X_\tau \mathbf{1}_{\{\tau < \infty\}} \right) | \mathcal{F}_t \right] \mathbf{1}_{\{t < \tau\}}$$
$$= \left(E_Q \left[\left(\int_0^\tau dD_u + X_\tau \right) \mathbf{1}_{\{\tau < \infty\}} | \mathcal{F}_t \right] + E_Q \left[\mathbf{1}_{\{\tau = \infty\}} \int_0^\infty dD_u | \mathcal{F}_t \right] \right) \mathbf{1}_{\{t < \tau\}}. \tag{6}$$

Substituting (6) into (5) and then into (4),

$$\lim_{t \to \infty} S^*_t = - \int_0^\infty dD_u \mathbf{1}_{\{\tau = \infty\}} + \mathbf{1}_{\{\tau = \infty\}} E_Q \left[\int_0^\tau dD_u + X_\tau \mathbf{1}_{\{\tau < \infty\}} | \mathcal{F}_\infty \right]$$
$$= - \int_0^\infty dD_u \mathbf{1}_{\{\tau = \infty\}} + \mathbf{1}_{\{\tau = \infty\}} E_Q \{ \int_0^\infty dD_u | \mathcal{F}_\infty \} = 0.$$

$\square$

Note that, in general, $\int_0^\infty dD_u + X_\tau \mathbf{1}_{\{\tau < \infty\}}$ need not be P-integrable. In this regard, Lemma 2 shows that the existence of Q implies that $\int_0^\infty dD_u + X_\tau \mathbf{1}_{\{\tau < \infty\}}$ is Q-integrable.

Lemma 3. *The fundamental wealth process $W^* = (W_t)_{t \geq 0}$ given by $W^*_t = S^*_t + \int_0^{\tau \wedge t} dD_u + X_\tau \mathbf{1}_{\{\tau \leq t\}}$ is a uniformly integrable martingale under $Q \in \mathcal{M}^e_{loc}(W)$ closed by*

$$W^*_\infty = \int_0^\tau dD_u + X_\tau \mathbf{1}_{\{\tau < \infty\}}.$$

Proof. By Lemma 2,

$$W_\infty^* := \lim_{t\to\infty} W_t^* = \lim_{t\to\infty} \left(S_t^* + \int_0^{t\wedge\tau} dD_u + X_\tau \mathbf{1}_{\{\tau\leq t\}} \right)$$

$$= \int_0^\tau dD_u + X_\tau \mathbf{1}_{\{\tau<\infty\}} \qquad \text{a.s.}$$

W_∞^* is in $L^1(Q)$ because $S_\infty \geq 0$, $W_\infty \in L^1$, and $W_\infty^* + S_\infty = W_\infty$. Observe that

$$E_Q\left[W_\infty^*|\mathcal{F}_t\right] = E_Q\left[\left(\int_t^\tau dD_u + X_\tau\right)|\mathcal{F}_t\right]\mathbf{1}_{\{t<\tau\}}$$

$$+ \left(-\int_\tau^t dD_u + X_\tau\right)\mathbf{1}_{\{\tau\leq t\}} + \left(\int_0^t dD_u\right)\mathbf{1}_{\{t<\tau\}}$$

$$= S_t^* + \int_0^{t\wedge\tau} dD_u + X_\tau \mathbf{1}_{\{\tau\leq t\}} = W_t^*.$$

It follows that W^* is a closable and hence uniformly integrable martingale. $\square$

The Asset Price Bubble

Definition 5 (Bubble). *The asset price bubble β_t for S_t is given by*

$$\beta_t = S_t - S_t^*.$$

As indicated, the asset price bubble is the asset's market price less the asset's fundamental price.

3 Properties of Bubbles

In this section, we analyze the properties of asset price bubbles, applying semimartingale theory and potential theory. We begin with a nonstandard definition.

Definition 6 (Strict Local Martingale). *A strict local martingale is a local martingale that is not a martingale.*

The term "strict local martingale" is not common in the literature, but it can be found in the recent book of Delbaen and Schachermayer [10], who in turn refer to a paper of Elworthy et al. [13]. We hasten to remark that their definition of a strict local martingale is different from our definition. Indeed, Delbaen and Schachermayer refer to a strict local martingale as being a local martingale that is not a uniformly integrable martingale. They allow a strict local martingale to be actually a martingale, as long as the martingale itself is not uniformly integrable. Our definition is more appropriate for the study of bubbles, as is made clear shortly.

3.1 Characterization of Bubbles

Theorem 2. *If there exists a nontrivial bubble $\beta := (\beta)_{t \geq 0} \neq 0$ in an asset's price, then we have three and only three possibilities:*

1. *β is a local martingale (which could be a uniformly integrable martingale) if $P(\tau = \infty) > 0$.*
2. *β is a local martingale but not a uniformly integrable martingale if it is unbounded, but with $P(\tau < \infty) = 1$.*
3. *β is a strict Q-local martingale, if τ is a bounded stopping time.*

Proof. Fix $Q \in \mathcal{M}_{loc}^e(W)$. Because W is a closable supermartingale (see proof of Lemma 2), there exists $W_\infty \in L^1(Q)$ such that W converges to W_∞ a.s. Let

$$\beta_t' = W_t - E_Q[W_\infty|\mathcal{F}_t]. \tag{7}$$

Then $(\beta_t')_{t \geq 0}$ is a (nonnegative) local martingale, because it is a difference of a local martingale and a uniformly integrable martingale. By Lemma 3,

$$E_Q[W_\infty|\mathcal{F}_t] = E_Q[W_\infty^*|\mathcal{F}_t] + E_Q[S_\infty|\mathcal{F}_t] = W_t^* + E_Q[S_\infty|\mathcal{F}_t]. \tag{8}$$

By the definition of wealth processes, and applying Equations (7) and (8),

$$\beta_t = S_t - S_t^* = W_t - W_t^*$$
$$= \left(E_Q[W_\infty|\mathcal{F}_t] + \beta_t^1\right) - \left(E_Q[W_\infty|\mathcal{F}_t] - E_Q[S_\infty|\mathcal{F}_t]\right) = \beta_t' + E_Q[S_\infty|\mathcal{F}_t].$$

If $\tau < T$ for $T \in \mathbb{R}^+$, then $S_\infty = 0$. A bubble $\beta_t = \beta_t' = 0$ for $t \geq \tau$ and in particular $\beta_T = 0$. If β_t is a martingale,

$$\beta_t = E_Q[\beta_T|\mathcal{F}_t] = 0 \qquad \forall t \leq T.$$

It follows that β is a strict local martingale. This proves Part 1. For Part 2, assume that β is a uniformly integrable martingale. Then by Doob's optional sampling theorem, for any stopping time $\tau_0 \leq \tau$,

$$\beta_{\tau_0} = E_Q[\beta_\tau|\mathcal{F}_{\tau_0}] = 0, \tag{9}$$

and because β is optional, it follows from (for example) the section theorems of P.A. Meyer that $\beta = 0$ on $[0, \tau]$. Therefore the bubble does not exist. For Part 3, $(E_Q[S_\infty|\mathcal{F}_t])_{t \geq 0}$ is a uniformly integrable martingale. $\qquad \square$

As indicated, there are three types of bubbles that can be present in an asset's price. Type 1 bubbles occur when the asset has infinite life with a payoff at $\{\tau = \infty\}$. Type 2 bubbles occur when the asset's life is finite, but unbounded. Type 3 bubbles are for assets whose lives are bounded. In a subsequent section, we provide an intuitive economic explanation for why these bubbles exist. Before that, however, we provide some examples.

3.2 Examples

This section presents simple examples of bubbles of Types 1, 2, and 3.

A Uniformly Integrable Martingale Bubble: Fiat Money (Type 1)

Example 2. Let S be fiat money given by $S_t = 1$ for all $t \geq 0$. Fiat money is money that the government declares to be legal tender, although it cannot be converted into standard specie. Because money never matures, $\tau = \infty$ and $X_\infty = 1$. Money pays no dividend and hence $D \equiv 0$. Therefore $S^* \equiv 0$ and

$$\beta_t = S_t - S_t^* = 1 \qquad \forall t \geq 0.$$

The entire value of money comes from the bubble, its payoff $X_\infty = 1$, and it is a trivial uniformly integrable martingale. Note that in our setting, fiat money is equivalent to our money market account (paying zero interest for all time).

A Martingale Bubble (Type 2)

Example 3. Let the asset's maturity τ be a positive random time with $P(\tau > t) > 0$ for all $t \geq 0$. Let the fundamental price process be $(S_t^*)_{t \geq 0} = (\mathbf{1}_{\{t < \tau\}})_{t \geq 0}$, with payoff 1 at time τ. Set a process β by

$$\beta_t = \frac{1 - \mathbf{1}_{\{\tau \leq t\}}}{Q(\tau > t)}.$$

Lemma 4 shows that β is a martingale that is not a uniformly integrable martingale, with $\beta_\infty = 0$. Then

$$S = S^* + \beta$$

is a price process with a nonuniformly integrable bubble.

Lemma 4. *Let τ_0 be a positive finite random variable such that $P(\tau_0 > t) > 0$ for all $t \geq 0$. Let $D = (D_t)_{t \geq 0} = (\mathbf{1}_{\{\tau_0 \leq t\}})_{t \geq 0}$ and $\mathbb{D} = (\mathcal{D}_t)_{t \geq 0}$ be a natural filtration generated by D. Then a process N defined by*

$$N_t = \frac{1 - D_t}{P(\tau_0 > t)}$$

is a martingale that is not a uniformly integrable martingale, and $N_\infty = 0$.

Proof. By the structure of $\mathbb{D}$ (e.g., see Protter [25, Lemma on p. 121]) for $s < t$,

$$P(\tau_0 > t | \mathcal{D}_s) = \mathbf{1}_{\{\tau_0 > t\}} P(\tau_0 > t | \tau_0 > s) = \mathbf{1}_{\{\tau_0 > s\}} \frac{P(\tau_0 > t)}{P(\tau_0 > s)}.$$

Therefore,

$$E\left[1 - D_t | \mathcal{D}_s\right] = (1 - D_s)\frac{P(\tau_0 > t)}{P(\tau_0 > s)}.$$

This shows that N is a martingale. Observe that $N_t = 0$ on $\{t > \tau_0\}$ and hence N converges to 0 a.s., because $\tau_0 < \infty$ a.s. If N is a uniformly integrable martingale, then N is closable by N_∞ and $N_t = E[N_\infty | \mathcal{D}_t] \equiv 0$, which is not true. Therefore N is not uniformly integrable. $\square$

This example has the asset's maturity τ having a positive probability of continuing past any given future time t. Although finite with probability one, the asset's life is unbounded.

A Strict Local Martingale Bubble (Type 3)

The following example is essentially that contained in Cox and Hobson [7, Example 3.5, p. 9 and 2.2.1, p. 4]. In this example, although the asset has finite maturity T, a bubble still exists, as the following lemma shows.

Example 4. Let $D = (D_t)_{t \geq 0} \equiv 0$ $\tau = T$ and $X_\tau = X_T = 1$. Then the fundamental price at $t \geq 0$ is $S_t^* = \mathbf{1}_{\{t < T\}}$. Define a process β by

$$\beta_t = 1 + \int_0^t \frac{\beta_u}{\sqrt{T-u}} dB_u, \tag{10}$$

where $B = (B_t)_{t \geq 0}$ is a standard Q-Brownian motion. Lemma 5 shows that β is a strict local martingale with $\beta_T = 0$. Then

$$S = S^* + \beta$$

is a price process with a strict local martingale bubble.

Lemma 5. *A process β defined by Equation (10) is a continuous local martingale on $[0, T]$.*

Proof. The stochastic integral $(\int_0^t dB_s / \sqrt{T-s})_{t \geq 0}$ is a local martingale but not a martingale on $[0, T)$ (because it is a stochastic integral of a predictable integrand w.r.t Brownian motion), such that

$$\left[\int_0^\cdot dB_s/\sqrt{T-s}, \int_0^\cdot dB_s/\sqrt{T-s}\right]_u = -\ln\left[1 - \frac{u-t}{T}\right] := A_u$$

and continuous on $[0, T)$. By the Dubins–Schwartz theorem, there exists a Brownian motion $\tilde{B}$ such that

$$d\beta_u = \beta_u d\tilde{B}_{A_u}$$

and

$$\beta_u = \beta_0 \mathcal{E}\left(\tilde{B}\right)_{A_u} = \beta_0 \exp\left(\tilde{B}_{A_u} - \frac{1}{2}A_u\right)$$

for all $u < T$. By the law of the iterated logarithm, we can show that $\lim_{t \to \infty} \mathcal{E}(B)_t = 0$. Inasmuch as A_u is monotonic and $\lim_{u \to \infty} A_u = \infty$,

$$\lim_{u \to T} \beta_u = 0 \qquad \text{a.s.}$$

Because we set $\beta_T = 0$, $\beta_{T-} = S_T$ and β is continuous on $[0, T]$, $E_Q[\beta_T] = 0 < E_Q[\beta_0]$ implies that β is not a martingale.   □

3.3 A Bubble Decomposition

In this section, we refine Theorem 2 to obtain a unique decomposition of an asset price bubble that yields some additional insights. The key tool is the decomposition of a positive supermartingale.

Theorem 3 (Riesz Decomposition I). *Let X be a right-continuous supermartingale such that $E[X_t^-] = \lim_{t \to \infty} E[X_t^-] < \infty$. Then the limit $X_\infty = \lim_{t \to \infty} X_t$ a.s. exists and $E[\|X_\infty\|] < \infty$. X has the decomposition $X = U + V$, where U is a right-continuous version of the uniformly integrable martingale $E[X_\infty | \mathcal{F}_t]$ and V is a right-continuous supermartingale that is zero a.s. at infinity. V is positive if $(X^-)_{t \geq 0}$ are uniformly integrable.*

Proof. See Dellacherie and Meyer [11, V.34, 35 and VI.8, p. 73].   □

Definition 7 (Potential). *A positive right-continuous supermartingale such that $\lim_{t \to \infty} E[Z_t] = 0$ is called a* potential.

Theorem 4 (Riesz Decomposition II). *Every right-continuous positive supermartingale X can be decomposed as a sum of $X = Y + Z$, where Y is a right-continuous martingale and Z is a potential. This decomposition is unique, except on an evanescent set, and Y is the greatest right-continuous martingale bounded above by X.*

Proof. See Dellacherie and Meyer [11, VI.9, p. 73].   □

Theorem 5. *S admits a unique (up to an evanescent set) decomposition*

$$S = S^* + \beta = S^* + (\beta^1 + \beta^2 + \beta^3), \qquad (11)$$

where $\beta = (\beta_t)_{t \geq 0}$ is a càdlàg local martingale and

- *β^1 is a càdlàg nonnegative uniformly integrable martingale with $\beta_t^1 \to X_\infty$ a.s.*
- *β^2 is a càdlàg nonnegative nonuniformly integrable martingale with $\beta_t^2 \to 0$ a.s.*
- *β^3 is a càdlàg nonnegative supermartingale (and strict local martingale) such that $E[\beta_t^3] \to 0$ and $\beta_t^3 \to 0$ a.s. That is, β_t^3 is a potential.*

Furthermore, $(S^* + \beta^1 + \beta^2)$ is the greatest submartingale bounded above by W.

Proof. Let β be a càdlàg process defined by $\beta_t^1 = E_Q[S_\infty | \mathcal{F}_t]$. Define

$$K_t = W_t - (W_t^* + \beta_t^1) = W_t - E_Q[W_t | \mathcal{F}_t].$$

By Theorem 3, $K = (K_t)_{t \geq 0}$ is a nonnegative supermartingale and K converges to 0 a.s. Let M be a uniformly integrable martingale such that

$$0 \leq M_t \leq W_t - K_t \qquad \forall t \geq 0.$$

Because $W - K$ converges to 0 a.s., M converges to 0 a.s. Then $M \equiv 0$. Therefore K is unique up to an evanescent set. By Theorem 4, K has a unique decomposition:

$$K = \beta^2 + \beta^3,$$

where β^2 is a martingale, β^3 is a nonnegative supermartingale such that $E[\beta_t^3] \to 0$, which implies β^3 converges to 0 a.s. Because K converges to 0 a.s., $\beta^2 = K - \beta^3$ converges to 0 a.s. Because β^2 is a càdlàg process defined as

$$\beta_t^2 = \lim_{u \to \infty} E_Q[K_{t+u} | \mathcal{F}_t],$$

and $K_s \geq 0$ for all $s \in [0, \infty)$, $\beta^2 \geq 0$. This completes the proof.   □

As in the previous Theorem 2, β^1, β^2, β^3 give the Type 1, 2, and 3 bubbles, respectively. First, for Type 1 bubbles with infinite maturity, we see that the Type 1 bubble component converges to the asset's value at time ∞, X_∞. This time ∞ value X_∞ can be thought of as analogous to fiat money, embedded as part of the asset's price process. Indeed, it is a residual value that pays zero dividends for all finite times. Second, this decomposition also shows that for finite maturity assets $\tau < \infty$, the critical threshold is that of uniform integrability. This is due to the fact that when $\tau < \infty$, the Type 2 and 3 bubble components of $\beta = (\beta_t)_{t \geq 0}$ have to converge to 0 a.s., whereas they need not converge in L^1.

As a direct consequence of this theorem, we obtain the following corollary.

Corollary 1. *Any asset price bubble β has the following properties.*

1. $\beta \geq 0$.
2. $\beta_\tau \mathbf{1}_{\{\tau < \infty\}} = 0$.
3. *If $\beta_t = 0$ then $\beta_u = 0$ for all $u \geq t$.*

Proof. Parts 1 and 2 hold by Theorem 5. A nonnegative supermartingale stays at 0 once it hits 0, which implies Part 3.   □

This is a key result. Condition 1 states that bubbles are always nonnegative; that is, the market price can never be less than the fundamental value. Condition 2 states that if the bubble's maturity is finite $\tau < \infty$, then the bubble must burst on or before τ. Finally, Condition 3 states that if the bubble ever bursts before the asset's maturity, then it can never start again. Alternatively stated, Condition 3 states that in the context of our model, bubbles must either exist at the start of the model, or they never will exist. And, if they exist and burst, then they cannot start again. The fact that this model does not include bubble birth is a weakness of the theory, due in part to the fact that the markets are complete and there is a unique martingale measure.

3.4 No-Dominance

In this section, we add Assumption 3 (the assumption of no-dominance) to the previous structure to see what additional insights can be obtained. We only consider assets whose maturities are finite (i.e., $\tau < \infty$ a.s). This means that we only consider bubbles of Types 2 and 3.

Let W be the wealth process generated by the asset with price S. Now, by our complete markets Assumption 2, we know that there exist $\pi^1, \pi^2 \in L(W)$ such that

$$W_t^* = W_0^* + \int_0^t \pi_u^1 dW_u, \qquad \beta_t = \beta_0 + \int_0^t \pi_u^2 dW_u,$$

where W^* is the fundamental wealth process and β is the asset price bubble. Let $\eta^i = (\eta_t^i)_{t \geq 0}$ be holdings in the money market account given by Equation (2) so that the trading strategies (π^i, η^i) are self-financing. Because $\beta_\infty = 0$, two portfolios represented by W and W^* have the same cash flows. Because $W_\infty^* \geq 0$, π^i represents an admissible trading strategy. This observation implies that there are two alternative ways of obtaining the asset's cash flows. The first is to buy and hold the asset, obtaining the wealth process W. The second is to hold the admissible trading strategy π^1, obtaining the wealth process W^* instead. The cost of obtaining the first position is $W_0 \geq W_0^*$, with strict inequality if a bubble exists. This implies that if there is a bubble, then the second method for buying the asset dominates the first, yielding the following proposition.

Proposition 1. *Under Assumption 3, Type 2 and Type 3 bubbles do not exist.*

Proof. For any admissible payoff function, there is an admissible trading strategy to replicate S^*. Under Assumption 3, $V_0(S_0^*) \geq V_0(S_0)$ and $V_0(S_0) \geq V_0(S_0^*)$. Hence $V_0(S_0) = V_0(S_0^*)$, because the cash flow of a synthetic asset S^* and an asset S_t are the same. It follows that $\beta \equiv 0$ and Type 2 or Type 3 bubbles do not exist. $\square$

This proposition implies that given both the NFLVR assumption 1 and the no-dominance assumption 3, the only possible asset price bubbles are those of Type 1. Essentially, under these two weak no-arbitrage assumptions, only infinite horizon assets can have bubbles in complete markets.

4 The Economic Intuition

This section provides the economic intuition underlying the existence of asset price bubbles of Types 1, 2, and 3.

4.1 Type 1 Bubbles

Type 1 bubbles are for assets with infinite lives, with positive probability. As argued after Theorem 5, Type 1 bubbles are due to a component of an asset's price process X_∞ that is obtained at time ∞. This component of the asset's price is analogous to fiat money, a residual value received at time ∞. As such, bubbles of Type 1 are uninteresting from an economic perspective because they represent a permanent (but stochastic) wedge between an asset's fundamental value and its market price, generated by an exogenously given value at time ∞.

4.2 Type 2 Bubbles

Type 2 bubbles are for assets with finite, but unbounded, lives. In a Type 2 bubble, the market price of the asset exceeds its fundamental value. To take advantage of this discrepancy, one would form a trading strategy that is long the fundamental value, and short the asset's price. This is possible because the market is complete. If held until the asset's maturity, when $\beta_\tau^2 \mathbf{1}_{\{\tau < \infty\}} = 0$, this would (if possible) create an arbitrage opportunity (FLVR). Unfortunately, for any sample path of the asset price process, the trading strategy must terminate at some finite time. And, there is a positive probability that the bubble exceeds this termination time, ruining the trading strategy, and making it "risky" and not an arbitrage. This situation enables asset price bubbles of Type 2 to exist in our economy.

4.3 Type 3 Bubbles

Type 3 bubbles are for assets with finite and bounded lives. In a Type 3 bubble, the market price of the asset exceeds its fundamental value. Just as for a Type 2 bubble, to take advantage of this discrepancy, one would form a trading strategy that is long the fundamental value, and short the asset's price. This is possible because the market is complete. If held until the asset's maturity, when $\beta_\tau^3 \mathbf{1}_{\{\tau < \infty\}} = 0$, this would (if possible) create an arbitrage opportunity (FLVR). Unfortunately, to be a FLVR trading strategy, the trading strategy must be admissible. Shorting the asset is an inadmissible trading strategy, because if the price of the asset becomes large enough, the value of the trading strategy will fall below any given lower bound. Hence, there are NFLVR with Type 3 bubbles.

Alternatively stated, the first fundamental theorem of asset pricing is formulated for admissible trading strategies. And, admissible trading strategies are used to exclude doubling strategies, which would be possible otherwise. Restricting the class of trading strategies to be admissible (to exclude doubling strategies) implies that it also excludes shorting the asset for a fixed time horizon (short and hold) as an admissible trading strategy. This removes downward selling pressure on the asset price process, allowing bubbles to exist in an arbitrage-free setting.

The question naturally arises, therefore, whether the class of admissible trading strategies can be relaxed further, to exclude both doubling strategies, but still allow shorting the stock over a fixed investment horizon. Unfortunately, the answer is no. To justify this statement, we briefly explore the concept of admissibility. The standard definition of admissibility, the one we adopted, yields the following set of possible trading strategy values,

$$\mathcal{W} = \bigcup_a \{W_u : W_u \geq -a, \forall u \in [0, T]\}.$$

As usual, W is the wealth process generated by a risky asset with price process S. The weakest notion of admissibility consistent with NFLVR (see Strasser [29]) yields the following set of trading strategy values,

$$\mathcal{W}^* = \left\{X = H \cdot W : H \in L(W) \wedge \lim_{n \to \infty} E_Q[(H \cdot W)_{\sigma_n}^- \mathbf{1}_{\{\sigma_n < \infty\}}] = 0\right\}, \quad (12)$$

where the notation Z^- for a random variable Z means $Z^- = -(Z \wedge 0)$, and

$$\sigma_n = \inf\{t \in [0, T] : X_t \leq -n\}.$$

Clearly $\mathcal{W} \subset \mathcal{W}^*$. Replacing our definition by this weaker notion of admissibility does not affect our analysis for Type 3 bubbles. Short-selling an asset with a Type 3 bubble is not admissible even in the sense of (12) as follows from Lemma 6.

Lemma 6. *Assume that S has a Type 3 bubble β. Then a trading strategy $H = (H_t)_{t \geq 0} = (-\mathbf{1}_{\{t \leq T\}})_{t \geq 0}$ is not an admissible strategy, and $W_0 - W_T \notin \mathcal{W}^*$.*

Proof. It suffices to show that if $W_0 - W_T \in \mathcal{W}^*$ then a Type 3 bubble does not exist. Observe that

$$(H \cdot W)_{\sigma_n}^- = (W_0 - W_{\sigma_n \wedge T})^- = (W_{\sigma_n \wedge T} - W_0)^+ \geq W_{\sigma_n \wedge T} - W_0. \quad (13)$$

By definition, σ_n takes a value in $[0, T] \cup \{\infty\}$ and $(\sigma_n \wedge T)\mathbf{1}_{\{\sigma_n < \infty\}} = \sigma_n$. By (13) and hypothesis,

$$\lim_{n\to\infty} E_Q[W_{\sigma_n}\mathbf{1}_{\{\sigma_n<\infty\}}] = \lim_{n\to\infty} E_Q[(W_{\sigma_n} - W_0)\mathbf{1}_{\{\sigma_n<\infty\}}]$$
$$\leq \lim_{n\to\infty} E_Q[(H\cdot W)_{\sigma_n}^{-}\mathbf{1}_{\{\sigma_n<\infty\}}] = 0.$$

Because $W \geq 0$, $\lim_{n\to\infty} E_Q[W_{\sigma_n}\mathbf{1}_{\{\sigma_n<\infty\}}] = 0$. Because W is a supermartingale and $W_0 \geq 0$,

$$E_Q[(H\cdot W)_T^{-}] = E_Q[(W_T - W_0)^{+}] \leq E_Q[W_T] \leq E_Q[W_0] < \infty.$$

By [29, Theorem 1.4], $\{(H\cdot W)_t\}_{t\geq 0} = (W_0 - W_t)_{t\geq 0}$ is a supermartingale. Then by [1, Theorem 3.3], there exists a martingale M such that $(W_0 - W_t)^{-} \leq M_t$ for $0 \leq t \leq T$. Then $W_t \leq M_t + W_0$ for $0 \leq t \leq T$. Because W is a local martingale and $(M_t + W_0)_{t\geq 0}$ is a martingale, W is a martingale. Because $0 \leq \beta \leq W$, β is also a martingale and a Type 3 bubble does not exist. □

This motivation for the existence of stock price bubbles is consistent with the rich literature on the question, "If stocks are overpriced, why aren't prices corrected by short sales?" To answer this question, two types of short-sales constraints were used. The first constraint is a structural limitation in the economy caused by a limited ability and/or costs to borrow an asset for a shortsale (see, e.g., [23; 12; 6; 8]). The second constraint is indirect and is caused by the risk associated with short sales (see, e.g., [3; 28]). Using Internet stock data from the alleged bubble period (1999 to 2000), Battalio and Schultz [2] argue that put–call parity holds and the constraint on shortsales was not the reason for the alleged Internet stock bubble.

5 Bubbles and Contingent Claims Pricing

This section studies the pricing of contingent claims in markets where the underlying asset price process has a bubble. Bubbles can have two impacts on a contingent claim's value. The first is that a bubble in the underlying asset price process influences the contingent claim's price. The second is that the contingent claim itself can have a bubble. This section explores these possibilities in our market setting. For the remainder of this section we assume that the risky asset S does not pay dividends, so that $W_t = S_t$. We restrict our attention to European contingent claims in this paper because under the NFLVR and no-dominance assumptions, American contingent claims provide no additional insights. However, this is not true for the incomplete market setting (cf. [21]). Following our analysis for the underlying asset price process, the first topic to discuss is the fundamental price for a contingent claim.

5.1 The Fundamental Price of a Contingent Claim

Definition 8. *The fundamental price $V(H)_t$ of a European contingent claim with payoff function H at maturity T is given by*

$$V_t^*(H) = E_Q[H(S)_T|\mathcal{F}_t],$$

where $H(S)_T$ denotes a functional of the path of S on the time interval $[0, T]$. That is, $H(S)_T = H(S_r; 0 \leq r \leq T)$.

Note that in this definition, the market price of the asset $S = (S_t)_{0 \leq t \leq T}$, and not its fundamental value, is used in the payoff function. This makes sense because the contingent claim is written on the market price of the asset, and not its fundamental value. As seen in Theorem 6 below, this definition is equivalent to the fair price as defined by Cox and Hobson [7]. We believe Definition 8 is more natural inasmuch as it is valid in an incomplete market setting as well.

Theorem 6 (Cox and Hobson Theorem 3.3). *If the market is complete, the fundamental price is equivalent to the smallest initial cost to finance a replicating portfolio of a contingent claim.*

Proof. Let $\theta = (\theta_u)_{0 \leq u \leq T}$ be an admissible trading strategy and $v = (v_t)_{t \geq 0}$ be a wealth process associated with θ with initial value v_0:

$$v_t = v_0 + \int_0^t \theta_u dS_u. \tag{14}$$

Let $\mathcal{V}$ be a subcollection of wealth processes defined by (14) such that

$$\mathcal{V} = \{v : v_T \geq H(S)_T, \text{ admissible, self-financing}\}.$$

Fix $v \in \mathcal{V}$. By the definition of risk-neutral measure, v is a local martingale. Because $H(S)_T \geq 0$, v is nonnegative and hence v is a supermartingale. Then there exists a decomposition on $[0, T]$:

$$v = M + C,$$

where M is a uniformly integrable martingale and C is a potential (a nonnegative supermartingale converging to 0). This decomposition is unique (up to an evanescent set). In addition, M is the greatest martingale dominated by v (see [11, V.34, 35 and VI.8, 9 on page 73] for a discussion of this decomposition). At option maturity date T,

$$v_T = H(S)_T = M_T + C_T,$$

and $v_t = M_t = C_t = 0$ on $t > T$, $C_T = 0$ and hence $M_T = H(S)_T$. Recall that M is a uniformly integrable martingale, whence:

$$M_t = E_Q[H(S)_T|\mathcal{F}_t] \qquad \text{a.s.}$$

Because we assume a complete market, there exists a predictable process θ such that

$$M_t = M_0 + \int_0^T \theta_u dS_u.$$

Because H is positive, $M \geq 0$ and hence this strategy is admissible. Therefore for any potential C, $v = M + C$ is a superreplicating portfolio. (M is a replicating portfolio. Adding C makes it superreplicating except for the case $C \equiv 0$). The fair price is the infimum of such v_ts:

$$V_t^*(H) = \inf_{v \in \mathcal{V}} v_t = M_t + \inf_{C_t:\text{potential with } C_T=0} C_t = M_t + 0 = E_Q[H(S)_T|\mathcal{F}_t],$$

which completes the proof.   □

Because contingent claims discussed here have a fixed maturity T, by Theorem 2, contingent claims cannot have Type 1 or Type 2 bubbles. The only possible bubbles in the contingent claim's price are of Type 3. We explore these bubbles below. However, this does not imply that the existence of Type 1 or 2 bubbles in the underlying asset's price does not affect the price of the contingent claim. Indeed, it appears within the payoff function H as a component of the asset price S_T.

5.2 A Contingent Claim's Price Bubble

Analogous to the underlying asset, a contingent claim's price bubble is defined by

$$\delta_t = V_t(H) - V_t^*(H),$$

where $V_t(H)$ is the market price of the contingent claim at time t.

5.3 Bubbles Under NFLVR

This section studies a contingent claim's price bubble under NFLVR. Assume that S_t is a nondividend-paying asset with $\tau > T$ a.s. for some $T \in \mathbb{R}^+$. Let $C_t(K)^*$, $P_t(K)^*$, $F_t(K)^*$ be the fundamental prices of a call option, put option, and forward contract on S.

Lemma 7 (Put–Call Parity for Fundamental Prices). *The fundamental prices satisfy put–call parity:*

$$C^*(K) - P^*(K) = F^*(K).$$

Proof. At maturity of an option with terminal time T,

$$(S_T - K)^+ - (K - S_T)^+ = S_T - K \qquad \forall K \geq 0.$$

Because a fundamental price of a contingent claim with payoff function H is $E_Q[H(S)_T|\mathcal{F}_t]$,

$$C_t^*(K) - P_t^*(K) = E_Q[(S_T - K)^+|\mathcal{F}_t] - E_Q[(K - S_T)^+|\mathcal{F}_t]$$
$$= E_Q[S_T - K|\mathcal{F}_t] = F_t^*(K).$$

□

However, the market prices of the call, put, and forward need not satisfy put–call parity.

Example 5. Let B^i, $i = \{1, 2, 3, 4, 5\}$ be independent Brownian motions. Define processes M^i by

$$M_t^1 = \exp\left(B_t^1 - \frac{t}{2}\right), \qquad M_t^i = 1 + \int_0^t \frac{M_s^i}{\sqrt{T-s}} dB_s^i, \quad 2 \leq i \leq 5.$$

Consider a market with a finite time horizon $[0, T]$. The market is complete with respect to the filtration generated by $\{(M_t^i)_{t \geq 0}\}_{i=1}^5$. M^1 is a uniformly integrable martingale on $[0, T]$. By Lemma 5, $(M_t^i)_{i=2}^5$ are nonnegative strict local martingales that converge to 0 a.s. as $t \to T$. Let $S_t^* = \sup_{s \leq t} M_s^1$. Suppose the market prices in this model are given by

- $S = S^* + M^2$.
- $C(K) = C^*(K) + M^3$.
- $P(K) = P^*(K) + M^4$.
- $F(K) = F^*(K) + M^5$.

All of the traded securities in this example have bubbles. To take advantage of any of these bubbles $\{M^i\}_{i=2}^4$ based on the time T convergence, an agent must short-sell at least one asset. However, as shown in Lemma 6, shorting an asset with a Type 3 bubble is not admissible. Therefore, such strategies are not a free lunch with vanishing risk.

In summary, this example shows that Assumption 1 is not strong enough to exclude bubbles in contingent claims. And, given the existence of bubbles in calls and puts, we get various possibilities for put–call parity in market prices.

- $C_t(K) - P_t(K) = F_t(K)$ if and only if $\delta_t^F = \delta_t^c - \delta_t^p$.
- $C_t(K) - P_t(K) = S_t - K$ if and only if $\delta_t^S = \delta_t^c - \delta_t^p$.

This example validates the following important observation. In the well-studied Black–Scholes economy (a complete market under the standard NFLVR structure), contrary to common belief, the Black–Scholes formula need not hold! Indeed, if there is a bubble in the market price of the option M^3, then the market price $C(K)$ can differ from the option's fundamental price $C^*(K)$, the Black–Scholes formula. This insight has numerous ramifications; for example, it implies that the implied volatility (from the Black–Scholes formula) does not have to equal the historical volatility. In fact, if there is a bubble, then the implied volatility should exceed the historical volatility, and there exist no arbitrage opportunities! (Note that this is with the market still being complete.) This possibility, at present, is not commonly understood. However, all is not lost. One additional assumption returns the Black–Scholes economy to normalcy, but an additional assumption is required. This is the assumption of no-dominance, which we discuss in the next section.

5.4 Bubbles Under No-Dominance

This section analyzes the behavior of the market prices of call and put options under Assumption 3. We start with a useful lemma.

Lemma 8. *Let H' be a payoff function of a contingent claim such that $V_t(H') = V_t^*(H')$. Then for every contingent claim with payoff H such that $H(S)_T \leq H'(S_T)$, $V_t(H) = V_t^*(H)$.*

Proof. Because contingent claims have bounded maturity, we only need to consider Type 3 bubbles. Let $\mathcal{L}$ be a collection of stopping times on $[0,T]$. Then for all $L \in \mathcal{L}$, $V_L(H) \leq V_L(H')$ by Assumption 3. Because $V(H')$ is a martingale on $[0,T]$ it is a uniformly integrable martingale and in class (D) on $[0,T]$. Then $V(H)$ is also in class (D) and it is a uniformly integrable martingale on $[0,T]$. (See Jacod and Shiryaev [19, Definition 1.46, Proposition 1.47, p. 11]). Therefore Type 3 bubbles do not exist for this contingent claim. ☐

    This lemma states that if we have a contingent claim with no bubbles, and this contingent claim dominates another contingent claim's payoff, then the dominated contingent claim will not have a bubble as well. Immediately, we get the following corollary.

Corollary 2. *If $H(S)_T$ is bounded, then $V(H) \equiv V(H^*)$. In particular, a put option does not have a bubble.*

Proof. Assume that $H(S)_T < \alpha$ for some $\alpha \in \mathbb{R}^+$. Then applying Lemma 8 for $H(x) = \alpha$, we have the desired result. ☐

Theorem 7. *$C_t(K) - C_t^*(K) = S_t - E_Q[S_T|\mathcal{F}_t]$ for all $K \geq 0$. This implies calls and forwards (with $K = 0$) can only have Type 3 bubbles and that they must be equal to the asset price Type 3 bubble.*

Proof. Let $C_t(K)$, P_t, and $F_t(K)$ denote market prices of call, put options with strike K and a forward contract with delivery price K. Then

$$F_t^*(K) = E_Q[S_T|\mathcal{F}_t] - K \leq S_t - K.$$

By Assumption 3, the price of two admissible portfolios with the same cash flow are the same. Thus,

$$F_t = S_t - K = F_t^*(K) + (S_t - E_Q[S_T|\mathcal{F}_t]), \qquad (15)$$

implying a forward contract has a Type 3 bubble of size $\beta_t^3 = S_t - E_Q[S_T|\mathcal{F}_t]$. To investigate put–call parity, take the conditional expectation on the identity: $(S_T - K)^+ - (K - S_T)^+ = S_T - K$.

$$C_t^*(K) - P_t^*(K) = F_t^*(K) \leq S_t - K. \qquad (16)$$

By Assumption 3 and Equation (15),

$$C_t(K) - P_t(K) = F_t(K) = S_t - K.$$

By subtracting Equation (16) from Equation (15),

$$[C_t(K) - C_t^*(K)] - [P_t(K) - P_t(K)^*] = \beta_t^3.$$

By Corollary 2, $P_t(K) - P_t(K)^* = 0$, so $C_t(K) - C_t^*(K) = \beta_t^3$. $\square$

This theorem states that a call option's bubble, if it exists, must equal the stock price's Type 3 bubble. But, we know from Proposition 1 that under the no-dominance assumption, asset prices have no Type 3 bubbles. Thus, call options have no bubbles under the no-dominance assumption, as well.

Inasmuch as both European calls and puts have no bubbles under the no-dominance assumption, put–call parity (as in [24]) holds, as well.

6 Conclusion

This paper reviews and extends the mathematical finance literature on bubbles in complete markets. We provide a new characterization theorem for bubbles under the standard no-arbitrage (NFLVR) framework, showing that bubbles can be of three types. Type 1 bubbles are uniformly integrable martingales, and these can exist for assets with infinite lifetimes. Type 2 bubbles are nonuniformly integrable martingales, and these can exist for assets with finite, but unbounded, lives. Last, Type 3 bubbles are strict local martingales, and these can exist for assets with finite lives. In addition, we show that bubbles can only be nonnegative, and must exist at the start of the model. Bubble birth cannot occur in the standard NFLVR, complete markets structure.

When one adds a no-dominance assumption (from [24]), we show that only Type 1 bubbles are possible. In addition, under Merton's no-dominance hypothesis, put–call parity holds and there are no bubbles in standard call and put options. Our analysis implies that if one believes asset price bubbles exist and are an important economic phenomenon, then asset markets must be incomplete. Incomplete market bubbles are studied in a companion paper, which is in preparation.

Acknowledgments

It is a privilege to be included in this volume honoring Dilip Madan on his 60th birthday. Dilip is one of those pioneers who carried the torch through the dark years when only a few people in the United States—and for that matter, the world—considered mathematical finance a subject worthy of study. Now that he is 60, and a leader in the field, he can look back with satisfaction on a job well done. More than just an academic researcher, Dilip is also an entertaining communicator. His presentations are always lucid, often sarcastic, and humorous. He is also one of the few prominent researchers with intimate ties to industry. He is loved by his students and his colleagues. We are proud to be considered his friends.

The authors wish to thank Nicolas Diener and Alexandre Roch for helpful discussions concerning this paper, and financial bubbles in general. Philip Protter gratefully acknowledges support from an NSF grant DMS-0604020 and an NSA grant H98230-06-1-0079.

References

1. J.P. Ansel and C. Stricker. Couverture des actifs contingents et prix maximum. *Ann. Inst. H. Poincaré Probab. Statist.*, 30:303–315, 1994.
2. R. Battalio and P. Schultz. Options and bubble. *Journal of Finance*, 59(5), 2017–2102, 2006.
3. J.B. DeLong, A. Shleifer, L.H. Summers, and R.J. Waldmann. Noise trader risk in financial markets. *Journal of Political Economy*, 98:703–738, 1990.
4. P. Brémaud and M. Yor. Changes of filtrations and of probability measures. *Z. Wahrsch. Verw. Gebiete*, 45:269–295, 1978.
5. G. Cassese. A note on asset bubbles in continuous time. *International Journal of Theoretical and Applied Finance*, 8:523–536, 2005.
6. J. Chen, H. Hong, and S.C. Jeremy. Breadth of ownership and stock returns. *Journal of Financial Economics*, 66:171–205, 2002.
7. A.M. Cox and D.G. Hobson. Local martingales, bubbles and option prices. *Finance and Stochastics*, 9:477–492, 2005.
8. G. D'Avolio. The market for borrowing stock. *Journal of Financial Economics*, 66:271–306, 2002.
9. F. Delbaen and W. Schachermayer. The fundamental theorem of asset pricing for unbounded stochastic processes. *Math. Ann.*, 312:215–250, 1998.
10. F. Delbaen and W. Schachermayer. *The Mathematics of Arbitrage*. Springer-Verlag, 2006.
11. C. Dellacherie and P.A. Meyer. *Probabilities and Potential B*. North-Holland, 1982.
12. D. Duffie, N. Gârleanu, and L. Pedersen. Securities lending, shorting and pricing. *Journal of Financial Economics*, 66:307-339, 2002.
13. D. Elworthy, X-M. Li, and M. Yor. The importance of strictly local martingales: Applications to redial Ornstein-Uhlenback processes. *Probability Theory and Related Fields*, 115:325–255, 1999.
14. M. Émery. Compensation de processus à variation finie nonlocalement intégrables. *Séminare de probabilités*, XIV:152–160, 1980.
15. C. Gilles. Charges as equilibrium prices and asset bubbles. *Journal of Mathematical Economics*, 18:155–167, 1988.
16. C. Gilles and S.F. LeRoy. Bubbles and charges. *International Economic Review*, 33:323–339, 1992.
17. K.X. Huang and J. Werner. Asset price bubbles in Arrow-Debreu and sequential equilibrium. *Economic Theory*, 15:253–278, March 2000.

18. J.M. Harrison and S. Pliska. Martingales and stochastic integrals in the theory of continuous trading. *Stochastic Processes and Their Applications*, 11:215–260, 1981.

19. J. Jacod and A.N. Shiryaev. *Limit Theorems for Stochastic Processes*, 2nd edition. Springer-Verlag, 2003.

20. R.A. Jarrow and D.B. Madan. Arbitrage, martingales, and private monetary value. *Journal of Risk*, 3:73–90, 2000.

21. R.A. Jarrow, P. Protter, and K. Shimbo. Bubbles in incomplete markets. In preparation, Cornell University, 2006.

22. M. Loewenstein and G.A. Willard. Rational equilibrium asset-pricing bubbles in continuous trading models. *Journal of Economic Theory*, 91:17–58, 2000.

23. E. Oftek and M. Richardson. Dotcom mania: The rise and fall of Internet stock prices. *Journal of Finance*, 58:1113-1137, 2003

24. R.C. Merton. Theory of rational option pricing. *Bell Journal of Economics*, 4:141–183, 1973.

25. P. Protter. *Stochastic Integration and Differential Equations*, 2nd edition. Springer-Verlag, 2005.

26. M. Schweizer. Martingale densities for general asset prices. *Journal of Mathematical Economics*, 21:363–378, 1992.

27. M. Schweizer. On the minimal martingale measure and the Föllmer-Schweizer decomposition. *Stochastic Anal. Appl.*, 13:573–599, 1995.

28. A. Shleifer and R.W. Vishny. The limit of arbitrage. *Journal of Finance*, 52:35–55, 1997.

29. E. Strasser. Necessary and sufficient conditions for the supermartingale property of a stochastic integral with respect to a local martingale. *Séminare de Probabilités XXXVII*, eds. J. Azéma, M. Émery, M. Ledoux, and M. Yor, Lecture Notes in Mathematics, 1832, Springer, 2004.

Taxation and Transaction Costs in a General Equilibrium Asset Economy

Xing Jin[1] and Frank Milne[2]

[1] Warwick Business School
University of Warwick
Coventry CV4 7AL, United Kingdom
Xing.Jin@wbs.ac.uk

[2] Department of Economics
Queen's University
Kingston, Ontario K7L 3N2, Canada
milnef@econ.queensu.ca

Summary. Most financial asset-pricing models assume frictionless competitive markets that imply the absence of arbitrage opportunities. Given the absence of arbitrage opportunities and complete asset markets, there exists a unique martingale measure that implies martingale pricing formulae and replicating asset portfolios. In incomplete markets, or markets with transaction costs, these results must be modified to admit nonunique measures and the possibility of imperfectly replicating portfolios. Similar difficulties arise in markets with taxation. Some theoretical research has argued that some taxation functions will imply arbitrage opportunities and the nonexistence of a competitive asset economy. In this paper we construct a multiperiod, discrete time/state general equilibrium model of asset markets with transaction costs and taxes. The transaction cost technology and the tax system are quite general, so that we can include most discrete time/state models with transaction costs and taxation. We show that a competitive equilibrium exists. Our results require careful modeling of the government budget constraints to rule out tax arbitrage possibilities.

Key words: Taxation; transaction costs; general equilibrium; asset economy.

1 Introduction

There is an extensive literature addressing the role of taxation and transaction costs in competitive financial markets. In an earlier paper [15], we proposed a general competitive asset economy with very general transaction technologies and provided sufficient conditions for the existence of a competitive equilibrium. This model allowed us to consider economies with brokers, dealers, and a wide range of market constraints on trade (e.g., short-sale constraints). We considered economies with convex and nonconvex transaction technologies. In a complementary paper, Milne and Neave [18] provide characterizations of agent and competitive equilibria in such an economy and demonstrate that most known discrete time/state models can be accommodated as special cases.

In this paper we address another major imperfection in asset markets: taxation. As pointed out by Schaefer [20] and Dammon and Green [6], there are difficulties in dealing with taxes in a general equilibrium one-period setting. To clear markets, when there are no restrictions on asset trading, relative prices must reflect the after-tax marginal rates of substitution of all agents simultaneously. When tax rates differ across investors, however, this condition can be impossible to achieve. Schaefer [20] and Dammon and Green [6] give examples showing that when tax rates between investors are sufficiently different, there may exist arbitrage opportunities for at least one investor: this is inconsistent with the existence of general competitive equilibrium. The main reason that investors can exploit tax arbitrage opportunities is that they can infinitely short sell assets, and exploit unlimited tax rebates on capital losses. An alternative and more realistic view, which can accommodate asset short selling and individuals in different tax brackets, is where potential tax arbitrage opportunities exist, but there are realistic restrictions on their exploitation.

Unlike Schaefer [20] and Dammon and Green [6], we investigate a closed model where the government is modeled explicitly, and tax rebates are limited (ultimately) by the government budget constraint. Long before government revenues and wealth are exhausted, sections of the tax code limit the ability of private agents or organizations to exploit unbounded loopholes in the tax law. For example, our modeling captures the essential features of the current U.S. tax code. The Tax Reform Act of 1986 has eliminated the net capital gains deduction. Prior to this date, individuals could deduct 60 percent of net long-term capital gains from income. A $3000 loss limitation was applied to net long-term capital losses. Under the new rules, the capital gains deduction is eliminated, but the $3000 loss limitation is retained. More generally, the government can always make contingent provisions in the application of tax laws to avoid large revenue losses from implausible (unbounded) claims. More realistically, its revenues from wages and salaries dwarf any revenue or drains from financial taxes: in other words, the government cannot, and will not, promise infeasible tax rebates to consumers and firms. We call this a No Ponzi Game (NPG) condition. We stress that this constraint is a weak bound and that more restrictive conditions could be introduced by appealing to detailed tax laws or financial regulations. These tighter constraints would not destroy our equilibrium existence result.

Conceptually, the forms of tax on capital gains and losses are very complicated to analyze because the investment strategies depend on the whole history of the investment (e.g., see Dammon and Spatt [7] and Dammon et al. [8]). Instead of giving concrete examples of capital gains and tax rebate rules, we impose an upper bound on rebates for each investor. The bound can be regarded as being exogenously determined by government and legal consider-ations. Realistically, it is far less than the bound that would exhaust govern-ment resources.

Rather than considering bounds on short positions, the upper bound on rebates is the first friction in the present paper and is an asymmetric treat-ment of taxes on long and short positions. If the asymmetry is sufficiently pronounced to eliminate arbitrage opportunities, the only motivation for in-dividuals to take short positions is to construct a profile of portfolio cash flows that is not feasible with entirely nonnegative portfolio weights. However, when there are a large number of securities, and tax asymmetries have eliminated arbitrage opportunities, individuals may have little motivation to take short positions. Consequently, prices may be similar to those that would result if short-sales were explicitly disallowed.

The second friction considered in this paper is transaction costs. Effec-tively, we have extended our earlier paper [15] on transaction costs to include taxation. This means short sales have additional costs over and above taxation asymmetries, and are consistent with the discussion of Allen and Gale [1]. In this paper, we do not impose any constraints on short selling, and short-sales are determined endogenously. Limited tax rebates on capital losses and trans-action costs are suggested as possible explanations for the lack of apparent arbitrage and why a general competitive equilibrium can exist.

Another difficult issue is to construct a model that is sufficiently general to deal with the complexities of the tax law and yet remain tractable. There are many papers that consider the implications for asset prices or asset al-locations in specialized models (for a small sample, see Constantinides [4], Dammon and Green [6], Dammon and Spatt [7], Dammon et al. [8], Dybvig and Ross [12], Ross [19], Green [14], and Zechner [22]). Before discussing the properties of an equilibrium, it is important to note that there are sufficient restrictions imposed on the economy to imply the existence of an equilibrium. As we indicated above, if agents face agent-specific tax functions that allow tax arbitrage possibilities, then an equilibrium may not exist. This issue has been addressed in a two-period exchange economy by Dammon and Green [6] and Jones and Milne [16]. Dammon and Green [6] consider restrictions on tax functions to eliminate arbitrage and define a set of arbitrage-free asset prices. Their paper considered tax functions of considerable generality and exploited the theory of recession (asymptotic) cones to determine arbitrage-free prices. Jones and Milne [16] argued that Dammon and Green [6] did not include the government sector explicitly, ignoring the feasibility constraints implicit in the government budget constraint. Once these constraints were introduced and recognized by the agents, there were natural bounds on the possible asset trades consistent with tax arbitrage. These restrictions allowed more general and complex tax functions to be consistent with equilibrium. Of course the differences in the two models could be resolved by assuming that Dammon and Green's tax functions included the implicit tax rules that come into play as soon as large tax arbitrages were claimed from the government by consumers. In that sense the two models could be made consistent.

The models in Dammon and Green [6] and Jones and Milne [16] were two-period exchange economies. In this paper, we extend the two-period model to a more general multiperiod economy with an explicit government sector, productive firms, brokers/dealers who operate the costly transaction technology, and spot commodity markets. The aim of the paper is to show how the ideas of earlier simpler models can be extended in a number of realistic directions. In particular, the introduction of firms allows us to accommodate models (see Zechner [22], Swoboda and Zechner [21], and Graham [13] for a sample) that discuss the interaction of personal and corporate taxation, the trade-off between corporate equity and debt taxation, and their impact on corporate financial structure. The introduction of government is more general than the modeling in Jones and Milne [16], in that we allow the government to choose commodity trades optimally, given its net tax revenues. We show that the introduction of spot commodity markets is easily accommodated, although we exclude commodity and wage income taxation for simplicity. We allow general tax functions on asset capital gains and dividends that can accommodate most properties of tax codes on financial assets. For example, the tax functions can be nonlinear, convex, or piecewise linear; they can depend upon dynamic asset strategies, so that capital gains or other complex tax systems can be incorporated; the tax functions can include financial subsidies as well as taxes; and the state contingent taxation functions can be interpreted to include (random) legal interpretations of the tax code where a dynamic asset position is deemed to violate the code and subsequent income or capital gains are taxed at a higher rate with possible penalties. In the body of the paper, we rule out nonconvex tax functions due to subsidy/tax thresholds. Later in the paper, we discuss how this (and other extensions) could be incorporated in a more extensive model.

The model can be used as a basic structure to discuss incomplete asset markets and government policies to use the tax code and subsidies to complete asset markets, or at least improve the welfare of some agents.

The remainder of the paper is organized as follows. In Section 2 we set out the model and present fundamental assumptions. Section 3 is devoted to the proof of the main theorem. In Section 4 we discuss extensions or variations of the model and how they could be incorporated in an extended or modified version of the model. Also we discuss equilibrium efficiency properties and the complexity of general characterizations of an equilibrium.

2 The Economic Setting

Consider an economy with uncertainty characterized by a event tree such as that depicted in Figure 1 of Duffie [11]. This tree consists of a finite set of nodes $\mathbf{E}$ and directed arcs $\mathbf{A} \subset \mathbf{E} \times \mathbf{E}$ such that $(\mathbf{E}, \mathbf{A})$ forms a tree with a distinguished root e_0. The number of immediate successor nodes of any $e \in \mathbf{E}$ is denoted $\#e$. A node $e \in \mathbf{E}$ is terminal if it has no successor node. Let $\mathbf{T}$ denote the set of all terminal nodes. The immediate successor nodes of any nonterminal node $e \in \mathbf{E}$ are labeled $e^{+1}, \ldots, e^{+K}$, where $K = \#e$. The subtree with root e is denoted $\mathbf{E}(e)$. In particular, $\mathbf{E} = \mathbf{E}(e_0)$. Suppose there are N securities and M commodities at any node $e \in \mathbf{E}' = \mathbf{E} - \mathbf{T}$. This assumption can be relaxed by assuming that the numbers of securities can vary across nodes, because we do not require every asset to be held by agents at node e_0. This relaxation covers the case where some securities are issued and some mature in interim periods. For the sake of simplicity, we assume that there is only one commodity at each terminal node (this is largely for expositional convenience and avoids some minor technical issues). At each node, all securities first distribute dividends, and then are available for trading; that is, all security prices are ex-dividend.

Let $p^C(e) = (p_1^C(e), \ldots, p_M^C(e))$ denote the spot price of commodities at node $e \in \mathbf{E}'$. At each node $e \in \mathbf{E}'$, asset n $(n = 1, \ldots, N)$ has a buying price $p_n^B(e)$ and a selling price $p_n^S(e)$ and a dividend $p_1^C(e)D_n(e)$, where we take the first commodity as numeraire. Moreover, let $D_n(e)$ be the dividend of asset n at the terminal node $e' \in \mathbf{T}$. We denote $p^B(e) = (p_1^B(e), \ldots, p_N^B(e))$, $p^S(e) = (p_1^S(e), \ldots, p_N^S(e))$ and $p(e) = (p^B(e), p^S(e))$ and $D(e) = (D_1(e), \ldots, D_N(e))$. And it is assumed that dividends are always nonnegative.

2.1 Intermediaries

Suppose that there are H brokers or intermediaries (indexed by h) with an objective (utility) function $U_h^B(\cdot)$ defined over the nonnegative orthant, and commodity endowment vector $\omega_h^B(e)$ at each node e. They are intermediaries specializing in the transaction technology that transforms bought and sold assets. Let $\phi_{h,n}^B(e)(\phi_{h,n}^S(e))$ be the number of bought (sold) asset n supplied by intermediary h at node $e \in \mathbf{E}'$ (denote $\phi_h^B(e) = (\phi_{h,1}^B(e), \ldots, \phi_{h,N}^B(e))^T$ and $\phi_h^S(e) = (\phi_{h,1}^S(e), \ldots, \phi_{h,N}^S(e))^T$, where T is the transpose transformation) and $z_h(e) = (z_{h,1}(e), \ldots, z_{h,M}(e))^T$ be the vector of contingent commodities used up in the activity of intermediation at node $e \in \mathbf{E}'$. Then the broker pays tax on capital gains via a general tax function $T_h^{BC}(e) = T_h^{BC}(e)((p^B(e'), p^S(e'), \phi_h(e'))_{e' \in \mathbf{PA}(e)})$, where $\mathbf{PA}(e)$ is a path from e_0 to e; and pays tax on dividends via a general tax function

$$T_h^{BD}(e) = T_h^{BD}(e)\left(D(e)\Big[\sum_{e' \in \mathbf{PA}(e)-\{e\}} (\phi_h^B(e') - \phi_h^S(e'))\Big]\right).$$

At any terminal node $e \in \mathbf{T}$, the broker receives dividends

$$D(e)\Big[\sum_{e' \in \mathbf{PA}(e)-\{e\}} (\phi_h^B(e') - \phi_h^S(e'))\Big]$$

and pays tax via a general tax function

$$T_h^{BD}(e) = T_h^{BD}(e)\left(D(e)\left[\sum_{e' \in \mathbf{PA}(e)-\{e\}} (\phi_h^B(e') - \phi_h^S(e'))\right]\right).$$

Denote $z_h = (\dots, z_h(e), \dots)_{e \in \mathbf{E}'} \in R_+^{|\mathbf{E}'| \times M}$, where $|\mathbf{E}'|$ denotes the number of nodes in the set $\mathbf{E}'$, and the portfolio plan by

$$\phi_h = (\phi_h(e))_{e \in \mathbf{E}'} = (\phi_h^B(e), \phi_h^S(e))_{e \in \mathbf{E}'} \in R_+^{2(|\mathbf{E}'| \times N)}.$$

And set

$$
\begin{aligned}
\gamma_h = \Bigg(& p^B(e_0)\phi_h^S(e_0) - p^S(e_0)\phi_h^B(e_0) - T_h^{BC}(e_0) \\
& + p(e_0)\left(\omega_h^B(e_0) - z_h(e_0)\right), \Big(p^B(e)\phi_h^S(e) - p^S(e)\phi_h^B(e) \\
& + p_1^C(e)D(e)\left[\sum_{e' \in \mathbf{PA}(e)-\{e\}} (\phi_h^B(e') - \phi_h^S(e'))\right] \\
& - T_h^{BC}(e) - T_h^{BD}(e) + p(e)\left(\omega_h^B(e) - z_h(e)\right)\Big)_{e \in \mathbf{E}'-\{e_0\}}, \\
& \left(D(e)\left[\sum_{e' \in \mathbf{PA}(e)-\{e\}} (\phi_h^B(e') - \phi_h^S(e'))\right] - T_h^{BD}(e)\right)_{e \in \mathbf{T}}\Bigg).
\end{aligned}
$$

For intermediary h, let $\mathbf{T}(h, e) \subseteq R_+^N \times R_+^N \times R^M$ denote his or her technology at node e.

The maximization problem of broker h can be stated as

$$\sup_{(\phi_h, z_h) \in \Gamma_h^B(\tilde{p})} U_h^B(\gamma_h),$$

where $\Gamma_h^B(\tilde{p})$ is the space of feasible trade-production plans $(\phi_h, z_h) = (\phi_h^B, \phi_h^S, z_h)$ given $\tilde{p} = (p^B, p^S, p^C)$, which satisfies:

(2.1) $(\phi_h^B(e), \phi_h^S(e), z_h(e)) \in \mathbf{T}(h, e)$ and $z_h(e) \geq 0$,

(2.2) $p^B(e_0)\phi_h^S(e_0) - p^S(e_0)\phi_h^B(e_0) + p(e_0)\left(\omega_h^B(e_0) - z_h(e_0)\right) - T_h^{BC}(e_0) \geq 0$,
and $p^B\phi_h^S(e) - p^S(e)\phi_h^B(e) + p^C(e)\left(\omega_h^B(e) - z_h(e)\right) +$
$p_1^C(e)D(e)\left[\sum_{e' \in \mathbf{PA}(e)-\{e\}} (\phi_h^B(e') - \phi_h^S(e'))\right] - T_h^{BC}(e) - T_h^{BD}(e) \geq 0$,
$\forall e \in \mathbf{E}' - \{e_0\}$,

(2.3) $D(e)\left[\sum_{e' \in \mathbf{PA}(e)-\{e\}} (\phi_h^B(e') - \phi_h^S(e'))\right] - T_h^{BD}(e) \geq 0, \forall e \in \mathbf{T}$,

(2.4) $\gamma_h \geq 0$.

Remark 1. The intermediary formulation allows the agent to trade on his own account; or by interpreting the transaction technology more narrowly, it can be restricted to a pure broker with direct pass-through of assets bought and sold (see Milne and Neave [18] for further discussion). Note: our tax functions are sufficiently general in formulation that our capital gains tax function could incorporate dividend taxes as well (apart from the final date).

2.2 Firms

Suppose there are J firms. At node e_0, the firm $j \in \mathbf{J} = \{1, \ldots, J\}$ has an initial endowment $\omega_j^F(e_0)$, which gives the firm a positive cash flow. The firm chooses an input plan $y_j^-(e_0) \in R_+^M$ and a trading strategy $\beta_j(e_0) = (\beta_j^B(e_0), \beta_j^S(e_0))^T = (\beta_{j,1}^B(e_0), \ldots, \beta_{j,N}^B(e_0), \beta_{j,1}^S(e_0), \ldots, \beta_{j,N}^S(e_0))^T \in R_+^{2N}$, where $\beta_{j,n}^B(e_0)(\beta_{j,n}^S(e_0))$ represents the purchase (sale) of asset n by firm j at node e_0. Then, the firm pays tax according to the tax function $T_j^F(e_0)$. At every node $e (\in \mathbf{E}')$ other than node e_0, the firm produces an output $y_j^+(e) \in R_+^M$ and receives a net dividend

$$p_1^C(e) D(e) \Big[\sum_{e' \in \mathbf{PA}(e) - \{e\}} (\beta_j^B(e') - \beta_j^S(e')) \Big],$$

then chooses an input plan $y_j^-(e) \in R_+^M$ and a trading strategy $\beta_j(e) = (\beta_j^B(e), \beta_j^S(e))^T = (\beta_{j,1}^B(e), \ldots, \beta_{j,N}^B(e), \beta_{j,1}^S(e), \ldots, \beta_{j,N}^S(e))^T \in R_+^{2N}$ and pays tax $T_j^{FC}(e)$ and $T_j^{FD}(e)$ on capital gains and ordinary income. It is assumed $(y_j^+(e) - y_j^-(e)) \in \mathbf{Y}_j(e) \subseteq R^M$, where $\mathbf{Y}_j(e)$ is a production set. At each terminal node e, the firm j produces $y_j^+(e) \in R_+^M$, gets its dividend, and then pays tax $T_j^{FD}(e)$. Set

$$\delta(e_0) = p^C(e_0)\left(\omega_j^F(e_0) - y_j^-(e_0)\right) + p^S(e_0)\beta_j^S(e_0) - p^B(e_0)\beta_j^B(e_0) - T_j^{FC}(e_0),$$

$$\delta(e) = p^C(e)\left(y_j^+(e) - y_j^-(e)\right) + p^S(e)\beta_j^S(e) - p^B(e)\beta_j^B(e)$$
$$+ p_1^C(e)D(e)\Big[\sum_{e' \in \mathbf{PA}(e) - \{e\}} (\beta_i^B(e') - \beta_i^S(e')) \Big] - T_j^{FD}(e) - T_j^{FC}(e),$$

$$e \in \mathbf{E}' - \{e_0\},$$

$$\delta(e) = p^C(e)y_j^+(e) + p_1^C(e)D(e)\Big[\sum_{e' \in \mathbf{PA}(e) - \{e\}} (\beta_i^B(e') - \beta_i^S(e')) \Big] - T_j^{FD}(e),$$

$$e \in \mathbf{T}.$$

Therefore, the problem of the firm j can be stated as

$$\max_{(y_j, \beta_j) \in \Gamma_j^F(\tilde{p})} U_j\left(\left(\delta_j(e)\right)_{e \in \mathbf{E}}\right),$$

where $\Gamma_j^F(\tilde{p})$ is the feasible set of production-portfolio plans of firm j given $\tilde{p}$, which satisfies:

$$\delta_j(e) \geq 0, e \in \mathbf{E}.$$

Remark 2. The modeling of the objective of the firm avoids the failure of profit to be well defined, and the nonapplicability of the Fisher separation theorem. We can think of the firm being either a sole proprietorship or that the utility function is derived from some group preference arrangement (for a more detailed discussion of these issues see Kelsey and Milne [17]). The utility function can incorporate a number of interpretations (e.g., it could include job perquisite consumption by a manager, where her optimal perquisite consumption is endogenously determined in equilibrium).

2.3 Consumers

In addition to brokers/intermediaries and firms, there are I consumers, indexed by $i \in \mathbf{I} = \{1, \ldots, I\}$. The consumer i has endowment $\omega_i^C(e)$ of commodity at any node $e \in \mathbf{E}$. At any node $e \in \mathbf{E}'$, the consumer i has a consumption $x_i(e) = (x_{i,1}(e), \ldots, x_{i,M}(e))^T \in R_+^M$ and chooses an asset portfolio $\alpha_i(e) = (\alpha_i^B(e), \alpha_i^S(e))^T = (\alpha_{i,1}^B(e), \ldots, \alpha_{i,N}^B(e), \alpha_{i,1}^S(e), \ldots, \alpha_{i,N}^S(e))^T \in R_+^{2N}$, where $\alpha_{i,n}^B(e)(\alpha_{i,n}^S(e))$ represents the purchase (sale) of asset n by consumer i at node e. The consumer pays capital gains tax via a tax function

$$T_i^{CC}(e)\left(\left(p^B(e'), p^S(e'), \alpha_i(e') \right)_{e' \in \mathbf{PA}(e)} \right)$$

and dividend taxes via a function defined similarly to the broker tax function, $T_i^{CD}(e)$. At any terminal node $e \in \mathbf{T}$, the consumer receives dividends

$$D(e)\left[\sum_{e' \in \mathbf{PA}(e) - \{e\}} (\alpha_i^B(e') - \alpha_i^S(e')) \right]$$

and pays tax via a general tax function $T_i^{CD}(e)$, and then consumes. We denote the consumption plan $x_i = (\ldots, x_i(e), \ldots)_{e \in \mathbf{E}'} \in R_+^{|\mathbf{E}'| \times M}$, and the portfolio plan $\alpha_i = (\alpha_i(e))_{e \in \mathbf{E}'} = (\alpha_i^B(e), \alpha_i^S(e))_{e \in \mathbf{E}'} \in R_+^{2(|\mathbf{E}'| \times N)}$.

The consumer i has a consumption set $X_i \subseteq R_+^{\mathbf{E}' \times M}$ and a utility function $U_i^C(\cdot)$ over the consumption plan x_i and the consumption

$$D(e)\left[\sum_{e' \in \mathbf{PA}(e) - \{e\}} (\alpha_i^B(e') - \alpha_i^S(e')) \right] - T_i^{CD}(e)$$

at terminal nodes. Denote

$$\tilde{x}_i = \left(x_i, \left(D(e)\left[\sum_{e' \in \mathbf{PA}(e)} (\alpha_i^B(e') - \alpha_i^S(e')) \right] - T_i^{CD}(e) \right)_{e \in \mathbf{T}} \right).$$

So the problem of consumer i can be expressed as

$$\max_{(x_i, \alpha_i) \in \Gamma_i^C(\tilde{p}, \gamma, \delta)} U_i^C(\tilde{x}_i),$$

where $\Gamma_i^C(\tilde{p}, \gamma, \delta)$ is the feasible set of portfolio consumption plans of consumer i, given $(\tilde{p}, \gamma, \delta)$, which satisfies

$$p^C(e)x_i(e) + p^B(e)\alpha_i^B(e) - p^S(e)\alpha_i^S(e) + T_i^{CC}(e) + T_i^{CD}(e)$$

$$\leq p^C(e)\omega_i(e) + p_1^C(e)D(e)\Big[\sum_{e' \in \mathbf{PA}(e)-\{e\}} (\alpha_i^B(e') - \alpha_i^S(e'))\Big]$$

$$+ \sum_j \eta_{i,j}\delta_j(e) + \sum_h \theta_{i,h}\gamma_h(e) \quad \forall e \in \mathbf{E'},$$

$$D(e)\Big[\sum_{e' \in \mathbf{pa}(e)} (\alpha_i^B(e') - \alpha_i^S(e'))\Big] - T_i^{CD}(e) \geq 0 \quad \forall e \in \mathbf{T},$$

where $\eta_{i,j}$ is the share of the profit of the jth producer owned by the ith consumer. The numbers $\eta_{i,j}$ are positive or zero, and for every j, $\sum_{i=1}^{I}\eta_{i,j} = 1$. The numbers $\theta_{i,h}$ have the same explanation.

2.4 Government

The government is also included as part of the economy. Rather than providing a detailed analysis of the operations of the government, we place weak restrictions on government activity. For simplicity, assume that the government has net resources $\omega_G(e) \in R_+^M$ at node e, sets tax rates ex ante, and then consumes $x_G(e)$ at node e. The government can always propose precommitted tax laws that avoid government bankruptcy due to errors or subtle legal interpretations; and its revenues from taxes on wages and salaries dwarf any revenue or drains from financial taxes. In other words, the government cannot promise infeasible tax rebates to consumers and firms. We call this a No Ponzi Game (NPG) condition, which has been introduced in macroeconomic models to eliminate unbounded borrowing positions by consumers and/or governments (cf. [3]). We stress that this constraint is a weak bound, and that more restrictive conditions could be introduced by appealing to detailed tax laws or financial regulations.

Suppose that the government has preferences over its consumption set $X_G = R_+^{|\mathbf{E'}| \times M}$, represented by a utility function U_G (this is simplistic but avoids more complex issues of government decision making). At node $e \in \mathbf{E'}$, it spends its income, which comes from its endowment $\omega_G(e)$ and tax revenue

$$T(e) = \sum_h \left(T_h^{BC}(e) + T_h^{BD}(e)\right) + \sum_j \left(T_j^{FC}(e) + T_j^{FD}(e)\right)$$

$$+ \sum_i \left(T_i^{CC}(e) + T_i^{CD}(e)\right).$$

The government's problem can be stated as

$$\max_{x_G \in \Gamma_G} U^G(x_G),$$

where Γ_G denote the set of government's consumption x_G, which satisfies

$$p^C(e)x_G(e) \leq p^C(e)\omega_G(e) + T(e).$$

2.5 Competitive Equilibrium and Assumptions

To conclude the section, we give the definition of a competitive equilibrium and assumptions made throughout the remainder of this paper.

Definition 1. *A competitive equilibrium with asset taxation and transaction costs is a nonnegative vector of prices* $\tilde{p}^* = (p^{*C}, p^{*B}, p^{*S})$ *and allocations* (x_i^*, α_i^*) *for all* $i \in \mathbf{I}$; (z_h^*, ϕ_h^*) *for all* $h \in \mathbf{H}$; (y_j^*, β_j^*), *for all* $j \in \mathbf{J}$; x_G^* *such that:*

(i) (z_h^*, ϕ_h^*) *solves the broker problem for each* $h \in \mathbf{H}$,
(ii) (y_j^*, β_j^*) *solves the firm's problem for each* $j \in \mathbf{J}$,
(iii) (x_i^*, α_i^*) *solves the consumer problem for each* $i \in \mathbf{J}$,
(iv) x_G^* *solves government's problem,*
(v) $\sum_i x_i^* + \sum_h z_h^* + \sum_j y_j^{*-} + x_G^* = \omega_G + \sum_j y_j^{*+} + \sum_i \omega_i^C + \sum_j \omega_j^F + \sum_h \omega_h^B$;
(vi) $\sum_i \alpha_{i,n}^{*B} + \sum_j \beta_{j,n}^{*B} = \sum_h \phi_{h,n}^{*S}$ *and* $\sum_i \alpha_{i,n}^{*S} + \sum_j \beta_{j,n}^{*S} = \sum_h \phi_{h,n}^{*B}$ *if* $p_n^{*B} > p_n^{*S}$; $\sum_i \alpha_{i,n}^{*B} + \sum_j \beta_{j,n}^{*B} + \sum_h \phi_{h,n}^{*B} = \sum_h \phi_{h,n}^{*S} + \sum_i \alpha_{i,n}^{*S} + \sum_j \beta_{j,n}^{*S}$ *if* $p_n^{*B} = p_n^{*S}$.

Notice that condition (vi) allows for the extreme case where the transaction technology is costless.

The following assumptions are made in the remainder of this paper. For consumer i:

(A1) $X_i = R_+^{|\mathbf{E}'| \times M}$,
(A2) U_i^C is a continuous, concave, and strictly increasing function,
(A3) T_i^{CC} and T_i^{CD} are continuous convex functions, and there exist positive constants $c_i^{CC}(e) < 1$, $c_i^{CD}(e) < 1$ such that $T_i^{CC}(e)[x] \leq c_i^{CC}(e)x$, $T_i^{CD}(e)[x] \leq c_i^{CD}(e)x$, for any $x \geq 0$; that is, the taxes can never be larger than revenues,
(A4) We assume that the tax rebates on capital losses and ordinary income on long and short positions have lower bounds; that is, $T_i^{CC}(e) + T_i^{CD}(e) > \alpha_i^C(e)$, where α_i^C is determined by the government,
(A5) $\omega_i^C(e) > 0 \; \forall e \in \mathbf{E}'$, where for $x = (x_1, \ldots, x_n) \in R^n, x \gg 0$ means $x_i > 0, i = 1, \ldots, n$; $x > 0$ means $x_i \geq 0, i = 1, \ldots, n$ but $x_{i_0} > 0$ for at least one i_0.

For broker h:

(A6) For each e, $\mathbf{T}(h, e)$ is a closed and convex set with $0 \in \mathbf{T}(h, e)$,
(A7) For any e and given $x = (x_1, \ldots, x_{2N}, z_1, \ldots, z_M) \in \mathbf{T}(h, e)$, if $y = \sum_{n=1}^{2N} x_n \longrightarrow \infty$, then $|(z_1, \ldots, z_M)| = \sum_{m=1}^{M} z_m \longrightarrow \infty$,
(A8) For each e, if $(\psi, z) \in \mathbf{T}(h, e)$ and $z' \geq z$, then $(\psi, z') \in \mathbf{T}(h, e)$ (free disposal),
(A9) $U_h^B(\cdot)$ is a continuous, concave, and strictly increasing function,
(A10) $\omega_h^B(e) > 0, \forall e \in \mathbf{E}'$,

(**A11**) T_h^{BC} and T_h^{BD} satisfy Assumptions (A3) and (A4) for some $c_h^{BC}(e)$, $c_h^{BD}(e)$, and $\alpha_h^B(e)$.

For firm j:

(**A12**) For each $e \in \mathbf{E}'$, $\mathbf{Y}_j(e)$ is a closed and convex set ,
(**A13**) For each $e \in \mathbf{E}'$, $\mathbf{Y}_j(e) \cap \mathbf{R}_+^{2M} = \{0\}$,
(**A14**) For each $e \in \mathbf{E}'$, $(\sum_j \mathbf{Y}_j(e)) \cap (-\sum_j \mathbf{Y}_j(e)) = \{0\}$,
(**A15**) $U_j^F(\cdot)$ is a continuous, concave, and strictly increasing function,
(**A16**) $\omega_j^F(e_0) > 0$,
(**A17**) T_j^{FC} and T_j^{FD} satisfies Assumptions (A3) and (A4) for some $c_j^{FC}(e)$, $c_j^{FD}(e)$, and $\alpha_j^F(e)$.

For government:

(**A18**) U^G is a continuous, concave, and strictly increasing function,
(**A19**) $\omega_G(e) > 0$,
(**A20**) For each security and any $e \in \mathbf{E}'$ there exists a terminal node $e' \in \mathbf{E}(e)$ such that the dividend of this security is positive at this node,
(**A21**) $\sum_i \alpha_i^C(e) + \sum_j \alpha_j^F(e) + \sum_h \alpha_h^B(e) \geq 0, \forall e \in \mathbf{E}'$.

All assumptions except (A7), (A15), (A4), (A11), (A17), and (A21) are standard. (A7) says that transactions must consume resources. Assumptions (A4), (A11), (A17), and (A21) say that the government has designed into the tax code, and in the application of tax laws, measures that avoid large revenue losses from implausible (unbounded) claims. In other words, the government cannot/will not promise infeasible tax rebates to consumers and firms, preventing consumers from exploiting unlimited arbitrage opportunities through tax rebates on capital losses. According to the Federal Tax Code, the government often recognizes realized capital gains and does not recognize realized capital losses. Observe that the tax functions and trading restrictions are sufficiently general to allow the government to levy taxes on capital gains, interest, and dividends, and delay paying tax rebates on capital losses. Also the tax functions are state contingent allowing for random audits or (random from the point of view of agents and the government) legal interpretations. Strictly speaking, a fully rational system would model the intricacies of the legal tax system; here we simply assume that is an exogenous mechanism that is precommitted by the government.

3 The Competitive Equilibrium Existence Theorem

The main theorem of this paper is the following.

Theorem 1. *Suppose that Assumptions (A1)–(A21) hold. Then there exists an equilibrium $(\xi^*, \widetilde{p}^*)$; that is, $(\xi^*, \widetilde{p}^*)$ satisfies (i)–(vi) of Definition 1, where $\xi^* = ((x_i^*, \alpha_i^*)_{i \in \mathbf{I}}, (z_h^*, \phi_h^*)_{h \in \mathbf{H}}, (y_j^*, \beta_j^*)_{j \in \mathbf{J}}, x_G^*)$.*

3.1 The Modified Economy

To show the existence of equilibrium, as in Arrow and Debreu [2], or Debreu [9], we construct a bounded commodity spot market. Unfortunately, this method does not apply directly because of the possible emptiness of the budget correspondence of agents in a multiperiod assets market. To overcome this problem, we approximate the original economy with a sequence of new economies with positive commodity prices, which has the property that the limit of the equilibria of the new economies is an equilibrium of the original economy. In the following, we first truncate the commodity spot market as in Arrow and Debreu [2] and then truncate the asset space.

First of all, we truncate the commodity space. For broker h, define

$\mathbf{Z}_h = \{z_h$: there exists $(\phi_h^B, \phi_h^S) \gg 0$ such that $(\phi_h^B(e), \phi_h^S(e), z_h(e)) \in \mathbf{T}(h, e), \forall e \in \mathbf{E}'\}$;

$\widehat{\mathbf{Z}}_h = \{z_h \in \mathbf{Z}_h$: there exist $x_i \in \mathbf{X}_i, i = 1, \ldots, I, y_j \in \mathbf{Y}_j, j = 1, \ldots, J,$
$z_{h'} \in \mathbf{Z}_{h'}, h' \neq h,$ and $x_G \in \mathbf{X}_G$ such that $\sum_i x_i(e) + \sum_j y_j^-(e) + \sum_h z_h(e) + x_G(e) - \omega_G(e) - \sum_j y_j^+(e) - \sum_i \omega_i^C(e) - \sum_j \omega_j^F(e) - \sum_h \omega_h^B(e) \ll 0,$
$\forall e \in \mathbf{E}'\}$.

For consumer i, define

$\widehat{\mathbf{X}}_i = \{x_i \in \mathbf{X}_i$: there exist $x_{i'} \in \mathbf{X}_{i'}, i' \neq i, y_j \in \mathbf{Y}_j, j = 1, \ldots, J,$
$z_h \in \mathbf{Z}_h, h = 1, \ldots, H,$ and $x_G \in \mathbf{X}_G$ such that $\sum_i x_i(e) + \sum_j y_j^-(e) + \sum_h z_h(e) + x_G(e) - \omega_G(e) - \sum_j y_j^+(e) - \sum_i \omega_i^C(e) - \sum_j \omega_j^F(e) - \sum_h \omega_h^B(e) \ll 0, \forall e \in \mathbf{E}'\}$.

For firm j, set

$\widehat{\mathbf{Y}}_j = \{y_j \in \mathbf{Y}_j$: there exist $x_i \in \mathbf{X}_i, i = 1, \ldots, I, y_{j'} \in \mathbf{Y}_{j'}, j' \neq j,$
$z_h \in \mathbf{Z}_h, h = 1, \ldots, H,$ and $x_G \in \mathbf{X}_G$ such that $\sum_i x_i(e) + \sum_j y_j^-(e) + \sum_h z_h(e) + x_G(e) - \omega_G(e) - \sum_j y_j^+(e) - \sum_i \omega_i^C(e) - \sum_j \omega_j^F(e) - \sum_h \omega_h^B(e) \ll 0, \forall e \in \mathbf{E}'\}$.

For the government, set

$\hat{\mathbf{X}}_G = \{x_G \in \mathbf{X}_G$: there exist $x_i \in \mathbf{X}_i, i = 1, \ldots, I, y_j \in \mathbf{Y}_j, j = 1, \ldots, J,$
$z_h \in \mathbf{Z}_h, h = 1, \ldots, H,$ such that $\sum_i x_i(e) + \sum_j y_j^-(e) + \sum_h z_h(e) + x_G(e) - \omega_G(e) - \sum_j y_j^+(e) - \sum_i \omega_i^C(e) - \sum_j \omega_j^F(e) - \sum_h \omega_h^B(e) \ll 0, \forall e \in \mathbf{E}'\}$.

As in Arrow and Debreu [2], we have the boundedness of sets $\widehat{\mathbf{X}}_i, \widehat{\mathbf{Y}}_j, \widehat{\mathbf{Z}}_h$, and $\hat{\mathbf{X}}_G$, which is stated in the following lemma. The proof is omitted, because it is standard.

Lemma 1. *The sets $\widehat{\mathbf{X}}_i, \widehat{\mathbf{Y}}_j, \widehat{\mathbf{Z}}_h$, and $\hat{\mathbf{X}}_G$ are all convex and compact.*

Now we turn to truncating trading sets of asset. For broker h, define $\mathbf{\Phi}_h = \{\phi_h = (\phi_h^B, \phi_h^S)$: there exists $z_h \in \widehat{\mathbf{Z}}_h$ such that $(\phi_h(e), z_h(e)) \in \mathbf{T}(h, e), \forall e \in \mathbf{E}'\}$.

In the following lemma, $\mathbf{\Phi}_h$ is shown to be bounded, which plays an important role in proving the existence of the general equilibrium.

Lemma 2. *The set $\mathbf{\Phi}_h$ is a compact and convex subset of $R^{|\mathbf{E}'| \times N}$, $h = 1, \ldots, H$.*

Proof. If Assumption (A6) holds, then the set $\mathbf{\Phi}_h$ is a closed convex set. The convexity of $\mathbf{\Phi}_h$ is obvious. It remains to show its closedness. To this end, suppose that for each k, there exist $\phi_h^{(k)} \in \mathbf{\Phi}_h$ and $z_h^k \in \widehat{\mathbf{Z}}_h$ such that $(\phi_h^{(k)}, z_h^k) \in \mathbf{T}(h, e), \forall e \in \mathbf{E}'$, and, in particular, by (A8), $(\phi_h^{(k)}, \bar{z}_h^k) \in \mathbf{T}(h, e), \forall e \in \mathbf{E}'$, where $\bar{z}_h^k = (\max_k z_{h,1}^k, \ldots, \max_k z_{h,M}^k)$. If $\{\phi_h^{(k)}\}$ is unbounded, we may suppose $(\phi^{(k)})_{h,1}^B \longrightarrow \infty$ without loss of generality. But by Assumption (A7),

$$\lim_{k \longrightarrow \infty} |\bar{z}|_h^k = \infty,$$

which provides a contradiction, because $\widehat{\mathbf{Z}}_h$ is bounded by Lemma 1, and proves the boundedness of $\{\phi_h^{(k)}\}$. Hence, we can choose a subsequence $\{\phi_h^{(k_n)}\}$ from $\{\phi_h^{(k)}\}$ such that

$$\lim_{n \longrightarrow \infty} \phi_h^{(k_n)} = \phi_h;$$

this implies, inasmuch as $\mathbf{T}(h, e)$ is closed, that $\mathbf{\Phi}_h$ is compact. $\square$

By Lemma 2, there exists a positive constant M_0 such that for any $(\phi_h^B, \phi_h^S) \in \mathbf{\Phi}_h$, $\phi_{h,n}^B(e) \le M_0$ and $\phi_{h,n}^S(e) \le M_0, n = 1, \ldots, N, h = 1, \ldots, H$, $\forall e \in \mathbf{E}'$. For consumer i, define

$$\widehat{\mathbf{\Psi}}_i = \left\{ \alpha_i = (\alpha_i^B, \alpha_i^S) : \alpha_{i,n}^B(e) \le HM_0, n = 1, \ldots, N, \forall e \in \mathbf{E}' \right\}.$$

For firm j, define

$$\widehat{\mathbf{\Theta}}_j = \left\{ \beta_j = (\beta_j^B, \beta_j^S) : \beta_{j,n}^B(e) \le HM_0, n = 1, \ldots, N, \forall e \in \mathbf{E}' \right\}.$$

Define

$$C^1 = \left\{ (c_1, \ldots, c_{2(|\mathbf{E}'| \times N)}) : |c_i| \le M_1, i = 1, \ldots, 2(|\mathbf{E}'| \times N) \right\} \left(\subseteq R_+^{2(|\mathbf{E}'| \times N)} \right),$$

where $M_1 > HM_0$. Let $\widehat{\mathbf{X}}_i'$ be the set of $x_i \in \mathbf{X}_i$ with property that

$$\sum_i x_i(e) + \sum_j y_j^-(e) + \sum_h z_h(e) + x_G(e)$$

$$- \omega_G(e) - \sum_j y_j^+(e) - \sum_i \omega_i^C(e) - \sum_j \omega_j^F(e) - \sum_h \omega_h^B(e)$$

$$\ll D(e) \left[\sum_{e' \in \mathbf{PA}(e) - \{e\}} M_1(e') \right], \forall e \in \mathbf{E}',$$

for some $x_{i'} \in \mathbf{X}_{i'}, i' \neq i, y_j \in \mathbf{Y}_j, j = 1, \ldots, J, z_h \in \mathbf{Z}_h, h = 1, \ldots, H$, and $x_G \in \mathbf{X}_G$. Here

$$M_1(e') = (3M_1, \ldots, 3M_1).$$

Likewise, we can define $\widehat{\mathbf{Y}}_j'$, $\widehat{\mathbf{Z}}_h'$, and $\widehat{\mathbf{X}}_G'$. As in Lemma 1, it can be shown that $\widehat{\mathbf{X}}_i'$, $\widehat{\mathbf{Y}}_j'$, $\widehat{\mathbf{Z}}_h'$, and $\widehat{\mathbf{X}}_G'$ are compact and convex, and therefore, there exists a cube $C^2 = \{(c_1, \ldots, c_{|\mathbf{E}'| \times M}) : |c_i| \leq M_2, i = 1, \ldots, |\mathbf{E}'| \times M\}$ ($\subseteq R^{|\mathbf{E}'| \times M}$) such that C^2 contains in its interior for all sets $\widehat{\mathbf{X}}_i'$, $\widehat{\mathbf{Y}}_j'$, $\widehat{\mathbf{Z}}_h'$, and $\widehat{\mathbf{X}}_G'$. Define $\widetilde{\mathbf{X}}_i = C^2 \cap X_i$, $\widetilde{\mathbf{X}}_G = C^2 \cap X_G$, $\widetilde{\mathbf{Y}}_j = C^2 \cap Y_j$, $\widetilde{\mathbf{Z}}_h = C^2 \cap Z_h$, and $\widetilde{\boldsymbol{\Phi}}_h = \widetilde{\boldsymbol{\Psi}}_i = \widetilde{\boldsymbol{\Theta}}_j = C^1$. And let $\widetilde{\Gamma}_i^C(\tilde{p}, \gamma, \delta)$, $\widetilde{\Gamma}_j^F(\tilde{p})$, $\widetilde{\Gamma}_h^B(\tilde{p})$, and $\widetilde{\Gamma}_G$ be the resultant modification of $\Gamma_i^C(\tilde{p}, \gamma, \delta)$, $\Gamma_j^F(\tilde{p})$, $\Gamma_h^B(\tilde{p})$, and Γ_G, respectively. Define

$$\triangle = \Big\{ \tilde{p} = (p^C, p^B, p^S) : \sum_{n=1}^{N} (p_n^B(e) + p_n^S(e)) + \sum_{m=1}^{M} p_m^C(e) = 1,$$

$$p_n^B(e) \geq p_n^S(e) \geq 0, p_m^C(e) \geq 0, m = 1, \ldots, M, n = 1, \ldots, N, \forall e \in \mathbf{E}' \Big\},$$

$$\triangle_k = \Big\{ \tilde{p} = (p^C, p^B, p^S) \in \triangle : p_m^C(e) \geq 1/k,$$

$$m = 1, \ldots, M, \forall e \in \mathbf{E}' \Big\},$$

where $k \geq M$. In this model, excess demand is not necessarily homogeneous of degree zero in prices; thus the equilibrium prices may depend on the normalization chosen. For example, this will be the case for specific taxes.

In the next section, we prove the existence of the modified economy, and then prove the existence of equilibrium of the original economy.

3.2 The Proof of Existence of General Equilibrium

We adopt the Arrow–Debreu technique to prove existence in the modified economy. Given $\tilde{p} \in \triangle_k$, let

$$\mu_i^C = \mu_i^C(\tilde{p}, \gamma, \delta) = \Big\{ (\psi_i, x_i) : U_i^C(x_i) = \sup_{(\bar{\psi}_i, \bar{x}_i) \in \widetilde{\Gamma}_i^C(\tilde{p}, \gamma, \delta)} U_i^C(\bar{x}_i) \Big\}.$$

$$\mu_j^F = \mu_j^F(\tilde{p}) = \Big\{ (\beta_j, y_j) : U_j^F(\delta_j) = \sup_{(\beta_j', y_j') \in \widetilde{\Gamma}_i^F(\tau)} U_j^F(\delta_j') \Big\}.$$

$$\mu_h^B = \mu_h^B(\tilde{p}) = \Big\{ (\phi_h, z_h) : U_h^B(\gamma_h) = \sup_{(\bar{\phi}_h, \bar{z}_h) \in \widetilde{\Gamma}_h^B(\gamma_h)} U_h^B(\bar{\gamma}_h) \Big\}.$$

$$\upsilon_G = \upsilon_G(\tilde{p}) = \Big\{ x_G : U^G(x_G) = \sup_{\bar{x}_G \in \widetilde{\Gamma}_G} U^G(\bar{x}_G) \Big\}.$$

$$\mu^M(e) = \Big\{ p \in \triangle_k | p^B(e) \Big(\sum_i \alpha_i^B(e) + \sum_j \beta_j^B(e) - \sum_h \phi_h^S(e) \Big)$$

$$- p^S(e) \Big(\sum_i \alpha_i^S(e) + \sum_j \beta_j^S(e) - \sum_h \phi_h^B(e) \Big)$$

$$+ \sum_{m=2}^M p_m^C(e) \Big[\sum_i x_{i,m}(e) + \sum_j y_{j,m}^-(e) + \sum_h z_{h,m}(e) + x_{G,m}(e)$$

$$- \Big(\omega_{G,m}(e) + \sum_j y_{j,m}^+(e) + \sum_i \omega_{i,m}^C(e) \Big)$$

$$+ \sum_j \omega_{j,m}^F(e) + \sum_h \omega_{h,m}^B(e) \Big) \Big]$$

$$+ p_1^C(e) \Big\{ D(e) \Big[\sum_{e' \in \mathbf{PA}(e)-\{e\}} (\alpha_i^B(e') - \alpha_i^S(e')) \Big]$$

$$+ D(e) \Big[\sum_{e' \in \mathbf{PA}(e)-\{e\}} (\beta_i^B(e') - \beta_i^S(e')) \Big]$$

$$+ D(e) \Big[\sum_{e' \in \mathbf{PA}(e)-\{e\}} (\phi_h^B(e') - \phi_h^S(e')) \Big] \Big\} \text{ is maximum} \Big\}.$$

Now we make the last assumption, the boundary condition for government.

(**A22**) If $\tilde{p}(k) = (p^C(k), p^B(k), p^S(k))$, $p_m^C(k) \to 0$ as k goes to infinity, then $x_{G,m}(\tilde{p}(k)) \to \infty$, where $x_{G,m}(\tilde{p}(k))$ is the government's optimal consumption of commodity m corresponding to price $\tilde{p}(k)$.

This assumption makes the plausible condition that the government can increase its consumption of a commodity to infinity when the price of that commodity declines to zero. A sufficient condition to imply this would be to assume that the government utility function has the property that the marginal utility of its consumption goes to infinity as consumption goes to zero. Here we merely assume the condition directly.

Now we turn to the proof of the lower hemi-continuity of $\tilde{\Gamma}_i^C(p, \gamma, \delta)$, $\tilde{\Gamma}_j^F(p)$, $\tilde{\Gamma}_h^B(\gamma_h)$, and $\tilde{\Gamma}^G$.

Lemma 3. $\tilde{\Gamma}_i^C(p, \gamma, \delta), \tilde{\Gamma}_j^F(p), \tilde{\Gamma}_h^B(\gamma_h)$, and $\tilde{\Gamma}^G$ are lower hemi-continuous on $\triangle_k$ for each k, and therefore, $\mu_i^C(p, \gamma, \delta), \mu_j^F(\tilde{p}), \mu_h^B(\tilde{p})$, and $\upsilon_G(\tilde{p})$ are upper hemi-continuous on $\triangle_k$ for each k.

Proof. It suffices to show that the correspondences $\tilde{\Gamma}_i^C(p, \gamma, \delta), \tilde{\Gamma}_j^F(p), \tilde{\Gamma}_h^B(\gamma_h)$, and $\tilde{\Gamma}^G(\tau)$ all have interior points for the given price $\tilde{p} \in \triangle_k$. From (A5), $\tilde{\Gamma}_i^C(p, \gamma, \delta)$ has an interior point; from (A10), $\tilde{\Gamma}_h^B(\gamma_h)$ has an interior point; from (A16), $\tilde{\Gamma}_j^F(p)$ has an interior point, because the firm can buy some security at the initial node and sell it gradually afterwards; and from (A19) and (A20), $\tilde{\Gamma}^G(\tau)$ has an interior point. By Berge's maximum theorem and standard methods, we can prove that the correspondences μ_i^C, μ_j^F, μ_h^B, and υ_G are upper hemi-continuous and convex-valued.   □

Define

$$\Psi(\xi, \tilde{p}) = \prod_{i=1}^{I} \mu_i^C \otimes \prod_{j=1}^{J} \mu_j^F \otimes \prod_{h=1}^{H} \mu_h^B \otimes \upsilon_G \otimes \prod_{e \in \mathbf{E}'} \mu^M.$$

Ψ is also upper hemi-continuous and convex-valued. Under these conditions, Ψ satisfies all the conditions of the Kakutani fixed point theorem. Thus there exist $(x_i^*(k), \alpha_i^*(k))$ for all $i \in \mathbf{I}$; $(z_h^*(k), \phi_h^*(k))$ for all $h \in \mathbf{H}$; $(y_j^*(k), \beta_j^*(k))$ for all $j \in \mathbf{J}$; $x_G^*(k)$ and $\tilde{p}^*(k)$ such that $(\xi^*(k), \tilde{p}^*(k)) \in \Psi(\xi^*(k), \tilde{p}^*(k))$, where $\xi^*(k) = ((x_i^*(k), \alpha_i^*(k))_{i \in \mathbf{I}}, (z_h^*(k), \phi_h^*(k))_{h \in \mathbf{H}}, (y_j^*(k), \beta_j^*(k))_{j \in \mathbf{J}}, x_G^*(k))$. In particular, for all $\tilde{p} \in \triangle_k$

$$p^B(e)\Big(\sum_i \alpha_i^{B*}(k, e) + \sum_j \beta_j^{B*}(k, e) - \sum_h \phi_h^{S*}(k, e)\Big)$$

$$- p^S(e)\Big(\sum_i \alpha_i^{S*}(k, e) + \sum_j \beta_j^{S*}(k, e) - \sum_h \phi_h^{B*}(k, e)\Big)$$

$$+ \sum_{m=2}^{M} p_m^C(e)\Big[\sum_i x_{i,m}^*(k, e) + \sum_j y_{j,m}^{-*}(k, e) + \sum_h z_{h,m}^*(k, e) + x_{G,m}^*(k, e)$$

$$- \Big(\omega_{G,m}(e) + \sum_j y_{j,m}^{+*}(k, e) + \sum_i \omega_{i,m}^C(e) + \sum_j \omega_{j,m}^F(e) + \sum_h \omega_{h,m}^B(e)\Big)\Big]$$

$$+ p_1^C(e)\Big\{D(e)\Big[\sum_{e' \in \mathbf{PA}(e)-\{e\}} \big(\alpha_i^{B*}(k, e') - \alpha_i^{S*}(k, e')\big)\Big]$$

$$+ D(e)\Big[\sum_{e' \in \mathbf{PA}(e)-\{e\}} \big(\beta_i^{B*}(k, e') - \beta_i^{S*}(k, e')\big)\Big]$$

$$+ D(e)\Big[\sum_{e' \in \mathbf{PA}(e)-\{e\}} \big(\phi_h^{B*}(k, e') - \phi_h^{S*}(k, e')\big)\Big]\Big\}$$

$$\leq 0, \ e \in \mathbf{E}', \tag{1}$$

where $(\alpha_i^{B*}(k, e))_{e \in \mathbf{E}'} = \alpha_i^{B*}(k)$, and the other variables are defined in exactly the same manner.

Because $(\xi^*(k), \tilde{p}^*(k))$ are bounded, we may assume that the sequence $(\xi^*(k), \tilde{p}^*(k))$ converges, say to $(\xi^*, \tilde{p}^*)$. By Lemma 3, in order to prove that $(\xi^*, \tilde{p}^*)$ is an equilibrium of the modified economy, we should show that $p_m^{C*}(e) > 0, m = 1, \ldots, M, e \in \mathbf{E}'$, and ξ^* satisfies market clearance.

Lemma 4. $p^{C*}(e) \gg 0, p^{B*}(e) \gg 0, e \in \mathbf{E}'$.

Proof. First of all, we prove that $p^{C*} \gg 0$. It is obvious that for all $\tilde{p} \in \triangle$, ξ^* satisfies (1), and in particular,

$$\sum_i \alpha_i^{B*}(e) + \sum_j \beta_j^{B*}(e) - \sum_h \phi_h^{S*}(e)$$

$$- \left(\sum_i \alpha_i^{S*}(e) + \sum_j \beta_j^{S*}(e) - \sum_h \phi_h^{B*}(e) \right) \ll 0; \tag{2}$$

$$\sum_i \alpha_i^{B*}(e) + \sum_j \beta_j^{B*}(e) - \sum_h \phi_h^{S*}(e) \ll 0; \tag{3}$$

$$\sum_i x_{i,m}^*(e) + \sum_j y_{j,m}^{-*}(e) + \sum_h z_{h,m}^*(e) + x_{G,m}^*(e)$$

$$- \left(\omega_{G,m}(e) + \sum_j y_{j,m}^{+*}(e) + \sum_i \omega_{i,m}^C(e) \right.$$

$$+ \sum_j \omega_{j,m}^F(e) + \sum_h \omega_{h,m}^B(e) \right)$$

$$\le 0, \quad m = 2, \ldots, M, \tag{4}$$

$$\sum_i x_{i,1}^*(e) + \sum_j y_{j,1}^{-*}(e) + \sum_h z_{h,1}^*(e) + x_{G,1}^*(e)$$

$$- \left(\omega_{G,1}(e) + \sum_j y_{j,1}^{+*}(e) + \sum_i \omega_{i,1}^C(e) \right.$$

$$+ \sum_j \omega_{j,1}^F(e) + \sum_h \omega_{h,1}^B(e) \right)$$

$$\le D(e) \left[\sum_{e' \in \mathbf{PA}(e) - \{e\}} \left(\alpha_i^{B*}(e') - \alpha_i^{S*}(e') \right) \right]$$

$$+ D(e) \left[\sum_{e' \in \mathbf{PA}(e) - \{e\}} \left(\beta_i^{B*}(e') - \beta_i^{S*}(e') \right) \right]$$

$$+ D(e) \left[\sum_{e' \in \mathbf{PA}(e) - \{e\}} \left(\phi_h^{B*}(e') - \phi_h^{S*}(e') \right) \right] \le 0, \tag{5}$$

where the last inequality follows from (2). This means that $x_i^* \in \widehat{\mathbf{X}}_i$, $y_j^* \in \widehat{\mathbf{Y}}_j$, $z_h^* \in \widehat{\mathbf{Z}}_h$, and $x_G^* \in \widehat{\mathbf{X}}_G$ and therefore, $(\phi_h^B, \phi_h^S) \in \mathbf{\Phi}_h$, $h = 1, \ldots, H$, $(\alpha_i^{B*}, \alpha_i^{S*}) \in \widehat{\mathbf{\Psi}}_i$, and $(\beta_i^{B*}, \beta_i^{S*}) \in \widehat{\mathbf{\Theta}}_j$. Thus, by Assumption (A22), $p_m^{C*}(e) > 0, m = 1, \ldots, M, \forall e \in \mathbf{E}'$, otherwise, $x_G^* \notin \widehat{\mathbf{X}}_G$.

Now we turn to proving $p^{B*}(e) \gg 0, \forall e \in \mathbf{E}'$. To the contrary, suppose $p_n^{B*}(e_0) = 0$ for some n and some $\tilde{e} \in \mathbf{E}'$. By considering the consumer i and noticing that $p^{C*}(e) \gg 0, \forall e \in \mathbf{E}'$, $(\alpha_i^{B*}, \alpha_i^{S*}) \ne \widehat{\mathbf{\Psi}}_i$, because, by Assumption (A20), the consumer can unlimitedly increase his wealth at one of the terminal nodes by buying asset n at the node $\tilde{e}$. Consequently, $p^{B*}(e) \gg 0, \forall e \in \mathbf{E}'$. $\square$

Lemma 5. ξ^* satisfies market clearance.

Proof. Note that from the proof of Lemma 4, $x_i^* \in \widehat{\mathbf{X}}_i$, $y_j^* \in \widehat{\mathbf{Y}}_j$, $z_h^* \in \widehat{\mathbf{Z}}_h$, and $x_G^* \in \widehat{\mathbf{X}}_G$. Hence,

$$p^{B*}(e)\Big(\sum_i \alpha_i^{B*}(e) + \sum_j \beta_j^{B*}(e) - \sum_h \phi_h^{S*}(e)\Big)$$

$$- p^{S*}(e)\Big(\sum_i \alpha_i^{S*}(e) + \sum_j \beta_j^{S*}(e) - \sum_h \phi_h^{B*}(e)\Big)$$

$$+ \sum_{m=2}^{M} p_m^{C*}(e)\Big[\sum_i x_{i,m}^*(e) + \sum_j y_{j,m}^{-*}(e) + \sum_h z_{h,m}^*(e) + x_{G,m}^*(e)$$

$$- \Big(\omega_{G,m}(e) + \sum_j y_{j,m}^{+*}(e) + \sum_i \omega_{i,m}^C(e) + \sum_j \omega_{j,m}^F(e) + \sum_h \omega_{h,m}^B(e)\Big)\Big]$$

$$+ p_1^{C*}(e)\Big\{D(e)\Big[\sum_{e' \in \mathbf{PA}(e)-\{e\}} (\alpha_i^{B*}(e') - \alpha_i^{S*}(e'))\Big]$$

$$+ D(e)\Big[\sum_{e' \in \mathbf{PA}(e)-\{e\}} (\beta_i^{B*}(e') - \beta_i^{S*}(e'))\Big]$$

$$+ D(e)\Big[\sum_{e' \in \mathbf{PA}(e)-\{e\}} (\phi_h^{B*}(e') - \phi_h^{S*}(e'))\Big]\Big\} = 0, \quad e \in \mathbf{E'}.$$

Suppose $p_n^{B*}(e) > p_n^{S*}(e)$. Define $p_n^S(e) = p_n^{S*}(e) + \varepsilon$, $p_M^C(e) = p_M^{C*}(e) - \varepsilon$ for ε sufficiently small, $p_l^S(e) = p_l^{S*}(e)$, $l \neq n$, $p_l^B(e) = p_l^{B*}(e)$, $l = 1, \ldots, N$, $p_m^C(e) = p_m^{C*}(e)$, $m = 1, \ldots, M-1$. Plugging this price into (1) and by the above equality and (5),

$$\varepsilon\Big(\sum_h \phi_{h,n}^{B*}(e) - \sum_i \alpha_{i,n}^{S*}(e) - \sum_j \beta_{j,n}^{S*}(e)\Big)$$

$$\leq \varepsilon\Big[\sum_i x_{i,M}^*(e) + \sum_j y_{j,M}^{-*}(e) + \sum_h z_{h,M}^*(e) + x_{G,M}^*(e)$$

$$- \Big(\omega_{G,M}(e) + \sum_j y_{j,M}^{+*}(e) + \sum_i \omega_{i,M}^C(e) + \sum_j \omega_{j,M}^F(e) + \sum_h \omega_{h,M}^B(e)\Big)\Big]$$

$$\leq 0, \quad e \in \mathbf{E'},$$

implying

$$\sum_h \phi_{h,n}^{B*}(e) - \sum_i \alpha_{i,n}^{S*}(e) - \sum_j \beta_{j,n}^{S*}(e) \leq 0, \quad e \in \mathbf{E'}. \tag{6}$$

On the other hand, as in the proof of $p^{B*}(e) \gg 0$, we can show that if $p_m^{S*}(e^0) = 0$ for some m and some $e^0 \in \mathbf{E'}$, then, by Assumption (A20), $\alpha_{i,n}^{S*}(e^0) = \beta_{j,n}^{S*}(e^0) = 0$, $\forall i,j$, and therefore, by (6), $\phi_{h,n}^{B*}(e^0) = 0$. Consequently, by noticing that $p^{C*}(e) \gg 0, p^{B*}(e) \gg 0, e \in \mathbf{E'}$, and combining (2), (3), (4), (5), and (6),

$$\sum_i x_i^*(e) + \sum_j y_j^{-*}(e) + \sum_h z_h^*(e) + x_G^*(e)$$

$$- \left(w_G(e) + \sum_j y_j^{+*}(e) + \sum_i w_i^C(e) + \sum_j w_j^F(e) + \sum_h w_h^B(e) \right) = 0;$$

$$\sum_i \alpha_{i,n}^{B*}(e) + \sum_j \beta_{j,n}^{B*}(e) - \sum_h \phi_{h,n}^{S*}(e)$$

$$- \left(\sum_i \alpha_{i,n}^{S*}(e) + \sum_j \beta_{j,n}^{S*}(e) - \sum_h \phi_{h,n}^{B*}(e) \right) = 0$$

if $p_n^{B*}(e) = p_n^{S*}(e)$; if $p_n^{B*}(e) > p_n^{S*}(e)$,

$$\sum_i \alpha_{i,n}^{B*}(e) + \sum_j \beta_{j,n}^{B*}(e) - \sum_h \phi_{h,n}^{S*}(e) = 0$$

and

$$\sum_i \alpha_{i,n}^{S*}(e) + \sum_j \beta_{j,n}^{S*}(e) - \sum_h \phi_{h,n}^{B*}(e) = 0$$

finishing the proof of Lemma 5.   □

Combining the above three lemmas, we have proved the existence of general equilibrium in the modified economy. Consequently, in exactly the same manner as in Arrow and Debreu [2], it can be shown that $(\xi^*, \tilde{p}^*)$ is an equilibrium of the original economy, completing the proof of Theorem 1.   □

4 Discussion

We have provided sufficient conditions for the existence of a competitive equilibrium for a multiperiod asset economy with taxes and transaction costs. The point of the paper is to show how common assumptions on taxes and transaction costs can be incorporated into a competitive asset economy in a consistent manner. For expositional reasons, we have introduced a number of assumptions to simplify the analysis and proofs. Below we sketch how one could relax these assumptions in a more general model and how one could modify the strategy of proof to deal with this added complexity.

Having proved the existence of a competitive equilibrium in this model, we should be interested in the efficiency properties of the equilibrium, uniqueness of the equilibria, and characterizations of the equilibrium. We argue that efficiency and uniqueness are problematic except under very strong conditions. The characterization of the asset prices and portfolios can be very complex; simple arbitrage pricing relations will occur under certain restrictive conditions, but one must be careful to check that the taxation functions and transaction costs are consistent with the standard results. Dynamic portfolios with transaction costs, trading constraints, and taxes will be highly sensitive to the specification of those functions. There are numerous simple models in the literature, but most rely on strong assumptions to obtain analytic results, or require numerical simulation.

4.1 Extensions and Generalizations of the Model

First, one could introduce multiple physical commodities in the final period. We assumed a single commodity in the last period to simplify the technical analysis. Apart from some additional technical restrictions, the addition of multiple commodities in the last period is a straightforward generalization.

Second, the brokers and firms have utility functions. We assumed this type of objective because profit maximization is no longer the unanimous objective for shareholders with incomplete markets or transaction costs. Similar problems arise with differing taxation across firms and shareholders in multiperiod economies. Rather than attempt to address this difficult issue here we have avoided it by assuming utility functions for brokers and firms. There have been attempts to discuss more abstract firm objectives (see Kelsey and Milne [17] and their references), but we do not pursue that line of argument here. Similar arguments can be applied to the government utility function. Clearly, assuming a utility function is a gross simplification of government decision making, but it suffices to provide sufficiently regular responses in the government demands for goods and services for our existence proof, and it closes our economy. It would be possible to introduce less regular preferences for government; see the techniques on voting outcomes for firm decision making in Kelsey and Milne [17] for some possible ideas that could apply to firms and governments. If the economy has a single physical commodity and the government cannot trade securities, then government consumption is specified by the budget constraint, and we can dispense with the government utility function.

Third, we have assumed that the transaction technology is convex. It is well known that transactions involve setup costs, or nonconvex technologies. In [15], we address that issue by considering approximate equilibria in economies with nonconvex transaction technologies. Clearly, these ideas could be introduced here for our transaction technologies. Similar arguments could be applied to nonconvex tax functions that arise from tax/subsidy thresholds that induce sharp nonconvexities in tax schedules. As long as the nonconvexities in these functions are not large in comparison to the overall economy, an approximate equilibrium could be obtained.

Fourth, we assumed that taxation over dividends and capital gains are additively separable. This is assumed for expositional convenience. We could have assumed that the capital gains tax function included dividend taxation and suppressed the dividend tax function. Notice that agent's asset demands and supplies will depend upon buying and selling prices, their utility functions, endowments, technology, and their tax functions. This allows considerable generality in taxation law. For example, our model allows for tax timing in buying or selling assets (for more discussion and examples of this type of behavior, see Dammon et al. [8]).

Fifth, although we consider the simple case of brokers/intermediaries, the model can be adapted quite easily to accommodate restrictions on consumer and/or firm asset positions that are imposed by regulation (see Milne and Neave [18] for either a reinterpretation of our model, or a more straightforward modification of the consumer model). Thus our model could be adapted to include discrete versions of Detemple and Murthy [10] and Cuoco and Liu [5] on trading restrictions.

Sixth, one can consider the government precommitting to a general taxation system of laws that specify contingent rates and rules depending upon the behavior of the agents. Given that the government could compute the equilibrium given the competitive actions of all other agents, then the government could simulate different tax codes and choose among the equilibria. The equilibria we have discussed here is merely one of that set.

Seventh, we have omitted taxation on commodities and services at each event. It is not difficult to add those taxation functions to accommodate the interaction of financial and income taxation and subsidies. This modification would be necessary to include standard economic discussions of the interaction of the taxation and social security systems.

4.2 Efficiency and Uniqueness

It is well known in the economics and finance literature that an equilibrium in incomplete asset markets is generically inefficient. This result carries over to the economy with transaction costs, because the incomplete market model is simply an economy with zero or infinite transaction costs on disjoint sets of assets. There are special cases where the economy is efficient; the representative agent model is a standard example. Another case is where the agents have identical hyperbolic absolute risk averse (HARA) von Neumann–Morgenstern preferences. Related to the efficiency property is the derivation of an objective function for the firm. With transaction costs and taxation, there are many cases where the objective function cannot be obtained by the Fisher separation theorem. We have assumed the existence of a firm utility function, but a more serious construction would introduce a welfare function that weighted the utility of the controlling agents. Or more abstractly, we could allow for an efficient outcome where a social-welfare function does not exist (see Kelsey and Milne [17]).

One strand of literature in public economics discusses incomplete markets and taxation with respect to the social security system. It suggests that the taxation system can help overcome asset market incompleteness. As the taxation and asset return functions enter in the budget constraints, it is possible to construct tax functions that would mimic the missing asset returns and prices. It is clear that altering the taxation system will also alter the actions of agents in equilibrium and the relative prices of assets, so that the design of the taxation system would require very detailed information about economic agents.

In general, the equilibria for the economy will not be unique. This is well known in the general equilibrium literature. Often in simple models with transaction costs and taxes, the modeler assumes HARA utilities (or identical risk-neutral preferences) to avoid the difficulties of nonuniqueness of equilibria. These restrictive assumptions greatly simplify the model, and in extreme cases imply a representative agent.

4.3 Characterizations of Equilibria

Our model includes the well-studied case where there are no transaction costs and taxes. This model implies martingale pricing results and the generalized Modigliani–Miller theorems for derivative or redundant assets. With complete markets, any equilibrium is efficient.

With transaction costs, or incomplete markets, it is well known that one can still derive martingale pricing results, but the martingale measure may not be unique. This type of model is consistent with a subcase of our model. Similar characterizations with bounds on martingale measures can be derived with the version of the model with different buying and selling prices for assets. A number of papers use this characterization to bound derivative prices in an equilibrium (see Milne and Neave [18] for a synthesis and discussion of the literature).

A more difficult issue arises with the introduction of a new derivative security that cannot be priced by arbitrage, and the derivation of pricing bounds. In a two-period setting, this model is identical to an industrial organization model with the introduction of a new commodity. The pricing of the commodity, and related commodities, will be very sensitive to the structure of the model and assumptions on strategic behavior by the agents. Clearly this type of model is not nested in our model, as we are restricted to an equilibrium model and do not consider comparative equilibria or strategic behavior.

Turning to the economy with taxation, there are a number of results that can be nested in our model. First, the introduction of taxation introduces distortions that destroy efficiency. But the purpose of government taxation is to raise revenue to provide public goods and redistribute wealth (we could have modified our model to allow for this extension). Second, the existence of transaction costs, or incomplete asset markets, raises the possibility of government taxation acting as a proxy for the missing or inactive asset markets. This type of argument is often used to justify social security systems to redistribute wealth and to act as an implied insurance market for the poor. Third, restricting our attention to asset pricing, it is easy to show that if taxation does not discriminate across assets and falls only on net income after asset trading, then the standard arbitrage pricing results continue to hold. As these results rely on zero arbitrage profits, and there is no profit in trading dynamic portfolios, the martingale pricing results follow. Fourth, when taxation discriminates across assets or dynamic asset portfolios, then arbitrage asset pricing must include taxes in the marginal conditions. These results are well known in corporate finance when dealing with corporate leverage, or dividend policy. In the case of capital gains taxation, characterizations will involve optimal stopping rules that will be very sensitive to the specification of the tax rules, asset price movements, and the preferences and income of the agent (for an example, see Dammon et al. [8]). Fifth, although we have only one government, it is easy to introduce other governments and their taxation systems. This modification would allow the model to deal with international taxation and tax arbitrage through financial markets.

5 Conclusion

We have constructed a model that is sufficiently general to include most known models of competitive asset economies with "frictions" in asset markets. We have omitted two important classes of frictions. The first is price-making behavior where agents' asset trades have an impact on prices. Clearly, this would violate our competitive assumption. The second friction involves asymmetric information between agents. Nearly all of the latter literature is restricted to partial equilibrium frameworks with strategic behavior, and is not consistent with our competitive assumptions.

References

1. F. Allen and D. Gale *Financial Innovation and Risk Sharing*. MIT Press, 1994.
2. K.J. Arrow and G. Debreu. Existence of an equilibrium for a competitive economy. *Econometrica*, 22:265–290, 1954.
3. O. Blanchard and S. Fischer. *Lectures in Macroeconomics*. MIT Press, 1989.
4. G. Constantinides. Capital market equilibrium with personal tax. *Econometrica*, 51:611–636, 1983.
5. D. Cuoco and H. Liu. A martingale characterization of consumption choices and hedging costs with margin requirements. *Mathematical Finance*, 10:355–385, 2000.
6. R.M. Dammon and R.C. Green. Tax arbitrage and the existence of equilibrium prices for financial assets. *Journal of Finance*, 42:1143–1166, 1987.
7. R. Dammon and C. Spatt. The optimal trading and pricing of securities with asymmetric capital gains taxes and transaction costs. *Review of Financial Studies*, 9:921–952, 1996.
8. R. Dammon, C. Spatt, and H. Zhang. Optimal asset allocation with taxable and tax-deferred investing. *Journal of Finance*, 59:999–1037, 2004.
9. G. Debreu. Existence of competitive equilibrium. *Handbook of Mathematical Economics*, Vol. II, eds. K. Arrow and M. Intriligator, North Holland, 1982.
10. J. Detemple and S. Murthy. Equilibrium asset prices and no-arbitrage with portfolio constraints. *Review of Financial Studies*, 10(4):1133–1174, 1997.

11. D. Duffie. Stochastic equilibria with incomplete financial markets. *Journal of Economic Theory*, 41:405–416, 1987.
12. P.H. Dybvig and S.A. Ross. Tax clienteles and asset pricing. *Journal of Finance*, 39:751–762, 1986.
13. J.R. Graham. Taxes and corporate finance: A review. *Review of Financial Studies*, 16, 4:1075–1130, 2003.
14. R.C. Green. A simple model of the taxable and tax-exempt yield curves. *Review of Financial Studies*, 6:233–264, 1993.
15. X. Jin and F. Milne. The existence of equilibrium in a financial market with transaction costs. *Quantitative Analysis in Financial Markets*, ed. M. Avellaneda, 1:323–343, 1999.
16. C. Jones and F. Milne. Tax arbitrage, existence of equilibrium, and bounded tax rebates. *Mathematical Finance*, 2:189–196, 1992.
17. D. Kelsey and F.Milne. The existence of equilibrium in incomplete markets and the objective function of the firm. *Journal of Mathematical Economics*, 25:229–245, 1996.
18. F. Milne and E. Neave. A general equilibrium financial asset economy with transaction costs and trading constraints. *WP. 1082*, Economics Dept., Queen's University, 2003.
19. S. Ross. Arbitrage and martingales with taxation. *Journal of Political Economy*, 95:371–393, 1987.
20. S. Schaefer. Taxes and security market equilibrium. *Financial Economics: Essays in Honor of Paul Cootner*, eds. W. Sharpe and C. Cootner, Prentice-Hall, 1982.
21. P. Swoboda and J. Zechner. Financial structure the tax system. Chapter 24 in *Finance: Handbooks in Operations Research and Management Science*, eds. R. Jarrow, V. Maksimovic, and W.T. Ziemba, North-Holland, 1995.
22. J. Zechner. Tax clienteles and optimal capital structure under uncertainty. *Journal of Business*, 63:465–491, 1990.

Calibration of Lévy Term Structure Models

Ernst Eberlein[1] and Wolfgang Kluge[2]

[1] Department of Mathematical Stochastics
University of Freiburg
79104 Freiburg, Germany
eberlein@stochastik.uni-freiburg.de

[2] BNP Paribas London Branch
London, NW1 6AA, United Kingdom
wolfgang.kluge@uk.bnpparibas.com

Summary. We review the Lévy-driven interest rate theory that has been developed in recent years. The intimate relations between the various approaches, as well as the differences, are outlined. The main purpose of this paper is to elaborate on calibration in the real world as well as in the risk-neutral setting.

Key words: Lévy processes; market models; forward rate model; calibration.

1 Introduction

Although the mathematical theory of Lévy processes in general originated in the first half of the last century, its use in finance started only in the last decade of that century. Because Brownian motion, itself a Lévy process, is so well understood, and also because a broad community is familiar with diffusion techniques, it is not surprising that this technology became the basis for the classical models in finance. On the other hand, it has been known for a long time that the normal distribution, which generates the Brownian motion and which is reproduced on any time horizon by this process, is only a poor approximation of the empirical return distributions observed in financial data. Of course, diffusion processes with random coefficients produce distributions different from the normal one; however, the resulting distribution on a given time horizon is not even known in general, and can only be determined approximately and visualized by Monte Carlo simulation. This remark holds for most of the extensions of the classical geometric Brownian motion model such as models with stochastic volatility or stochastic interest rates.

In [21] and [20], a genuine Lévy model for the pricing of derivatives was introduced by Madan and Seneta and Madan and Milne. Based on a three-parameter Variance-Gamma (VG) process as the driving process, they derived a pricing formula for standard call options. This approach was extended and refined in a series of papers by Madan and coauthors. We only mention the extension to the four-parameter CGMY-model in [4], which added more flexibility to the initial VG model. Based on an extensive empirical study of stock price data, in an independent line of research Eberlein and Keller introduced the hyperbolic Lévy model in [7]. Both processes, the VG as well as the hyperbolic Lévy process, are purely discontinuous and, therefore, in a sense opposite to the Brownian motion. Starting from empirical results, the basic concern in [7] was to develop a model which produces distributions that fit the observed empirical return distributions as closely as possible. This led to exponential Lévy models

$$S_t = S_0 \exp(L_t), \qquad t \geq 0, \tag{1}$$

to describe stock prices or indices. The log returns $\log S_t - \log S_{t-1}$ derived from model (1) are the increments of length 1 of the driving Lévy process L. Therefore, by feeding in the Lévy process L which is generated by an (infinitely divisible) empirical return distribution, at least on the time horizon 1, this model reproduces exactly the distribution that one sees in the data. This is not the case if one starts with a model for prices given by the stochastic differential equation

$$dS_t = S_{t-}d\widetilde{L}_t, \tag{2}$$

or equivalently by the stochastic exponential

$$S_t = \mathcal{E}(\widetilde{L})_t \tag{3}$$

of a Lévy process $\widetilde{L} = (\widetilde{L}_t)_{t \geq 0}$.

A model based on normal inverse Gaussian (NIG) distributions was added by Barndorff-Nielsen in [2]. Normal inverse Gaussian Lévy processes have nice analytic properties. As the class of hyperbolic distributions, NIG distributions constitute a subclass of the class of generalized hyperbolic distributions. The stock price model based on this five-parameter class was developed in [13] and [5]. VG distributions turned out to be another subclass. A further interesting class of Lévy models based on Meixner processes was introduced by Schoutens (see [23; 24]).

Calibration of the exponential Lévy model (1), at least with respect to the real-world (or historical) measure, is conceptually straightforward, because— as pointed out above—the return distribution is the one that generates the driving Lévy process. See [7] for calibration results in the case of the hyperbolic model. In this paper we study calibration of Lévy interest rate models. The corresponding theory has been developed in a series of papers starting with [14] and continuing with [11; 6; 12; 8; 18; 10]. During the extensions of the initial model, it turned out that the natural driving processes for interest rate models are time-inhomogeneous Lévy processes. They are described in the next section. Section 3 is a brief review of the three basic approaches: the Lévy forward rate model (HJM-type model), the Lévy forward process model, and the Lévy LIBOR (London Interbank Offered Rate) model.

In each of these approaches, a different quantity is modeled: the instantaneous forward rate $f(t,T)$, the forward process $F(t,T,U)$ corresponding to time points T and U, and the δ-(forward) LIBOR rate $L(t,T)$. The relation between the latter quantities is obvious, because $1 + \delta L(t,T) = F(t,T,T+\delta)$. Although $L(t,T)$ and $F(t,T,T+\delta)$ differ only by an additive and multiplicative constant, the two specifications lead to models that behave quite differently. The reason is that the changes of the driving process have a different impact on the forward LIBOR rates. In the Lévy LIBOR model, forward LIBOR rates change by an amount that is relative to their current level whereas the change in the Lévy forward process model does not depend on the actual level. Let us note that by construction the forward process model is easier to handle and implement. On the other hand, this model—as does the classical HJM and therefore also the Lévy forward rate model—produces negative rates with some (small) probability. Negative rates are excluded in the Lévy LIBOR model.

It is shown in Section 4 that there is also a close relation to the forward rate model. We prove that the forward process model can be seen as a special case of the forward rate model. In Section 5 we describe how the Lévy forward rate model can be calibrated with respect to the real-world measure as well as with respect to the risk-neutral martingale measure. Some explicit calibration results for driving generalized hyperbolic Lévy processes are given.

In Section 6 calibration of the Lévy forward process and the Lévy LIBOR model is discussed. Again, generalized hyperbolic Lévy processes, in particular NIG processes, are considered in the explicit results.

2 The Driving Process

Let $(\Omega, \mathcal{F}, \mathbb{F}, P)$ be a complete stochastic basis, where $\mathbb{F} = (\mathcal{F}_t)_{t \in [0, T^*]}$, the filtration, satisfies the usual conditions, $T^* \in \mathbb{R}^+$ is a finite time horizon, and $\mathcal{F} = \mathcal{F}_{T^*}$. The driving process $L = (L_t)_{t \in [0, T^*]}$ is a *time-inhomogeneous Lévy process*, that is, an adapted process with independent increments and absolutely continuous characteristics, which is abbreviated by PIIAC in [16]. We can assume that the paths of the process are right-continuous with left-hand limits. We also assume that the process starts in 0. The law of L_t is given by its characteristic function

$$E[e^{i\langle u, L_t \rangle}] = \exp \int_0^t \left[i\langle u, b_s \rangle - \frac{1}{2}\langle u, c_s u \rangle \right.$$
$$\left. + \int_{\mathbb{R}^d} \left(e^{i\langle u, x \rangle} - 1 - i\langle u, x \rangle \mathbf{1}_{\{|x| \leq 1\}} \right) F_s(dx) \right] ds. \quad (4)$$

Here, $b_s \in \mathbb{R}^d$, c_s is a symmetric nonnegative-definite $d \times d$-matrix, and F_s is a *Lévy measure*, that is, a measure on $\mathbb{R}^d$ that integrates $(|x|^2 \wedge 1)$ and satisfies

$F_s(\{0\}) = 0$. By $\langle \cdot, \cdot \rangle$, we denote the Euclidean scalar product on $\mathbb{R}^d$, and $|\cdot|$ is the corresponding norm. We assume that

$$\int_0^{T^*} \left(|b_s| + \|c_s\| + \int_{\mathbb{R}^d} (|x|^2 \wedge 1) F_s(dx) \right) ds < \infty, \tag{5}$$

where $\|\cdot\|$ denotes any norm on the $d \times d$-matrices. The triplet $(b, c, F) = (b_s, c_s, F_s)_{s \in [0, T^*]}$ represents the local characteristics of the process L. We impose a further moment assumption.

Assumption EM: *There are constants $M, \varepsilon > 0$, such that for every $u \in [-(1+\varepsilon)M, (1+\varepsilon)M]^d$,*

$$\int_0^{T^*} \int_{\{|x|>1\}} \exp\langle u, x \rangle F_s(dx) ds < \infty. \tag{6}$$

EM is a very natural assumption. It is equivalent to $E[\exp\langle u, L_t \rangle] < \infty$ for all $t \in [0, T^*]$ and all u as above. In the interest rate models that we consider, the underlying processes are always exponentials of stochastic integrals with respect to the driving processes L. In order to allow pricing of derivatives, these underlying processes have to be martingales under the risk-neutral measure and, therefore, a priori have to have finite expectations, which is exactly assumption EM.

In particular, under EM the random variable L_t itself has finite expectation, and consequently we do not need a truncation function. The representation (4) simplifies to

$$E[e^{i\langle u, L_t \rangle}] = \exp \int_0^t \Big[i\langle u, b_s \rangle - \frac{1}{2}\langle u, c_s u \rangle$$
$$+ \int_{\mathbb{R}^d} \left(e^{i\langle u, x \rangle} - 1 - i\langle u, x \rangle \right) F_s(dx) \Big] ds, \tag{4'}$$

where the characteristic b_s is now different from the one in (4). We always use the local characteristics (b, c, F) derived from (4').

Another consequence of assumption EM is that L is a special semimartingale, and thus its canonical representation has the simple form

$$L_t = \int_0^t b_s ds + \int_0^t \sqrt{c_s} dW_s + \int_0^t \int_{\mathbb{R}^d} x(\mu^L - \nu)(ds, dx), \tag{7}$$

where $W = (W_s)_{s \geq 0}$ is a standard d-dimensional Brownian motion, $\sqrt{c_s}$ is a measurable version of the square root of c_s, and μ^L is the random measure of jumps of L with compensator $\nu(ds, dx) = F_s(dx)ds$.

Assumption EM, which is assumed throughout the following sections, holds for all processes we are interested in, in particular for processes generated by generalized hyperbolic distributions. It excludes processes generated by stable distributions in general, but these processes are a priori not appropriate for developing a martingale theory to price derivative products.

We denote θ_s the cumulant associated with a process L as given in (7) with local characteristics (b_s, c_s, F_s); that is,

$$\theta_s(z) = \langle z, b_s \rangle + \frac{1}{2}\langle z, c_s z \rangle + \int_{\mathbb{R}^d}(e^{\langle z,x \rangle} - 1 - \langle z, x \rangle)F_s(dx). \qquad (8)$$

3 Lévy Term Structure Models

We give a short review of the three basic interest rate models that are driven by time-inhomogeneous Lévy processes. Although the focus is on different rates in these three approaches, the models are closely related. All three of them are appropriate for pricing standard interest rate derivatives.

3.1 The Lévy Forward Rate Model

Modeling the dynamics of instantaneous forward rates is the starting point in the Heath–Jarrow–Morton approach ([15]). The forward rate model driven by Lévy processes was introduced in [14] and developed further in [11], where in particular a risk-neutral version was identified. The model was extended to driving time-inhomogeneous Lévy processes in [6] and [8]. In the former reference, a complete classification of all equivalent martingale measures was achieved. As an unexpected consequence of this analysis, it turned out that under the standard assumption of deterministic coefficients for one-dimensional driving processes there is a single martingale measure, and thus—as in the Black–Scholes option pricing theory—there is a unique way to price interest rate derivatives. Explicit pricing formulae for caps, floors, swaptions, and other derivatives, as well as efficient algorithms to evaluate these formulae, are given in [8] and [9].

Denote $B(t, T)$ the price at time t of a zero coupon bond with maturity T. Obviously $B(T, T) = 1$ for any maturity date $T \in [0, T^*]$. Because zero coupon bond prices can be deduced from instantaneous forward rates $f(t, T)$ via $B(t, T) = \exp\left(-\int_t^T f(t, u)du\right)$ and vice versa, the term structure can be modeled by specifying either of them. Here we specify the forward rates. Its dynamics is given for any $T \in [0, T^*]$ by

$$f(t, T) = f(0, T) + \int_0^t \alpha(s, T)ds - \int_0^t \sigma(s, T)dL_s, \qquad (9)$$

where $L = (L_t)_{t\in[0,T^*]}$ is a PIIAC with local characteristics (b, c, F). For details concerning assumptions on the coefficients $\alpha(t, T)$ and $\sigma(s, T)$, we refer to [6] and [8]. The simplest case and at the same time the most important one for the implementation of the model is the case where α and σ are deterministic functions. Defining

$$A(s, T) := \int_{s \wedge T}^T \alpha(s, u)du \quad \text{and} \quad \Sigma(s, T) = \int_{s \wedge T}^T \sigma(s, u)du, \qquad (10)$$

one can derive the corresponding zero coupon bond prices in the form

$$B(t,T) = B(0,T)\exp\left(\int_0^t (r(s) - A(s,T))ds + \int_0^t \Sigma(s,T)dL_s\right), \quad (11)$$

where $r(s) := f(s,s)$ denotes the short rate. Choosing $T = t$ in (11), the risk-free savings account $B_t = \exp\left(\int_0^t r(s)ds\right)$ can be written as

$$B_t = \frac{1}{B(0,t)}\exp\left(\int_0^t A(s,t)ds - \int_0^t \Sigma(s,t)dL_s\right). \quad (12)$$

Now assume that $\Sigma(s,T)$ is deterministic and

$$0 \le \sigma^i(s,T) \le M, \quad i \in \{1,\ldots,d\}, \quad (13)$$

where M is the constant from assumption EM. From (11) one sees immediately that discounted bond prices $B(t,T)/B_t$ are martingales for all $T \in [0,T^*]$ if we choose

$$A(s,T) := \theta_s(\Sigma(s,T)), \quad (14)$$

because $\int_0^t \theta_s(\Sigma(s,T))ds$ is the exponential compensator of $\int_0^t \Sigma(s,T)dL_s$. Thus we are in an arbitrage-free market. Another useful representation of zero coupon bond prices that follows from (11) and (12) is

$$B(t,T) = \frac{B(0,T)}{B(0,t)}\exp\left(-\int_0^t A(s,t,T)ds + \int_0^t \Sigma(s,t,T)dL_s\right), \quad (15)$$

where we used the abbreviations

$$A(s,t,T) := A(s,T) - A(s,t) \quad \text{and} \quad \Sigma(s,t,T) = \Sigma(s,T) - \Sigma(s,t). \quad (16)$$

3.2 The Lévy Forward Process Model

This model was introduced in [12]. The advantage of this approach is that the driving process remains a time-inhomogeneous Lévy process during the backward induction that is done to get the rates in uniform form. Thus one can avoid any approximation, and the model is easy to implement.

Let $0 = T_0 < T_1 < \cdots < T_N < T_{N+1} = T^*$ denote a discrete tenor structure and set $\delta_k = T_{k+1} - T_k$. For zero coupon bond prices $B(t,T_k)$ and $B(t,T_{k+1})$, the forward process is defined by

$$F(t,T_k,T_{k+1}) = \frac{B(t,T_k)}{B(t,T_{k+1})}. \quad (17)$$

Therefore, modeling forward processes means specifying the dynamics of ratios of successive bond prices.

Let L^{T^*} be a time-inhomogeneous Lévy process on a complete stochastic basis $(\Omega, \mathcal{F}_{T^*}, \mathbb{F}, P_{T^*})$. The probability measure P_{T^*} can be interpreted as the forward measure associated with the settlement date T^*. The moment condition $\mathbb{EM}$ is assumed as before. The local characteristics of L^{T^*} are denoted (b^{T^*}, c, F^{T^*}). Two parameters, c and F^{T^*}, are free parameters, whereas the drift characteristic b^{T^*} is chosen to guarantee that the forward process is a martingale. Because we proceed by backward induction, let us use the notation $T_i^* := T_{N+1-i}$ and $\delta_i^* = \delta_{N+1-i}$ for $i \in \{0, \dots, N+1\}$. The following ingredients are needed.

(FP.1) For any maturity T_i there is a bounded, continuous, deterministic function $\lambda(\cdot, T_i) : [0, T^*] \to \mathbb{R}^d$ that represents the volatility of the forward process $F(\cdot, T_i, T_{i+1})$. We require for all $k \in \{1, \dots, N\}$,

$$\left| \sum_{i=1}^{k} \lambda^j (s, T_i) \right| \le M \qquad (s \in [0, T^*], j \in \{1, \dots, d\}), \qquad (18)$$

where M is the constant in assumption $\mathbb{EM}$ and $\lambda(s, T_i) = 0$ for $s > T_i$.

(FP.2) The initial term structure of zero coupon bond prices $B(0, T_i)$, $1 \le i \le N+1$, is strictly positive. Consequently, the initial values of the forward processes are given by

$$F(0, T_i, T_{i+1}) = \frac{B(0, T_i)}{B(0, T_{i+1})}. \qquad (19)$$

We begin to construct the forward process with the longest maturity and postulate

$$F(t, T_1^*, T^*) = F(0, T_1^*, T^*) \exp\left(\int_0^t \lambda(s, T_1^*) dL_s^{T^*} \right). \qquad (20)$$

Now we choose b^{T^*} such that $F(\cdot, T_1^*, T^*)$ becomes a P_{T^*}-martingale. This is achieved via the following equation,

$$\int_0^t \langle \lambda(s, T_1^*), b_s^{T^*} \rangle ds = -\frac{1}{2} \int_0^t \langle \lambda(s, T_1^*), c_s \lambda(s, T_1^*) \rangle ds \qquad (21)$$

$$- \int_0^t \int_{\mathbb{R}^d} \left(e^{\langle \lambda(s, T_1^*), x \rangle} - 1 - \langle \lambda(s, T_1^*), x \rangle \right) \nu^{T^*}(ds, dx),$$

where $\nu^{T^*}(ds, dx) = F_s^{T^*}(dx)ds$ is the P_{T^*}-compensator of the random measure of jumps μ^L given by the process L^{T^*}. Using Lemma 2.6 in [17], one can express the ordinary exponential (20) as a stochastic exponential, namely,

$$F(t, T_1^*, T^*) = F(0, T_1^*, T^*) \mathcal{E}_t(H(\cdot, T_1^*)),$$

where

$$H(t, T_1^*) = \int_0^t \sqrt{c_s}\lambda(s, T_1^*)dW_s^{T^*}$$
$$+ \int_0^t \int_{\mathbb{R}^d} \left(e^{\langle \lambda(s,T_1^*),x \rangle} - 1 \right)(\mu^L - \nu^{T^*})(ds, dx). \qquad (22)$$

Because $F(\cdot, T_1^*, T^*)$ is a martingale, we can define the forward martingale measure associated with the date T_1^* by setting

$$\frac{dP_{T_1^*}}{dP_{T^*}} = \frac{F(T_1^*, T_1^*, T^*)}{F(0, T_1^*, T^*)} = \mathcal{E}_{T_1^*}(H(\cdot, T_1^*)). \qquad (23)$$

Using Girsanov's theorem for semimartingales (see [16, Theorem III.3.24]), we can identify from (22) the predictable processes β and Y that describe the measure change, namely,

$$\beta(s) = \lambda(s, T_1^*) \quad \text{and} \quad Y(s, x) = \exp\langle \lambda(s, T_1^*), x \rangle.$$

Consequently $W_t^{T_1^*} := W_t^{T^*} - \int_0^t \sqrt{c_s}\lambda(s, T_1^*)ds$ is a standard Brownian motion under $P_{T_1^*}$, and $\nu^{T_1^*}(dt, dx) := \exp\langle \lambda(s, T_1^*), x \rangle \nu^{T^*}(dt, dx)$ is the $P_{T_1^*}$-compensator of μ^L.

Now we construct the forward process $F(\cdot, T_2^*, T_1^*)$ by postulating

$$F(t, T_2^*, T_1^*) = F(0, T_2^*, T_1^*)\exp\left(\int_0^t \lambda(s, T_2^*)dL_s^{T_1^*} \right),$$

where

$$L_t^{T_1^*} = \int_0^t b_s^{T_1^*} ds + \int_0^t \sqrt{c_s}dW_s^{T_1^*} + \int_0^t \int_{\mathbb{R}^d} x(\mu^L - \nu^{T_1^*})(ds, dx).$$

The drift characteristic $b^{T_1^*}$ can be chosen in an analogous way as in (21), and we define the next measure change from the resulting equation. Proceeding this way, we get all forward processes in the form

$$F(t, T_i^*, T_{i-1}^*) = F(0, T_i^*, T_{i-1}^*)\exp\left(\int_0^t \lambda(s, T_i^*)dL_s^{T_{i-1}^*} \right), \qquad (24)$$

with

$$L_t^{T_{i-1}^*} = \int_0^t b_s^{T_{i-1}^*} ds + \int_0^t \sqrt{c_s}dW_s^{T_{i-1}^*} + \int_0^t \int_{\mathbb{R}^d} x(\mu^L - \nu^{T_{i-1}^*})(ds, dx). \qquad (25)$$

$W^{T_{i-1}}$ is here a $P_{T_{i-1}^*}$-standard Brownian motion, and $\nu^{T_{i-1}}$ is the $P_{T_{i-1}^*}$-compensator of μ^L given by

$$\nu^{T_{i-1}}(dt, dx) = \exp\left(\sum_{j=1}^{i-1} \langle \lambda(t, T_j^*), x \rangle \right) F_t^{T^*}(dx)dt. \qquad (26)$$

The drift characteristic $b^{T^*_{i-1}}$ satisfies

$$\int_0^t \langle \lambda(s, T^*_i), b_s^{T^*_{i-1}} \rangle ds = -\frac{1}{2} \int_0^t \langle \lambda(s, T^*_i), c_s \lambda(s, T^*_i) \rangle ds \qquad (27)$$
$$- \int_0^t \int_{\mathbb{R}^d} \left(e^{\langle \lambda(s, T^*_i), x \rangle} - 1 - \langle \lambda(s, T^*_i), x \rangle \right) \nu^{T^*_{i-1}}(ds, dx).$$

All driving processes L^{T_i} remain time-inhomogeneous Lévy processes under the corresponding forward measures, because they differ only by deterministic drift terms.

3.3 The Lévy LIBOR Model

This approach has been described in full detail in [12]; therefore, we just list some of the properties. As in Section 3.2 the model is constructed by backward induction along the discrete tenor structure and is driven by a time-inhomogeneous Lévy process L^{T^*}, which is given on a complete stochastic basis $(\Omega, \mathcal{F}_{T^*}, \mathbb{F}, P_{T^*})$. As in the Lévy forward process model, P_{T^*} should be regarded as the forward measure associated with the settlement day T^*. L^{T^*} is required to satisfy assumption $\mathbb{EM}$ and can be written in the form

$$L_t^{T^*} = \int_0^t b_s^{T^*} ds + \int_0^t \sqrt{c_s} dW_s^{T^*} + \int_0^t \int_{\mathbb{R}^d} x(\mu^L - \nu^{T^*})(ds, dx), \qquad (28)$$

where $\nu^{T^*}(dt, dx) = F_s^{T^*}(dx)dt$ is the compensator of μ^L. The ingredients needed for the model are as follows.

(LR.1) For any maturity T_i, there is a bounded, continuous, deterministic function $\lambda(\cdot, T_i) : [0, T^*] \to \mathbb{R}^d$ that represents the volatility of the forward LIBOR rate process $L(\cdot, T_i)$. In addition,

$$\sum_{i=1}^N |\lambda^j(s, T_i)| \le M, \quad s \in [0, T^*], j \in \{1, \dots, d\},$$

where M is the constant from assumption $\mathbb{EM}$ and $\lambda(s, T_i) = 0$ for $s > T_i$.

(LR.2) The initial term structure $B(0, T_i)$, $1 \le i \le N+1$, is strictly positive and strictly decreasing (in i). Consequently, the initial term structure $L(0, T_i)$ of forward LIBOR rates is given by

$$L(0, T_i) = \frac{1}{\delta_i} \left(\frac{B(0, T_i)}{B(0, T_{i+1})} - 1 \right) > 0.$$

Now we can start the induction by postulating that

$$L(t, T^*_1) = L(0, T^*_1) \exp \left(\int_0^t \lambda(s, T^*_1) dL_s^{T^*} \right). \qquad (29)$$

The drift characteristic b^{T^*} is chosen as in (21) to make this process a martingale. Writing (29) as a stochastic exponential and exploiting the relation $F(t, T_1^*, T^*) = 1 + \delta_1^* L(t, T_1^*)$, one gets the dynamics in terms of the forward process $F(\cdot, T_1^*, T^*)$. From this, the measure change can be done as in Section 3.2. As a result of the backward induction, one gets for each tenor time point the forward LIBOR rates in the form

$$L(t, T_j^*) = L(0, T_j^*) \exp\left(\int_0^t \lambda(s, T_j^*) dL_s^{T_{j-1}^*} \right) \tag{30}$$

under the corresponding forward martingale measure $P_{T_{j-1}^*}$. The successive forward measures are related by the following equation,

$$\frac{dP_{T_j^*}}{dP_{T_{j-1}^*}} = \frac{1 + \delta_j L(T_j^*, T_j^*)}{1 + \delta_j L(0, T_j^*)}. \tag{31}$$

The driving process $L^{T_{j-1}^*}$ in (30) has the canonical representation

$$L_t^{T_{j-1}^*} = \int_0^t b_s^{T_{j-1}^*} ds + \int_0^t \sqrt{c_s} dW_s^{T_{j-1}^*} + \int_0^t \int_{\mathbb{R}^d} x(\mu^L - \nu^{T_{j-1}^*})(ds, dx). \tag{32}$$

$W^{T_{j-1}^*}$ is a $P_{T_{j-1}^*}$-Brownian motion via

$$W_t^{T_{j-1}^*} = W_t^{T_{j-2}^*} - \int_0^t \sqrt{c_s} \alpha(s, T_{j-1}^*, T_{j-2}^*) ds,$$

where

$$\alpha(t, T_k^*, T_{k-1}^*) = \frac{\delta_k^* L(t-, T_k^*)}{1 + \delta_k^* L(t-, T_k^*)} \lambda(t, T_k^*). \tag{33}$$

Similarly $\nu^{T_{j-1}^*}$ is the $P_{T_{j-1}^*}$-compensator of μ^L which is related to the $P_{T_{j-2}^*}$-compensator via

$$\nu^{T_{j-1}^*}(ds, dx) = \beta(s, x, T_{j-1}^*, T_{j-2}^*) \nu^{T_{j-2}^*}(ds, dx),$$

where

$$\beta(t, x, T_k^*, T_{k-1}^*) = \frac{\delta_k^* L(t-, T_k^*)}{1 + \delta_k^* L(t-, T_k^*)} \left(e^{\langle \lambda(t, T_k^*), x \rangle} - 1 \right) + 1. \tag{34}$$

The backward induction guarantees that zero coupon bond prices $B(\cdot, T_j)$ discounted by $B(\cdot, T_k)$ (i.e., ratios $B(\cdot, T_j)/B(\cdot, T_k)$) are P_{T_k}-martingales for all $j, k \in \{1, \ldots, N+1\}$, and thus we have an arbitrage-free market. This follows directly for successive tenor time points from the relation $1 + \delta L(t, T_j) = B(t, T_j)/B(t, T_{j+1})$, because $L(t, T_j)$ is by construction a $P_{T_{j+1}}$-martingale. Expanding ratios with arbitrary tenor time points T_j and T_k into products of ratios with successive time points, one gets the result from this special case. To see this, one has to use Proposition 3.8 in [16, p. 168], which is a fundamental result for the analysis of all interest rate models where forward martingale measures are used.

Note that the driving processes $L_t^{T_{j-1}^*}$ that are derived during the backward induction are no longer time-inhomogeneous Lévy processes. This is clear from (34), because due to the random term $\beta(s, x, T_{j-1}^*, T_{j-2}^*)$, the compensator $\nu^{T_{j-1}^*}$ is no longer deterministic. One can force the process $\beta(\cdot, x, T_{j-1}^*, T_{j-2}^*)$ to become deterministic by replacing $L(t-, T_k^*)/(1 + \delta_k^* L(t-, T_k^*))$ by its starting value $L(0, T_k^*)/(1 + \delta_k^* L(0, T_k^*))$ in (34). This approximation is convenient for the implementation of the model, because then all driving processes are time-inhomogeneous Lévy processes. Because the process $Y(\cdot, x)$, which is used in the change from one compensator to the next in the forward process approach, is nonrandom, one can implement the model from Section 3.2 without any approximation.

4 Embedding of the Forward Process Model

In this section we show that the Lévy forward process model can be seen as a special case of the Lévy forward rate model. We choose the parameters of the latter in such a way that we get the forward process specification as shown in (24)–(27). In the martingale case, which is defined by (14), according to (15), zero coupon bond prices can be represented in the form

$$B(t, T) = \frac{B(0, T)}{B(0, t)} \exp\left(\int_0^t \Big(\tilde{\theta}_s(\Sigma(s, t)) - \tilde{\theta}_s(\Sigma(s, T)) \Big) ds + \int_0^t \Sigma(s, t, T) d\tilde{L}_s \right),$$
(35)

where $\tilde{L}$ is a time-inhomogeneous Lévy process with characteristics $(\tilde{b}, \tilde{c}, \tilde{F})$ under the (spot martingale) measure P. $\tilde{L}$ satisfies assumption $\mathbb{EM}$ and $\tilde{\theta}_s$ denotes the cumulant associated with the triplet $(\tilde{b}_s, \tilde{c}_s, \tilde{F}_s)$. Recall that $\Sigma(s, T) = \int_{s \wedge T}^T \sigma(s, u) du$.

The forward martingale measure $P_{T_i^*}$ associated with the settlement date T_i^* is related to the spot martingale measure P via the Radon–Nikodym derivative

$$\frac{dP_{T_i^*}}{dP} = \frac{1}{B_{T_i^*} B(0, T_i^*)} \qquad P\text{-a.s.}$$

Choosing $T = t = T_i^*$ in (11), one gets immediately the representation

$$\frac{dP_{T_i^*}}{dP} = \exp\left(-\int_0^{T_i^*} \tilde{\theta}_s(\Sigma(s, T_i^*)) ds + \int_0^{T_i^*} \Sigma(s, T_i^*) d\tilde{L}_s \right). \tag{36}$$

Because σ and, therefore, Σ are deterministic functions, $\tilde{L}$ is also a time-inhomogeneous Lévy process with respect to $P_{T_i^*}$, and its $P_{T_i^*}$-characteristics $(\tilde{b}^{T_i^*}, \tilde{c}^{T_i^*}, \tilde{F}^{T_i^*})$ are given by

$$\widetilde{b}_s^{T_i^*} = \widetilde{b}_s + \widetilde{c}_s \Sigma(s, T_i^*) + \int_{\mathbb{R}^d} \left(e^{\langle \Sigma(s, T_i^*), x \rangle} - 1 \right) x \, \widetilde{F}_s(dx),$$

$$\widetilde{c}_s^{T_i^*} = \widetilde{c}_s, \tag{37}$$

$$\widetilde{F}_s^{T_i^*}(dx) = e^{\langle \Sigma(s, T_i^*), x \rangle} \widetilde{F}_s(dx).$$

Because $\widetilde{L}$ is also a $P_{T_i^*}$-special semimartingale, it can be written in its $P_{T_i^*}$-canonical representation as

$$\widetilde{L}_t = \int_0^t \widetilde{b}_s^{T_i^*} ds + \int_0^t \sqrt{\widetilde{c}_s} dW_s^{T_i^*} + \int_0^t \int_{\mathbb{R}^d} x (\mu^{\widetilde{L}} - \widetilde{\nu}^{T_i^*})(ds, dx), \tag{38}$$

where W^{T_i} is a $P_{T_i^*}$-standard Brownian motion and where $\widetilde{\nu}^{T_i^*}(ds, dx) := \widetilde{F}_s^{T_i^*}(dx) ds$ is the $P_{T_i^*}$-compensator of $\mu^{\widetilde{L}}$, the random measure associated with the jumps of the process $\widetilde{L}$.

Using this representation in (35), we derive the forward process

$$F(t, T_{i+1}^*, T_i^*) = \frac{B(t, T_{i+1}^*)}{B(t, T_i^*)}$$

$$= \frac{B(0, T_{i+1}^*)}{B(0, T_i^*)} \exp\left(\int_0^t \left(\widetilde{\theta}_s(\Sigma(s, T_i^*)) - \widetilde{\theta}_s(\Sigma(s, T_{i+1}^*)) \right) ds \right.$$

$$\left. + \int_0^t \Sigma(s, T_i^*, T_{i+1}^*) d\widetilde{L}_s \right)$$

$$= F(0, T_{i+1}^*, T_i^*) \exp\left(I_t^1 + I_t^2 + \int_0^t \sqrt{\widetilde{c}_s} \Sigma(s, T_i^*, T_{i+1}^*) dW_s^{T_i^*} \right.$$

$$\left. + \int_0^t \int_{\mathbb{R}^d} \langle \Sigma(s, T_i^*, T_{i+1}^*), x \rangle (\mu^{\widetilde{L}} - \widetilde{\nu}^{T_i^*})(ds, dx) \right).$$

Here

$$I_t^1 := \int_0^t \left(\widetilde{\theta}_s(\Sigma(s, T_i^*)) - \widetilde{\theta}_s(\Sigma(s, T_{i+1}^*)) \right) ds$$

$$= \int_0^t \left[- \langle \Sigma(s, T_i^*, T_{i+1}^*), \widetilde{b}_s \rangle \right.$$

$$+ \frac{1}{2} \langle \Sigma(s, T_i^*), \widetilde{c}_s \Sigma(s, T_i^*) \rangle - \frac{1}{2} \langle \Sigma(s, T_{i+1}^*), \widetilde{c}_s \Sigma(s, T_{i+1}^*) \rangle$$

$$+ \int_{\mathbb{R}^d} \left(e^{\langle \Sigma(s, T_i^*), x \rangle} - e^{\langle \Sigma(s, T_{i+1}^*), x \rangle} + \langle \Sigma(s, T_i^*, T_{i+1}^*), x \rangle \right) \widetilde{F}_s(dx) \right] ds,$$

and making use of the first equation in (37),

$$I_t^2 := \int_0^t \langle \Sigma(s, T_i^*, T_{i+1}^*), \widetilde{b}_s^{T_i^*} \rangle ds$$

$$= \int_0^t \left[\langle \Sigma(s, T_i^*, T_{i+1}^*), \widetilde{b}_s \rangle + \langle \Sigma(s, T_i^*, T_{i+1}^*), \widetilde{c}_s \Sigma(s, T_i^*) \rangle \right.$$

$$\left. + \int_{\mathbb{R}^d} \langle \Sigma(s, T_i^*, T_{i+1}^*), x \rangle \left(e^{\langle \Sigma(s, T_i^*), x \rangle} - 1 \right) \widetilde{F}_s(dx) \right] ds.$$

Summing up I^1 and I^2 yields

$$I_t^1 + I_t^2 = -\frac{1}{2} \int_0^t \langle \Sigma(s, T_i^*, T_{i+1}^*), \widetilde{c}_s \Sigma(s, T_i^*, T_{i+1}^*) \rangle ds$$

$$- \int_0^t \int_{\mathbb{R}^d} \left(e^{\langle \Sigma(s, T_i^*, T_{i+1}^*), x \rangle} - 1 - \langle \Sigma(s, T_i^*, T_{i+1}^*), x \rangle \right) \widetilde{F}_s^{T_i^*}(dx) ds.$$

Hence, the forward process is given by

$$F(t, T_{i+1}^*, T_i^*)$$

$$= F(0, T_{i+1}^*, T_i^*) \exp\left(-\frac{1}{2} \int_0^t \langle \Sigma(s, T_i^*, T_{i+1}^*), \widetilde{c}_s \Sigma(s, T_i^*, T_{i+1}^*) \rangle ds \right.$$

$$- \int_0^t \int_{\mathbb{R}^d} \left(e^{\langle \Sigma(s, T_i^*, T_{i+1}^*), x \rangle} - 1 - \langle \Sigma(s, T_i^*, T_{i+1}^*), x \rangle \right) \widetilde{F}_s^{T_i^*}(dx) ds$$

$$+ \int_0^t \sqrt{\widetilde{c}_s} \Sigma(s, T_i^*, T_{i+1}^*) dW_s^{T_i^*}$$

$$\left. + \int_0^t \int_{\mathbb{R}^d} \langle \Sigma(s, T_i^*, T_{i+1}^*), x \rangle \left(\mu^{\widetilde{L}} - \widetilde{\nu}^{T_i^*} \right) (ds, dx) \right).$$

Now we shall specify the model parameters, that is, the volatility σ and the characteristics $(\widetilde{b}, \widetilde{c}, \widetilde{F})$ of $\widetilde{L}$, in such a way that the forward process dynamics match the dynamics given in (24)–(27). First, we choose

$$\Sigma(s, T_i^*, T_{i+1}^*) = \lambda(s, T_{i+1}^*).$$

This can be reached by setting

$$\sigma(s, u) := -\sum_{i=0}^{N} \frac{1}{\delta_{i+1}^*} \lambda(s, T_{i+1}^*) \mathbf{1}_{[T_{i+1}^*, T_i^*)}(u),$$

because

$$\Sigma(s, T_i^*, T_{i+1}^*) = -\int_{T_{i+1}^*}^{T_i^*} \sigma(s, u) du = \lambda(s, T_{i+1}^*).$$

Of course there are many other possibilities to specify σ. It could also be chosen to be continuous or smooth in the second variable.

Next, we specify the triplet $(\widetilde{b}, \widetilde{c}, \widetilde{F})$. $\widetilde{b}_s$ can be chosen arbitrary. We set $\widetilde{c}_s = c_s$ and

$$\widetilde{F}_s(dx) = \exp\langle -\Sigma(s, T^*), x\rangle F_s^{T^*}(dx), \qquad (39)$$

where F^{T^*} is the third characteristic of the driving process L^{T^*} in the Lévy forward process model. Then using the third equation in (37),

$$\widetilde{F}_s^{T_i^*}(dx) = \exp\langle \Sigma(s, T_i^*) - \Sigma(s, T^*), x\rangle F_s^{T^*}(dx)$$

$$= \exp\left(\sum_{j=1}^{i} \langle \Sigma(s, T_j^*) - \Sigma(s, T_{j-1}^*), x\rangle \right) F_s^{T^*}(dx)$$

$$= \exp\left(\sum_{j=1}^{i} \langle \lambda(s, T_j^*), x\rangle \right) F_s^{T^*}(dx),$$

and we arrive at the forward process

$$F(t, T_{i+1}^*, T_i^*) = F(0, T_{i+1}^*, T_i^*) \exp\left(\int_0^t \lambda(s, T_{i+1}^*) d\widetilde{L}_s^{T_i^*} \right),$$

where

$$\widetilde{L}_t^{T_i^*} = \int_0^t b_s^{T_i^*} ds + \int_0^t \sqrt{c_s} dW_s^{T_i^*} + \int_0^t \int_{\mathbb{R}^d} x(\mu^{\widetilde{L}} - \widetilde{\nu}^{T_i^*})(ds, dx).$$

The $P_{T_i^*}$-compensator $\widetilde{\nu}^{T_i}$ of $\mu^{\widetilde{L}}$ is given by

$$\widetilde{\nu}^{T_i^*}(dt, dx) = \exp\left(\sum_{j=1}^{i} \langle \lambda(t, T_j^*), x\rangle \right) F_t^{T^*}(dx) dt,$$

and finally $(b_s^{T_i^*})$ satisfies

$$\int_0^t \langle \lambda(s, T_{i+1}^*), b_s^{T_i^*}\rangle ds$$

$$= -\frac{1}{2} \int_0^t \langle \lambda(s, T_{i+1}^*), \widetilde{c}_s \lambda(s, T_{i+1}^*)\rangle ds$$

$$- \int_0^t \int_{\mathbb{R}^d} \left(e^{\langle \lambda(s, T_{i+1}^*), x\rangle} - 1 - \langle \lambda(s, T_{i+1}^*), x\rangle \right) \widetilde{\nu}^{T_i^*}(ds, dx).$$

Remark 1. This embedding works only for driving processes that are time-inhomogeneous Lévy processes. If both models are driven by a process with stationary increments (i.e., $F_s^{T^*}$ and $\widetilde{F}_s$ do not depend on s), in general we cannot embed the forward process model in the forward rate model.

5 Calibration of the Lévy Forward Rate Model

5.1 The Real-World Measure

In this section we consider the Lévy forward rate model with a time-homogeneous driving process L; that is, L has stationary increments. The goal is to estimate the parameters of the driving process under the real-world measure. For this purpose, we use market data of discount factors (zero coupon bond prices) for one year up to ten years, quoted between September 17, 1999, and September 17, 2001 (i.e., for 522 trading days).

The parameter estimation in the forward rate model is substantially more difficult than in a stock price model. The reason is that we have a number of different assets, namely ten bonds in the case of our data set (in theory of course an infinite number), but only one driving process. Therefore, we have to find a way to extract the parameters of the driving process from the log returns of all ten bond prices.

Let us start by considering the logarithm of the ratio between the bond price and its forward price on the day before; that is,

$$\mathrm{LR}(t, T) := \log \frac{B(t+1, t+T)}{B(t, t+1, t+T)}.$$

Here, $B(t, t+1, t+T)$ is the forward price of $B(t+1, t+T)$ at time t; that is,

$$B(t, t+1, t+T) := \frac{B(t, t+T)}{B(t, t+1)}.$$

We call LR the *daily log return*. Using (15), we get

$$\mathrm{LR}(t, T) = \log B(t+1, t+T) - \log B(t, t+T) + \log B(t, t+1)$$
$$= -\int_t^{t+1} A(s, t+1, t+T)ds + \int_t^{t+1} \Sigma(s, t+1, t+T)dL_s. \quad (40)$$

In what follows, we consider for simplicity the Ho–Lee volatility structure, that is, $\Sigma(s, T) = \hat{\sigma}(T-s)$ for a constant $\hat{\sigma}$, which we set equal to one without loss of generality. Similar arguments can be carried out for other stationary volatility structures, such as the Vasiček volatility function. By *stationary*, we mean that $\Sigma(s, T)$ depends only on $(T-s)$. We assume that the drift term also satisfies some stationarity condition, namely,

$$A(s, T) = A(0, T-s) \qquad \text{for } s \le t.$$

In the risk-neutral case given by (14), this stationarity follows from the stationarity of the volatility function $\Sigma(s, T)$. We get

$$-\int_t^{t+1} A(s, t+1, t+T)ds = -\int_0^1 A(s, 1, T)ds =: d(T), \qquad (41)$$

independent of t and

$$\int_t^{t+1} \Sigma(s, t+1, t+T)dL_s = (T-1)(L_{t+1} - L_t).$$

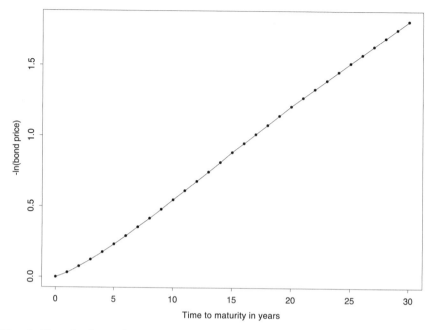

Fig. 1. Negative logarithms of bond prices on September 17, 1999 and interpolating cubic spline.

Consequently,

$$\mathrm{LR}(t, T) = d(T) + c(T) Y_{t+1}, \tag{42}$$

where $c(T) := (T-1)$ is deterministic and

$$Y_{t+1} := L_{t+1} - L_t \sim L_1$$

is $\mathcal{F}_{t+1}$-measurable and does not depend on T.

To estimate the parameters of the driving process, we first determine the daily log returns; that is, for $k \in \{0, 1, \ldots, 520\}$, $n \in \{1, \ldots, 10\}$,

$$\mathrm{LR}(k, k + (n \text{ years})) = \log B(k+1, k + (n \text{ years})) + \log \frac{B(k, k+1)}{B(k, k + (n \text{ years}))}.$$

Unfortunately, we can only get $B(k, k + (n \text{ years}))$ directly from our data set. To determine $B(k+1, k+(n \text{ years}))$ and $B(k, k+1)$, we use an idea developed in [22, Section 5.3] and interpolate the negative of the logarithm of the bond prices with a cubic spline. We do this procedure separately for each day of the data set. Figure 1 shows the interpolation for the first day (even for maturities up to 30 years).

Table 1. Estimated parameters of the distribution L_1 under the measure P.

α	β	δ	μ	λ	Method
1474224	-34659	1.1892e-12	7.5642e-08	2.37028	Max-Likelihood (GH)
590033	-14	1.3783e-06	3.2426e-11	-0.5	Max-Likelihood (NIG)
1195475	-114855	2.5614e-06	2.4723e-07	-0.5	Moments (NIG)

Because $E[L_1] = 0$, we know that

$$\mathrm{LR}(t, T) - E[\mathrm{LR}(t, T)] = (T-1)Y_{t+1}; \tag{43}$$

that is, the centered log returns are affine linear in T. Moreover, Y_1, Y_2, ..., Y_{521} are independent and equal to L_1 in distribution. The corresponding samples $y_1, y_2, \ldots, y_{521}$ could be calculated for a fixed $n \in \{1, 2, \ldots, 10\}$ via

$$y_{k+1} := \frac{\mathrm{LR}(k, k + (n \text{ years})) - \bar{x}_n}{(n \text{ years}) - 1}, \qquad \text{with} \tag{44}$$

$$\bar{x}_n := \frac{1}{521} \sum_{k=0}^{520} \mathrm{LR}(k, k + (n \text{ years})). \tag{45}$$

However, inasmuch as the centered empirical log returns $\mathrm{LR}(k, k+(n \text{ years})) - \bar{x}_n$ are not exactly affine linear in n (compare Figure 2), the y_{k+1} in (44) would then depend on n. Remember that the distribution of L_1 in the Lévy forward rate model does not depend on the time to maturity of the bonds. Therefore, we take a different approach and use the points

$$((1 \text{ year})-1, \mathrm{LR}(k, k+(1 \text{ year}))-\bar{x}_1), \ldots, ((10 \text{ years})-1, \mathrm{LR}(k, k+(10 \text{ years}))-\bar{x}_{10})$$

for a linear regression through the origin. The gradient of the straight line yields the value for y_{k+1}. Figure 2 shows the centered empirical log returns and the regression line for the first day of the data set. Repeating this procedure for each day provides us with the samples $y_1, y_2, \ldots, y_{521}$, which can now be used to estimate the parameters of L_1 by using maximum likelihood estimation.

The parametric class of distributions we use here are generalized hyperbolic distributions (see, e.g., [5]) or subclasses such as hyperbolic [7] or normal inverse Gaussian (NIG) distributions [2]. The resulting densities for our data set are shown in Figure 3. Figure 4 shows the same densities on a log-scale, which allows us to see the fit in the tails. The estimated distribution parameters corresponding to the densities in Figure 3 are given in Table 1.

One of the densities in Figures 3 and 4 was estimated using the method of moments. This is a somewhat simpler approach where one exploits the relation between moments of order i and the ith cumulant (see [1, 26.1.13 and 26.1.14]). Because for generalized hyperbolic distributions the cumulants are explicitly known, one can express the moments (up to order 4) as functions of the distribution parameters λ, α, β, δ, and μ. Applying the usual estimators for moments, one obtains the parameters.

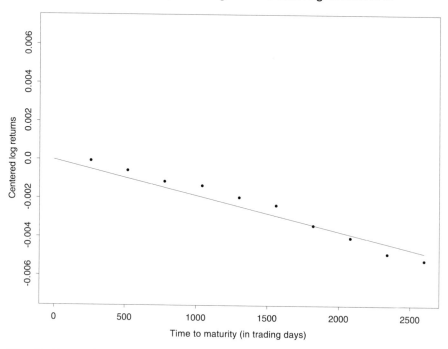

Fig. 2. Centered empirical log returns and regression line for the first day of the data set.

5.2 The Risk-Neutral Measure

There are very liquid markets for the basic interest rate derivatives such as caps, floors, and swaptions. Therefore, market prices for these instruments— typically quoted in terms of their (implied) volatilities with respect to the standard Gaussian model—contain a maximum of information. A cap is a series of call options on subsequent variable interest rates, namely LIBOR rates. Each option is called a caplet. It is easy to see that the payoff of a caplet can be expressed as the payoff of a put option on a zero coupon bond. In the same way a floor is a series of floorlets, and each floorlet is equivalent to a call option on a bond. Thus to price a floorlet one has to price a call on a bond. According to the general no-arbitrage valuation theory, the time-0-value of a call with strike K and maturity t on a bond with maturity T is

$$C_0(t, T, K) := E\left[\frac{1}{B_t}(B(t, T) - K)^+\right], \tag{46}$$

where the expectation is taken with respect to the risk-neutral measure (*spot martingale measure*). Because one would need the joint distribution of B_t

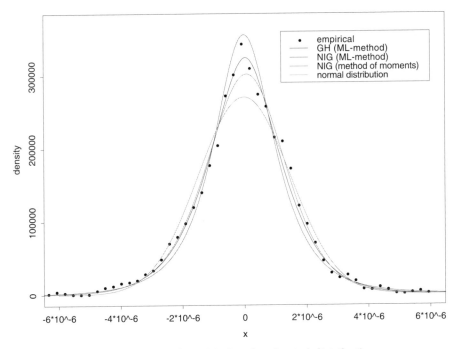

empirical
GH (ML-method)
NIG (ML-method)
NIG (method of moments)
normal distribution

Fig. 3. Densities of empirical and estimated distributions.

and $B(t, T)$ to evaluate this expectation, it is more efficient to express this expectation with respect to the forward measure associated with time t. One then gets

$$C_0(t, T, K) = B(0, t)E_{P_t}[(B(t, T) - K)^+]. \quad (47)$$

Once one has numerically efficient algorithms to compute these expectations, one can calibrate the model by minimizing the differences between model prices and market quotes simultaneously across all available option maturities and strikes. This has been described in detail in [8].

Let us mention that in the stationary case, one can derive risk-neutral parameters from the real-world parameters, which we estimated in Section 5.1. Because of the martingale drift condition (14), we get from (41)

$$d(T) = \int_0^1 (A(s, 1) - A(s, T))ds = \int_0^1 (\theta(\Sigma(\sigma, 1)) - \theta(\Sigma(s, T)))ds, \quad (48)$$

where θ is the logarithm of the moment generating function of $\mathcal{L}(L_1)$ under the risk-neutral measure. By stationarity $\mathcal{L}(Y_{n+1}) = \mathcal{L}(L_1)$ and $E[L_1] = 0$, therefore, (42) implies

$$E[\mathrm{LR}(t, T)] = d(T). \quad (49)$$

The arithmetic mean of the empirical log returns (45) is an estimator for the expectation on the left side. By a minimization procedure, one can now extract

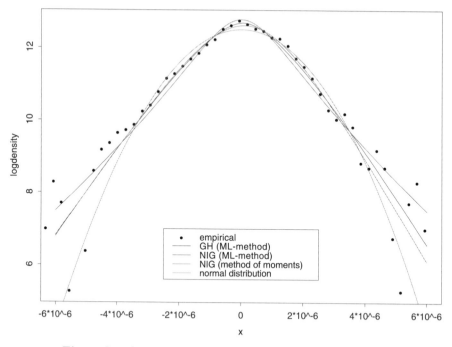

Fig. 4. Log-densities of empirical and estimated distributions.

the distribution parameters of $\mathcal{L}(L_1)$, which are implicit on the right-hand side of (48) from this equation.

In the case of generalized hyperbolic Lévy processes as driving processes, it is known from [22] that the parameters μ and δ do not change when one switches from the real-world to the risk-neutral distribution. Therefore, one has only to extract the remaining parameters λ, α, β via minimization of the distance between average log returns and the integral on the right side in (48).

It is clear that due to the highly liquid market in interest rate derivatives, the direct fit to market quotes as described at the beginning of this section provides more reliable calibration results than the derivation from real-world parameters. It will, therefore, be preferred by practitioners.

6 Calibration of Forward Process and LIBOR Models

Recall that a cap is a sequence of call options on subsequent LIBOR rates. Each single option is called a caplet. Let a discrete tenor structure $0 = T_0 < T_1 < \cdots < T_{n+1} = T^*$ be given as before. The time-T_j payoff of a caplet that is settled in arrears is

$$N\delta_{j-1}(L(T_{j-1}, T_{j-1}) - K)^+, \tag{50}$$

where K is the strike and N the notional amount, which we assume to be 1. The corresponding payoff of a floorlet settled in arrears is

$$N\delta_{j-1}(K - L(T_{j-1}, T_{j-1}))^+. \tag{51}$$

The time-t price of the cap is then

$$C_t(K) = \sum_{j=1}^{N+1} B(t, T_j) E_{P_{T_j}} \left[\delta_{j-1}(L(T_{j-1}, T_{j-1}) - K)^+ | \mathcal{F}_t\right]. \tag{52}$$

Given this formula, the LIBOR model is the natural approach to price caps, because then the LIBOR rates $L(T_{j-1}, T_{j-1})$ are given in the simple form (30) with respect to the forward martingale measure P_{T_j}.

Numerically efficient ways based on bilateral Laplace transforms to evaluate the expectations in (52) are described in [12]. For numerical purposes, the nondeterministic compensators that arise during the backward induction in the Lévy LIBOR model can be approximated by deterministic ones. Concretely, the stochastic ratios $\delta_j L(s-, T_j)/(1 + \delta_j L(s-, T_j))$ are replaced by their deterministic initial values $\delta_j L(0, T_j)/(1 + \delta_j L(0, T_j))$. An alternative approximation method, which is numerically much faster, is described in [18, Section 3.2.1].

Instead of basing the pricing on the LIBOR model, one can use the forward process approach outlined in Section 3.2. It is then more natural to write the caplet payoff (50) in the form

$$\left(1 + \delta_{j-1} L(T_{j-1}, T_{j-1}) - \widetilde{K}_{j-1}\right)^+, \tag{53}$$

where $\widetilde{K}_{j-1} = 1 + \delta_{j-1} K$. Because $1 + \delta_{j-1} L(T_{j-1}, T_{j-1}) = F(T_{j-1}, T_{j-1}, T_j)$, the pricing formula is then instead of (52)

$$C_t(K) = \sum_{j=1}^{N+1} B(t, T_j) E_{P_{T_j}} \left[(F(T_{j-1}, T_{j-1}, T_j) - \widetilde{K}_{j-1})^+ | \mathcal{F}_t\right]. \tag{54}$$

The implementation of this approach leads to a much faster algorithm, inasmuch as the backward induction is more direct in the case of the forward process model. Also, any approximation can be avoided here.

In the implementations, we use mildly time-inhomogeneous Lévy processes, namely those that are piecewise homogeneous Lévy processes. In order to catch the term structure of smiles, which one sees in implied volatility surfaces (see [8, Figure 1]), with sufficient accuracy, typically three Lévy parameter sets are needed: one Lévy process corresponding to short maturities up to one year roughly, a second one for maturities between one and five years, and a third Lévy process corresponding to long maturities from five to ten years. Instead of predetermining the breakpoints where the Lévy parameters change, one can actually include the choice of the breakpoints in the estimation procedure. These random breakpoints improve the calibration results further.

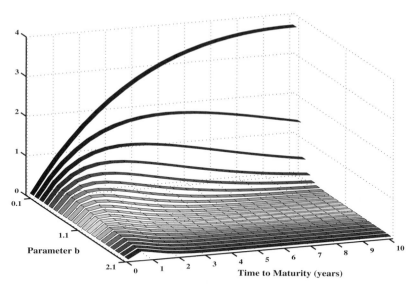

Fig. 5. A variety of shapes for the instantaneous volatility curve produced by (55) with $a = 1$, $b \in [0.1, 2.1]$, $c = 0.1$ (source: [19]).

According to ($\mathbb{FP}$.1) and ($\mathbb{LR}$.1), a volatility structure has to be chosen in both models. In [3], a broad spectrum of suitable volatility structures is discussed, which can be used in the forward process or the LIBOR model. A sufficiently flexible structure is given by

$$\lambda(t, T_j) = a(T_j - t) \exp(-b(T_j - t)) + c, \tag{55}$$

with three parameters a, b, c. Note that we consider one-dimensional processes and consequently also scalar volatility functions in all calibrations. Figure 5 shows a variety of shapes produced by formula (55).

Without loss of generality one can set $a = 1$, because this parameter can be included in the specification of the Lévy process. To see this, take for example a Lévy process L generated by a normal inverse Gaussian distribution; that is, $\mathcal{L}(L_1) = \mathrm{NIG}(\alpha, \beta, \delta, \mu)$ and $a \neq 1$. Define $\widetilde{L} = aL$ and $\widetilde{\lambda}(t, T) = \lambda(t, T)/a$; then (cf., e.g., [5]) $\mathcal{L}(\widetilde{L}_1) = \mathrm{NIG}(\alpha/|a|, \beta/|a|, |a|\delta, a\mu)$ and $\int_0^t \lambda(s, T) dL_s = \int_0^t \widetilde{\lambda}(s, T) d\widetilde{L}_s$. Thus, the model $(\widetilde{\lambda}, \widetilde{L})$ is exactly of the same type with parameter $a = 1$.

For the driving process L, we consider one-dimensional processes generated by generalized hyperbolic distributions $\mathrm{GH}(\lambda, \alpha, \beta, \delta, \mu)$, or subclasses such as the hyperbolic, where $\lambda = 1$, or normal inverse Gaussian distributions, where $\lambda = -\frac{1}{2}$. The generalized hyperbolic distribution is given by its characteristic function

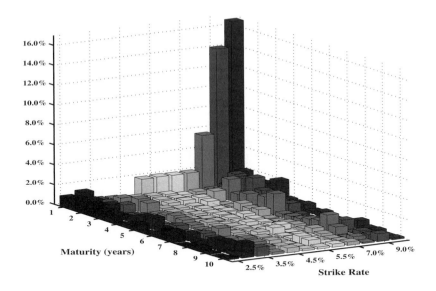

Fig. 6. Data set I. Absolute errors of EUR caplet calibration: forward process model (source: [19]).

$$\Phi_{\mathrm{GH}}(u) = e^{i\mu u}\left(\frac{\alpha^2 - \beta^2}{\alpha^2 - (\beta + iu)^2}\right)^{\lambda/2} \frac{K_\lambda(\delta\sqrt{\alpha^2 - (\beta + iu)^2})}{K_\lambda(\delta\sqrt{\alpha^2 - \beta^2})}, \qquad (56)$$

where K_λ denotes the modified Bessel function of the third kind with index λ. For normal inverse Gaussian distributions, this simplifies to

$$\Phi_{\mathrm{NIG}}(\mu) = e^{i\mu u}\frac{\exp(\delta\sqrt{\alpha^2 - \beta^2})}{\exp(\delta\sqrt{\alpha^2 - (\beta + iu)^2})}. \qquad (57)$$

The class of generalized hyperbolic distributions is so flexible that one does not have to consider higher-dimensional driving processes. Note that it would not be appropriate to classify a model driven by a one-dimensional Lévy process as a one-factor model, because the driving Lévy process itself is already a high-dimensional object. The notion of an x-factor model ($x = 1, 2, \ldots, n$) should be reserved for the world of classical Gaussian models.

To extract the model parameters from market quotes, one considers for each caplet the (possibly squared) difference between the market and model price. The objective function to be minimized is then the weighted sum over all strikes and over all caplets along the tenor structure. A possible choice of weights would be the at-the-money prices for the respective maturity. Figure 6 shows how close one gets to the empirical volatility surface on the basis of the forward process model. The figure shows the absolute difference between model and market prices expressed in volatilities. For the relevant part of the

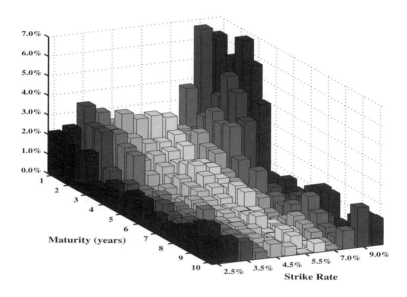

Fig. 7. Data set I. Absolute errors of EUR caplet calibration: LIBOR rate model (source: [19]).

moneyness–maturity plane the differences are below 1 %. The large deviations at the short end are of no importance, because the one-year caplet prices are of the order of magnitude 10^{-9} for the strike rates 8, 9, and 10 %. The underlying data set consists of cap prices in the Euro market on February 19, 2002. The differences one gets for the same data set on the basis of the Lévy LIBOR model are shown in Figure 7.

Comparing the two calibration results, one sees that the forward process approach yields a more accurate fit than the LIBOR approach. In [10], both the Lévy LIBOR as well as the Lévy forward process approach have been extended to a multicurrency setting, which takes the interplay between interest rates and foreign exchange rates into account. This model is also driven by a single time-inhomogeneous Lévy process, namely the process that drives the most distant forward LIBOR rate (or forward process) in the domestic market. Implementation of this sophisticated model was tested for up to three currencies (EUR, USD, and GBP).

Acknowledgments

Wolfgang Kluge has been partially supported by a grant from the Deutsche Forschungsgemeinschaft (DFG). We would like to thank N. Koval for providing some of the figures in Section 6.

References

1. M. Abramowitz and I.A. Stegun. *Handbook of Mathematical Functions*. Dover, 1968.
2. O.E. Barndorff-Nielsen (1998). Processes of normal inverse Gaussian type. *Finance and Stochastics*, 2:41–68, 1998.
3. D. Brigo and F. Mercurio. *Interest Rate Models: Theory and Practice*. Springer, 2001.
4. P. Carr, H. Geman, D. Madan, and M. Yor. The fine structure of asset returns: An empirical investigation. *Journal of Business*, 75:305–332, 2002.
5. E. Eberlein. Application of generalized hyperbolic Lévy motions to finance. *Lévy Processes: Theory and Applications*, eds. O.E. Barndorff-Nielsen, T. Mikosch, and S. Resnick, Birkhäuser Verlag, 319–337, 2001.
6. E. Eberlein, J. Jacod, and S. Raible. Lévy term structure models: No-arbitrage and completeness. *Finance and Stochastics*, 9:67–88, 2005.
7. E. Eberlein and U. Keller (1995). Hyperbolic distributions in finance. *Bernoulli*, 1:281–299, 1995.
8. E. Eberlein and W. Kluge. Exact pricing formulae for caps and swaptions in a Lévy term structure model. *Journal of Computational Finance*, 9(2):99–125, 2006.
9. E. Eberlein and W. Kluge. Valuation of floating range notes in Lévy term structure models. *Mathematical Finance*, 16:237–254, 2006.
10. E. Eberlein and N. Koval. A cross-currency Lévy market model. *Quantitative Finance*, 6(6):465–480, 2006.
11. E. Eberlein and F. Özkan. The defaultable Lévy term structure: Ratings and restructuring. *Mathematical Finance*, 13:277–300, 2003.
12. E. Eberlein and F. Özkan. The Lévy LIBOR model. *Finance and Stochastics*, 9:327–348, 2005.
13. E. Eberlein and K. Prause. The generalized hyperbolic model: Financial derivatives and risk measures. *Mathematical Finance – Bachelier Congress 2000*, eds. H. Geman, D. Madan, S. Pliska, and T. Vorst, Springer, 2002.
14. E. Eberlein and S. Raible. Term structure models driven by general Lévy processes. *Mathematical Finance*, 9:31–53, 1999.
15. E. Heath, R.A. Jarrow, and A. Morton. Bond pricing and the term structure of interest rates: A new methodology for contingent claims valuation. *Econometrica*, 60:77–105, 1992.
16. J. Jacod and A.N. Shiryaev. *Limit Theorems for Stochastic Processes*, 2nd edition, Springer, 2003.
17. J. Kallsen and A.N. Shiryaev. The cumulant process and Esscher's change of measure. *Finance and Stochastics*, 6:397–428, 2002.
18. W. Kluge. *Time-Inhomogeneous Lévy Processes in Interest Rate and Credit Risk Models*. Ph.D. thesis, University of Freiburg, Germany, 2005.
19. N. Koval. *Time-Inhomogeneous Lévy processes in Cross-currency Market Models*. Ph.D. thesis, University of Freiburg, Germany, 2005.
20. D.B. Madan and F. Milne. Option pricing with VG martingale components. *Mathematical Finance*, 1(4):39–55, 1991.
21. D.B. Madan and E. Seneta. The variance gamma (V.G.) model for share market returns. *Journal of Business*, 63:511–524, 1990.
22. S. Raible. *Lévy Processes in Finance: Theory, Numerics, and Empirical Facts*. Ph.D. thesis, University of Freiburg, Germany, 2000.

23. W. Schoutens. Meixner processes in finance. Report 2001-002, EURANDOM, Eindhoven, 2001.
24. W. Schoutens. *Lévy Processes in Finance: Pricing Financial Derivatives*. Wiley, 2003.

Pricing of Swaptions in Affine Term Structures with Stochastic Volatility

Massoud Heidari,[1] Ali Hirsa,[1] and Dilip B. Madan[2]

[1] Caspian Capital Management, LLC
745 Fifth Avenue
New York, NY 10151, USA
massoud.heidari@ixis-ccm.com; ali.hirsa@ixis-ccm.com

[2] Robert H. Smith School of Business
Department of Finance
University of Maryland
College Park, MD 20742, USA
dbm@rhsmith.umd.edu

Summary. In an affine term structure framework with stochastic volatility, we derive the characteristic function of the log swap rate. Having the characteristic function, we employ the fast Fourier transform (FFT) to price swaptions. Using ten years of swap rates and swaption premiums, model parameters are estimated using a square-root unscented Kalman filter. We investigate the relationship between model premiums and interest rate factors, as well as between market premiums and interest factors, to conclude that long-dated swaptions are highly correlated to the shape of the curve.

Key words: Affine term structure models; characteristic function; fast Fourier transform (FFT); swaptions; straddles.

1 Introduction

Although there exist closed-form and/or efficient approximate solutions for pricing swaptions in Heath–Jarrow–Morton (HJM [8]) and market (e.g., BGM [2]) models, these models have the following shortcomings: (a) the yield curve is exogenous to the model and the volatility structure constrains the shape of the yield curve, and (b) they are generally overparameterized. These classes of models are typically used for marking and risk management, and are great tools for interpolation–extrapolation purposes. In current application, these models are calibrated on a daily or a weekly basis.

In contrast to HJM or BGM models, affine term structure models (ATSMs) [6] are parsimonious, and the yield curve is endogenous to the model. Affine parameters can be estimated via filtering methods, and the need for updating is infrequent. This makes ATSMs ideal for analyzing the cross-properties of swap rates and swaptions.

Pricing swaptions in an affine framework with constant volatility dates back to the work by Munk [9], based on the assumption that the price of a European option on a coupon bond (European swaption) is approximately proportional to the price of a European option on a zero coupon bond with maturity equal to the stochastic duration of the coupon bond. In [5], the authors perform an approximation based on an Edgeworth expansion of the density of the coupon bond price. This requires the calculation of the moments of the coupon bond through the joint moments of the individual zero coupon bonds. Singleton and Umanstev [11] price swaptions based on approximation of the exercise region in the space of the underlying factors by line segments. Finally, Schrager and Pelsser's approximation [10] is based on the derivation of approximate dynamics in which the volatility of the forward swap rate is itself an affine function of the factors.

To check the approximation method for swaption premiums in [10] for constant volatility, we implemented the methodology and compared the approximate premiums against simulated ones for 12 different swaptions (1×2, 1×5, 1×10, 2×2, 2×5, 2×10, 5×2, 5×5, 5×10, 10×2, 10×5, 10×10). For 1,000,000 simulated paths, the absolute value of the largest relative error was less than 0.35%.

Our focus in this paper is on pricing swaptions in an affine term structure setting with stochastic volatility. Using the Schrager and Pelsser approximation [10], we derive the characteristic function of the log swap rate under the swap measure, and then employ the fast Fourier transform (FFT) to price swaptions. This efficient method enables us to price ATSM swaps and swaptions simultaneously and consistently. We use ten years of swap rates and swaption premiums to estimate model parameters via a square-root unscented Kalman filter. Time series of the interest rate factors are a byproduct of the estimation procedure. We then look more closely at the relationship between model premiums and interest rate factors, as well as between market premiums and interest factors, to conclude that long-dated swaptions are highly correlated with the slope of the yield curve.

2 Affine Term Structure Models with Stochastic Volatility

We are interested in a stochastic interest rate model with stochastic volatility. Using an affine structure for interest rates, the short rate is specified as

$$r_t = a_r + b_r^\top x_t,$$

where r_t and x_t are n-dimensional vectors, with n the number of factors in the model, and x_t follows the matrix Ornstein–Uhlenbeck equation driven by

Brownian noise, that is,

$$dx_t = (-b_\gamma - Bx_t)dt + \Sigma dW_t,$$

where $\{W_t\}$ is an n-dimensional vector of independent standard Brownian motion and Σ is the (constant) $n \times n$ volatility matrix. In general, the $n \times n$ matrix B may be full, with all eigenvalues having positive real part.

To incorporate stochastic volatility, we introduce the square root equation for (scalar) variance v_t, and write

$$dv_t = \kappa(\theta - v_t)dt + \lambda\sqrt{v_t}dZ_t,$$

where $\{Z_t\}$ is a standard Brownian motion, with the correlation between $\{W_t\}$ and $\{Z_t\}$ denoted by the vector ρ. We now reformulate the equation for x_t as

$$dx_t = (-b_\gamma - Bx_t)dt + \Sigma\sqrt{v_t}dW_t.$$

3 Swaption Problem Formulation

Let S denote the start time of the interest rate swap, T the end time of the swap, and t denote a time between the present time 0 and S, so $t < S < T$. Suppose fixed payments occur at times T_j, $n < j \le N$, where $T_N = T$. Define $T_n = S$, and let $\Delta_j = T_j - T_{j-1}$ be the interval between swap times. For simplicity, assume that the Δs are constant; that is, $\Delta_j = \Delta$ for all j.

The forward swap rate at time t is

$$k_t^{S,T} = \frac{P(t,S) - P(t,T)}{\Delta\sum_{j=n+1}^{N} P(t,T_j)}. \tag{1}$$

The present value of a swaption struck at the rate K is

$$V(K) = \left(\Delta\sum_{j=n+1}^{N} P(0,T_j)\right) E[(k_S^{S,T} - K)^+],$$

under the swap measure. The swaption pricing problem is to evaluate the expectation under the given interest rate model.

4 The Characteristic Function Pricing Method

The characteristic function pricing method values options using the characteristic function of the underlying process. For ease of notation, define the constant $c = \Delta\sum_{j=n+1}^{N} P(0,T_j)$, let X denote the random variable $\ln k_S^{S,T}$, let f be the probability density function of X under the swap measure, and $k = \ln K$. With this new notation,

$$V(k) = cE[(e^X - e^k)^+].$$

Denote the Fourier transform of $e^{\alpha k}V(k)$ by $\Psi(\cdot)$, where the term $\alpha > 0$ is a damping factor, which is included so that the Fourier integral is well defined. Thus,

$$\Psi(\nu) = \int_{-\infty}^{\infty} V(k)e^{(\alpha+i\nu)k}dk$$

$$= c\int_{-\infty}^{\infty}\int_{k}^{\infty} (e^x - e^k)e^{(\alpha+i\nu)k}f(x)dxdk$$

$$= c\int_{-\infty}^{\infty}\int_{-\infty}^{x} (e^x - e^k)e^{(\alpha+i\nu)k}dkf(x)dx$$

$$= \frac{c}{(\alpha+i\nu)(\alpha+1+i\nu)}\int_{-\infty}^{\infty} e^{i(\nu-i(\alpha+1))x}f(x)dx.$$

Let $c_2 = c/[(\alpha+i\nu)(\alpha+1+i\nu)]$ and $u = \nu - i(\alpha+1)$ to get

$$\Psi(\nu) = c_2 E[e^{iuX}] = c_2 E\left[e^{iu\ln k_S^{S,T}}\right],$$

where $E[e^{iu\ln k_S^{S,T}}]$ is the characteristic function of $\ln k_S^{S,T}$. If the characteristic function is known, then so is $\Psi(\nu)$, and the price of the swaption can be obtained by taking the inverse Fourier transform

$$V(K) = e^{-\alpha \ln K}\frac{1}{2\pi}\int_{-\infty}^{\infty} \Psi(\nu)e^{-i\nu \ln K}d\nu.$$

(Note that we have interchangeably denoted the argument of V as k and K.)

4.1 Fast Fourier Transform

The inverse transform can be evaluated numerically using the FFT, where the resulting integral can be discretized using Simpson's rule. Choose $B >> 1$, fix the number of time steps m, and let $h = B/m$. Then

$$\psi(t) \equiv \frac{1}{2\pi}\int_{-\infty}^{\infty} \Psi(\nu)e^{-i\nu k}d\nu \approx \text{Re}\left\{\frac{1}{\pi}\int_0^B \Psi(\nu)e^{-i\nu k}d\nu\right\}$$

$$\approx \text{Re}\left\{\frac{h}{3\pi}\sum_{l=1}^{m} \Psi((l-1)h)e^{-i(l-1)hk}(3+(-1)^l - \delta_{l-1})\right\}.$$

Next, consider the discrete times t_j, $0 \le j \le m-1$, where

$$t_j = (j-1)\lambda + A.$$

The parameter λ is chosen so that $\lambda h = 2\pi/m$, and A is chosen so that $t_{m-1} = t = \ln K$, namely $A = \ln K - (m-1)\lambda$. Then

$$\psi(t_j) = \text{Re}\left\{\sum_{k=1}^{m} e^{-i(2\pi/m)(j-1)(k-1)} X(k)\right\},$$

where we have let

$$X(k) = \frac{h}{3\pi}\Psi((k-1)h)e^{-i(k-1)hA}(3 + (-1)^k - \delta_{k-1}).$$

Thus, $\{\psi(t_j)\}$ (which includes $\psi(t)$) can be obtained by taking the real part of the FFT of $\{X(k)\}$.

5 Computing the Characteristic Function

One of the main contributions of this paper is the derivation of the characteristic function $E[e^{iu \ln k_S^{S,T}}]$ under the swap measure. The paper proceeds in two major steps. First, the bond prices $\{P(t, T_j)\}$ are derived. This is necessary because $k_t^{S,T}$ is a function of the bond prices. Then the main task of finding the characteristic function is undertaken.

5.1 Bond Prices

First, define two new variables, y_t and w_t, in terms of x_t and v_t:

$$y_t = \int_0^t x_s ds,$$

$$w_t = \int_0^t v_s ds.$$

Next, define the joint characteristic function of (x, v, y, w) at time 0 for fixed time T as

$$\phi(0, \xi, \zeta, \omega, \varphi) = E\left[\exp\left(i\xi^\top x_T + i\zeta^\top y_T + i\omega v_T + i\varphi w_T\right) \mid x_0 = x, v_0 = v\right],$$

where $(\xi, \zeta, \omega, \varphi)$ are the transform variables. More generally, define the characteristic function at time t, $0 < t < T$, as

$$\phi(t, \xi, \zeta, \omega, \varphi) = E\left[\exp\left(i\xi^\top x_t + i\zeta^\top \int_t^T x_s ds + i\omega v_t + i\varphi \int_t^T v_s ds\right)\right. \left|\, x_t = x, v_t = v\right]. \quad (2)$$

It is conjectured that the characteristic function at time 0 has a solution of the form

$$\phi(0, \xi, \zeta, \omega, \varphi) = \exp\left(-a(T) - b(T)^\top x - c(T)v\right).$$

Therefore, by the Markov property, it must be the case that

$$\phi(t, \xi, \zeta, \omega, \varphi) = \exp\left(-a(T-t) - b(T-t)^\top x - c(T-t)v\right). \tag{3}$$

In the next subsection, it is shown how to compute this characteristic function, or $a(\cdot)$, $b(\cdot)$, and $c(\cdot)$.

To compute bond prices, note that the bond price $P(t, T)$ is

$$P(t, T) = E\left[\exp\left(-\int_t^T r_s ds\right)\right] = E\left[\exp\left(-a_r(T-t) - b_r^\top \int_t^T x_s ds\right)\right]$$

$$= e^{-a_r(T-t)} E\left[\exp\left(-b_r^\top \int_t^T x_s ds\right)\right] = e^{-a_r(T-t)} \phi(t, 0, ib_r, 0, 0).$$

This corresponds to Equation (2) with $(\xi, \zeta, \omega, \varphi) = (0, ib_r, 0, 0)$. Thus,

$$P(t, T) = e^{-a_r(T-t)} \exp\left(- a(T-t, 0, ib_r, 0, 0) - b(T-t, 0, ib_r, 0, 0)^\top x\right.$$

$$\left. - c(T-t, 0, ib_r, 0, 0)v\right). \tag{4}$$

Here we are explicitly showing the dependence of $a(\cdot)$, $b(\cdot)$, and $c(\cdot)$ on $(\xi, \zeta, \omega, \varphi)$; that is, $a(t) = a(t, \xi, \zeta, \omega, \varphi)$ and so on.

Computing $a(\cdot)$, $b(\cdot)$, and $c(\cdot)$

The first thing to note is that evaluating Equation (3) at $t = T$ gives the following boundary conditions,

$$a(0) = 0,$$
$$b(0) = -i\xi,$$
$$c(0) = -i\omega.$$

The second is that the function $\mathcal{G}(t)$ defined as

$$\mathcal{G}(t) = \exp\left(i\zeta^\top \int_0^t x_s ds + i\varphi \int_0^t v_s ds\right) \phi(t, \xi, \zeta, \omega, \varphi),$$

is equal to

$$E[H(T)|\mathcal{F}_t],$$

with

$$H(T) = \exp\left(i\xi^\top x_t + i\zeta^\top \int_0^T x_s ds + i\omega v_t + i\varphi \int_0^T w_s ds\right).$$

Thus, for any $t_1 < t_2$,

$$E\left[\mathcal{G}(t_2)|\mathcal{F}_{t_1}\right] = E\left[E(H(T)|\mathcal{F}_{t_2})|\mathcal{F}_{t_1}\right] = E\left[H(T)|\mathcal{F}_{t_1}\right] = \mathcal{G}(t_1);$$

that is, $\mathcal{G}(t)$ is a martingale.

Because $\mathcal{G}(t)$ is a martingale, its derivative with respect to time (the dt term) must be identically 0. This leads to the expression

$$0 = (i\zeta^T x + i\varphi v)\phi + \phi_t + \phi_x(-b_\gamma - Bx) + \phi_v(\kappa(\theta - v))$$
$$+ \frac{1}{2}\text{Trace}(\phi_{xx}\Sigma\Sigma^T v) + \frac{1}{2}\phi_{vv}\lambda^2 v + \lambda\rho^T\Sigma^T\phi_{vx}v.$$

We can use Equation (2) to solve for the derivatives of ϕ to get

$$0 = \phi\Big\{i\zeta^T x + i\varphi v + a'(T-t) + b'(T-t)^T x + c'(T-t)v$$
$$- b^T(T-t)(-b_\gamma - Bx) - (\kappa(\theta - v))c(T-t)$$
$$+ \frac{1}{2}b^T(T-t)\Sigma\Sigma^T b(T-t)v + \frac{1}{2}\lambda^2 vc(T-t)^2 + \lambda\rho^T\Sigma^T b(T-t)c(T-t)v\Big\}.$$

Because this equation must hold for all (x, v), we get three simpler equations by grouping the x terms, the v terms, and the remaining terms:

$$a'(\tau) = -b^T(\tau)b_\gamma + \kappa\theta c(\tau), \tag{5}$$
$$b'(\tau) = -i\zeta - B^T b(\tau), \tag{6}$$
$$c'(\tau) = -i\varphi - (\kappa + \lambda\rho^T\Sigma^T b(\tau))c(\tau) - \frac{1}{2}b^T(\tau)\Sigma\Sigma^T b(\tau) - \frac{\lambda^2}{2}c^2(\tau), \tag{7}$$

where $\tau = T - t$.

We now have a set of equations, which are satisfied by $a(\cdot)$, $b(\cdot)$, and $c(\cdot)$. These ordinary differential equations (ODEs) are referred to as Riccati equations. The Riccati equations together with the boundary conditions completely characterize $a(\cdot)$, $b(\cdot)$, and $c(\cdot)$. Note that the ODE for $b(\cdot)$ is linear and has a known analytical solution. In stock models, $b' = 0$ and b is a constant, as there is no dependence of log stock on the state variable of log stock. Here we have a dependence of state variable drifts on the state variables, and this introduces $-B^T b$ and the associated additional structure to the equation. Also in stock, we have integrated vol but no interest in integrated log stock, but now we are interested in the integral of x_t, and this introduces $i\zeta$.

$$b(\tau) = -\exp(-B^T\tau)i\xi - (B^T)^{-1}(I - \exp(-B^T\tau))i\zeta,$$

where we employ a matrix exponential of the full matrix B^T.

We next have to solve the following ODEs for c and a.

$$c'(\tau) = -i\varphi - \left(\kappa + \lambda\rho^T\Sigma^T b(\tau)\right)c(\tau) - \frac{1}{2}b^T(\tau)\Sigma\Sigma^T b(\tau) - \frac{\lambda^2}{2}c^2(\tau),$$
$$a'(\tau) = -b^T(\tau)b_\gamma + \kappa\theta c(\tau).$$

The two ODEs can be solved numerically.

5.2 The Characteristic Function of the Log Swap Rate

The derivation of the characteristic function of $\ln k_S^{S,T}$ is done in multiple steps. First, the dynamics of $k_t^{S,T}$ are derived with respect to the Q-measure Brownian processes $\{W_t\}$ and $\{Z_t\}$. Second, the change of measure to the swap measure results in two new Brownian processes $\{\tilde{W}_t\}$ and $\{\tilde{Z}_t\}$. In Section 5.2, we show how these processes are derived. Third, the dynamics of x_t, v_t, and then k_t with respect to $\tilde{W}_t$ and $\tilde{Z}_t$ are derived. With these results in place, $\ln k_S^{S,T}$ and its characteristic function can be computed.

Dynamics of k_t

Applying the chain rule to Equation (1) gives

$$\Delta dk_t^{S,T} \sum P(t,T_i) + \Delta k_t^{S,T} \sum dP(t,T_i) + \Delta dk_t^{S,T} \sum dP(t,T_i)$$

$$= dP(t,S) - dP(t,T).$$

Because $k_t^{S,T}$ is a martingale under the swap measure, we can ignore the "dt" terms to get

$$\Delta dk_t^{S,T} \sum P(t,T_i) + \Delta k_t^{S,T} \sum dP(t,T_i) = dP(t,S) - dP(t,T).$$

Rearranging terms and substituting for $k_t^{S,T}$ gives

$$dk_t^{S,T} = \frac{dP(t,S) - dP(t,T) - \frac{P(t,S)-P(t,T)}{\sum P(t,T_i)} \sum dP(t,T_i)}{\Delta \sum P(t,T_i)},$$

or

$$\frac{dk_t^{S,T}}{k_t^{S,T}} = \frac{dP(t,S) - dP(t,T)}{P(t,S) - P(t,T)} - \frac{\sum dP(t,T_i)}{\sum P(t,T_i)}.$$

Differentiating Equation (4) and ignoring the "dt" terms gives

$$\frac{dP(t,T)}{P(t,T)} = -b^\top(T-t)dx - c(T-t)dv.$$

Thus

$$\frac{dk_t^{S,T}}{k_t^{S,T}} = \frac{P(t,S)}{P(t,S) - P(t,T)}[-b^\top(S-t)dx - c(S-t)dv]$$

$$-\frac{P(t,T)}{P(t,S) - P(t,T)}[-b^\top(T-t)dx - c(T-t)dv]$$

$$-\sum \frac{P(t,T_i)}{\sum P(t,T_j)}[-b^\top(T_i-t)dx - c(T_i-t)dv].$$

At this point, we make use of an approximation. It is argued in [1], that the $(P(t, T_j))/(\sum P(t, T_j))$ terms are low-variance martingales in the context of the market models (e.g., BGM). In [10], it is conjectured that this is also true in ATSMs.

As mentioned earlier, to check the approximation in [10], for the constant volatility case, we implemented the methodology and compared the approximated premiums against simulated ones for 12 different swaptions (1×2, 1×5, 1×10, 2×2, 2×5, 2×10, 5×2, 5×5, 5×10, 10×2, 10×5, 10×10). For 1,000,000 simulated paths, the absolute value of the largest relative error was less than 0.35%.

Making use of the same approximation in the stochastic volatility case, we replace the $(P(t, T_j))/(\sum P(t, T_j))$ terms by their conditional expected values, under the swap measure $(P(0, T_j))/(\sum P(0, T_j))$. This approximation yields the following.

$$
\frac{dk_t^{S,T}}{k_t^{S,T}} \approx \frac{P(0,S)}{P(0,S) - P(0,T)}[-b^T(S-t)dx - c(S-t)dv]
$$
$$
- \frac{P(0,T)}{P(0,S) - P(0,T)}[-b^T(T-t)dx - c(T-t)dv]
$$
$$
- \sum \frac{P(0,T_i)}{\sum P(0,T_j)}[-b^T(T_i-t)dx - c(T_i-t)dv].
$$

Letting

$$
\eta_t = -\frac{P(0,S)b(S-t) - P(0,T)b(T-t)}{P(0,S) - P(0,T)} + \sum \frac{P(0,T_i)}{\sum P(0,T_j)}b(T_i - t),
$$
$$
\delta_t = -\frac{P(0,S)c(S-t) - P(0,T)c(T-t)}{P(0,S) - P(0,T)} + \sum \frac{P(0,T_i)}{\sum P(0,T_j)}c(T_i - t),
$$

we have

$$
\frac{dk_t^{S,T}}{k_t^{S,T}} = \eta_t^T dx + \delta_t dv.
$$

Change of Measure

The random variable, which changes the Q measure to the swap measure is $z = z_T$, where $z_t = \Lambda_t/\Lambda_0$ and

$$
\Lambda_t = e^{-\int_0^t r_s ds} \Delta \sum P(t, T_i). \tag{8}
$$

The process Λ_t is a martingale under the Q measure, thus

$$
\frac{d\Lambda_t}{\Lambda_t} = \Phi_{1,t}^T dW + \Phi_{2,t} dZ, \tag{9}
$$
$$
= (\Phi_{1,t}^T + \Phi_{2,t}\rho^T)dW + \sqrt{1 - \|\rho\|_2^2}\Phi_{2,t}dY,
$$

for some $\Phi_{1,t}$, $\Phi_{2,t}$, and univariate Brownian process Y that is independent of W. Letting

$$\hat{\Phi}_{1,t} = \Phi_{1,t} + \Phi_{2,t}\rho,$$
$$\hat{\Phi}_{2,t} = \sqrt{1 - \|\rho\|_2^2}\Phi_{2,t},$$

gives

$$\frac{d\Lambda_t}{\Lambda_t} = \hat{\Phi}_{1,t}^T dW + \hat{\Phi}_{2,t}dY. \tag{10}$$

Solving this SDE gives

$$\frac{\Lambda_t}{\Lambda_0} = \exp\left(\int_0^t \hat{\Phi}_{1,t}^T dW + \hat{\Phi}_{2,t}dY - \frac{1}{2}\int_0^t \|\hat{\Phi}_{1,t}\|_2^2 + \hat{\Phi}_{2,t}^2 dt\right).$$

Thus the processes $\tilde{W}$ and $\tilde{Y}$, which are given by

$$d\tilde{W} = -\hat{\Phi}_{1,t} + dW,$$
$$d\tilde{Y} = -\hat{\Phi}_{2,t} + dY,$$

are independent univariate Brownian processes under the swap measure. Define the Brownian process $\tilde{Z}$ by

$$d\tilde{Z} = \rho^T d\tilde{W} + \sqrt{1 - \|\rho\|_2^2}d\tilde{Y}$$
$$= -(\hat{\Phi}_{1,t}^T\rho + \sqrt{1 - \|\rho\|_2^2}\hat{\Phi}_{2,t}) + dZ$$
$$= -(\Phi_{1,t}^T\rho + \Phi_{2,t}) + dZ.$$

Note that the correlation between $\tilde{W}$ and $\tilde{Z}$ is the vector ρ. Furthermore, it follows from Equations (9) and (10) that

$$dW = d\tilde{W} + < dW, \frac{d\Lambda_t}{\Lambda_t} >,$$
$$dZ = d\tilde{Z} + < dZ, \frac{d\Lambda_t}{\Lambda_t} > .$$

Dynamics of x and v Under the Swap Measure

From Equation (8)

$$\frac{d\Lambda_t}{\Lambda_t} = (\cdots)dt + \sum \frac{P(t, T_i)}{\sum P(t, T_j)}[-b^\top(T - t)dx - c(T - t)dv]$$
$$\approx (\cdots)dt + \sum \frac{P(0, T_i)}{\sum P(0, T_j)}[-b^\top(T - t)dx - c(T - t)dv]$$
$$= (\cdots)dt + \sum \frac{P(0, T_i)}{\sum P(0, T_j)}(-b^\top(T_i - t)\Sigma\sqrt{v}dW - c(T_i - t)\lambda\sqrt{v}dZ),$$

so

$$< dW, \frac{d\Lambda_t}{\Lambda_t} > = -\sum \frac{P(0,T_i)}{\sum P(0,T_j)} [\Sigma^\top b(T_i - t)\sqrt{v} + c(T_i - t)\lambda\sqrt{v}\rho]dt,$$

$$< dZ, \frac{d\Lambda_t}{\Lambda_t} > = -\sum \frac{P(0,T_i)}{\sum P(0,T_j)} [\sqrt{v}b^\top (T_i - t)\Sigma\rho + c(T_i - t)\lambda\sqrt{v}]dt.$$

So now we have

$$dx = (-b_\gamma - Bx + \mu_t v)dt + \Sigma\sqrt{v}d\tilde{W},$$
$$dv = \tilde{\kappa}_t(\tilde{\theta}_t - v)dt + \lambda\sqrt{v}d\tilde{Z},$$

where

$$\mu_t = -\sum \frac{P(0,T_i)}{\sum P(0,T_j)} [\Sigma\Sigma^\top b(T_i - t) + c(T_i - t)\lambda\Sigma\rho],$$

$$\pi_t = -\sum \frac{P(0,T_i)}{\sum P(0,T_j)} [\lambda b^\top (T_i - t)\Sigma\rho + c(T_i - t)\lambda^2],$$

$$\tilde{\kappa}_t = \kappa - \pi_t,$$

$$\tilde{\theta}_t = \frac{\kappa\theta}{\tilde{\kappa}_t}.$$

Dynamics of $k_t^{S,T}$ Revisited

Plugging into the expression for dk_t using the latest expressions for dx and dv and noting that k_t is a martingale under the swap measure and eliminating the "dt" terms gives

$$\frac{dk_t}{k_t} = \eta_t^T \Sigma\sqrt{v}d\tilde{W} + \delta_t\lambda\sqrt{v}d\tilde{Z}.$$

Solving for $\ln k_S^{S,T}$

We have that

$$d\ln k_t^{S,T} = \frac{dk_t^{S,T}}{k_t^{S,T}} - \frac{1}{2}\left(\frac{dk_t^{S,T}}{k_t^{S,T}}\right)^2$$

$$= \eta_t^T \Sigma\sqrt{v}d\tilde{W} + \delta_t\lambda\sqrt{v}d\tilde{Z} - \frac{1}{2}\left(\eta_t^T \Sigma\Sigma^T \eta_t + \delta_t^2\lambda^2 + 2\delta_t\lambda\eta_t^T \Sigma\rho\right)vdt$$

$$= \eta_t^T (dx + (b_\gamma + Bx - \mu_t v)dt) + \delta_t(dv + (-\kappa\theta + \tilde{\kappa}_t v)dt)$$

$$- \frac{1}{2}\left(\eta_t^T \Sigma\Sigma^T \eta_t + \delta_t^2\lambda^2 + 2\delta_t\lambda\eta_t^T \Sigma\rho\right)vdt.$$

Integrating and collecting terms gives

$$\ln k_S^{S,T} = \ln k_0^{S,T} + \int_0^S \eta_t^T b_\gamma dt - \int_0^S \delta_t \kappa \theta dt$$

$$+ \int_0^S \eta_t^T dx_t + \int_0^S \delta_t dv_t + \int_0^S \eta_t^T B x_t dt$$

$$+ \int_0^S \left(-\eta_t^T \mu_t + \delta_t \tilde{\kappa}_t - \frac{1}{2} \left(\eta_t^T \Sigma \Sigma^T \eta_t + \delta_t^2 \lambda^2 + 2\delta_t \lambda \eta_t^T \Sigma \rho \right) \right) v_t dt.$$

Noting that

$$\int_0^S \eta_t^T dx_t = \eta_S^T x_S - \eta_0^T x_0 - \int_0^S \frac{\partial \eta_t^T}{\partial t} x_t dt,$$

$$\int_0^S \delta_t dv_t = \delta_S v_S - \delta_0 v_0 - \int_0^S \frac{\partial \delta_t}{\partial t} v_t dt,$$

we have

$$\ln k_S^{S,T} = \ln k_0^{S,T} - \eta_0^T x_0 - \delta_0 v_0 + \int_0^S \eta_t^T b_\gamma dt - \int_0^S \delta_t \kappa \theta dt$$

$$+ \eta_S^T x_S + \delta_S v_S + \int_0^S (\eta_t^T B - \frac{\partial \eta_t^T}{\partial t}) x_t dt$$

$$+ \int_0^S \left(-\eta_t^T \mu_t + \delta_t \tilde{\kappa}_t - \frac{1}{2} \left(\eta_t^T \Sigma \Sigma^T \eta_t + \delta_t^2 \lambda^2 + 2\delta_t \lambda \eta_t^T \Sigma \rho \right) - \frac{\partial \delta_t}{\partial t} \right) v_t dt.$$

Letting

$$p_t = B^T \eta_t - \frac{\partial \eta_t}{\partial t},$$

$$q_t = -\eta_t^T \mu_t + \delta_t \tilde{\kappa}_t - \frac{1}{2} \left(\eta_t^T \Sigma \Sigma^T \eta_t + \delta_t^2 \lambda^2 + 2\delta_t \lambda \eta_t^T \Sigma \rho \right) - \frac{\partial \delta_t}{\partial t},$$

gives

$$\ln k_S^{S,T} = \ln k_0^{S,T} - \eta_0^T x_0 - \delta_0 v_0 + \int_0^S \eta_t^T b_\gamma dt - \int_0^S \delta_t \kappa \theta dt$$

$$+ \eta_S^T x_S + \int_0^S p_t^T x_t dt + \delta_S v_S + \int_0^S q_t v_t dt.$$

Note that $p_t = 0$. We can see this as follows. From Equation (6) (and the fact that $\zeta = ib_r$),

$$B^T b(\tau) - \frac{\partial b(\tau)}{\partial t} = B^T b(\tau) + b'(\tau) = b_r,$$

where $\tau = S - t$, $\tau = T - t$, or $\tau = T_i - t$. It now follows from the definition of η_t that

$$p_t = -\frac{P(0,S) - P(0,T)}{P(0,S) - P(0,T)} b_r + \sum \frac{P(0,T_i)}{\sum P(0,T_j)} b_r = 0.$$

Characteristic Function of $\ln k_S^{S,T}$

The approach is similar to that used in Section 5.1 to price bonds. Define the function

$$\phi(t, \hat{\xi}, \hat{\zeta}, \hat{\omega}, \hat{\varphi}) = E \left[\exp\left(i\hat{\xi}^T x_S + i\hat{\zeta} \int_t^S p_u^T x_u du + i\hat{\omega} v_S + i\hat{\varphi} \int_t^S q_u v_u du \right) \right.$$

$$\left. \Big|\, x_t = x, v_t = v \right], \qquad (11)$$

where the expectation is taken under the swap measure, and $(\hat{\xi}, \hat{\zeta}, \hat{\omega}, \hat{\varphi})$ are transform variables. Note that the characteristic function of $\ln k_S^{S,T}$ is

$$\Psi(u) = E \left[e^{iu \ln k_S^{S,T}} \right]$$

$$= \exp\left[iu \left(\ln k_0 - \eta_0^T x_0 - \delta_0 v_0 + \int_0^S \eta_t^T b_\gamma dt - \int_0^S \delta_t \kappa \theta dt \right) \right]$$

$$\times E \left[\exp\left(iu\eta_S^T x_S + iu\delta_S v_S + iu \int_0^S q_t v_t dt \right) \right]$$

$$= \exp\left[iu \left(\ln k_0 - \eta_0^T x_0 - \delta_0 v_0 + \int_0^S \eta_t^T b_\gamma dt - \int_0^S \delta_t \kappa \theta dt \right) \right]$$

$$\times \phi(0, \hat{\xi}, \hat{\zeta}, \hat{\omega}, \hat{\varphi}),$$

where $(\hat{\xi}, \hat{\zeta}, \hat{\omega}, \hat{\varphi}) = (u\eta_S, 0, u\delta_S, u)$. To solve for the characteristic function of $\ln k_S^{S,T}$, we must solve for $\phi(0, \hat{\xi}, \hat{\zeta}, \hat{\omega}, \hat{\varphi})$ and $\phi(t, \hat{\xi}, \hat{\zeta}, \hat{\omega}, \hat{\varphi})$. As before, it is conjectured that $\phi(t, \hat{\xi}, \hat{\zeta}, \hat{\omega}, \hat{\varphi})$ has a solution of the form

$$\phi(t, \hat{\xi}, \hat{\zeta}, \hat{\omega}, \hat{\varphi}) = \exp\left(\hat{a}(S-t) + \hat{b}^T(S-t)x + \hat{c}(S-t)v \right). \qquad (12)$$

The problem is then to solve for $\hat{a}(\cdot)$, $\hat{b}(\cdot)$, and $\hat{c}(\cdot)$.

Define the function

$$\hat{G}(t) = \exp\left(i\hat{\zeta} \int_0^t p_u^T x_u du + i\hat{\varphi} \int_0^t q_u v_u du \right) \phi(t, \hat{\xi}, \hat{\zeta}, \hat{\omega}, \hat{\varphi}).$$

This function is a martingale under the swap measure (same type of argument as employed in Section 5.1), and so its derivative with respect to time (the "dt" term) must be identically 0 leading to the expression

$$0 = \left(i\hat{\zeta} p_t^T x + i\hat{\varphi} q_t v \right) \phi + \phi_t + \phi_x \left(-b_\gamma - Bx + \mu_t v \right) + \phi_v \left(\tilde{\kappa}_t (\tilde{\theta}_t - v) \right)$$

$$+ \frac{1}{2} \text{Trace} \left(\phi_{xx} \Sigma \Sigma^T v \right) + \frac{1}{2} \phi_{vv} \lambda^2 v + \lambda v \phi_{vx}^T \Sigma \rho.$$

We can use Equation (12) to solve for the derivatives of ϕ to get

$$
\begin{aligned}
0 = {} & \phi\Big\{\Big(i\hat{\zeta}p_t^\top x + i\hat{\varphi}q_t v\Big) - \Big(\hat{a}'(S-t) + \hat{b}'(S-t)^\top x + \hat{c}'(S-t)v\Big)\Big\} \\
& + \hat{b}^\top(S-t)(-b_\gamma - Bx + \mu_t v) + \hat{c}(S-t)\tilde{\kappa}_t(\tilde{\theta}_t - v) \\
& + \frac{1}{2}\hat{b}^\top(S-t)\Sigma\Sigma^\top\hat{b}(S-t)v + \frac{1}{2}\hat{c}(S-t)^2\lambda^2 v + \lambda\hat{c}(S-t)v\hat{b}(S-t)^\top\Sigma\rho.
\end{aligned}
$$

Because this has to hold for all (x, v), we get the following three equations by grouping the x terms, v terms, and the remaining terms.

$$
\hat{a}'(\tau) = -\hat{b}^\top(\tau)b_\gamma + \hat{c}(\tau)\kappa\theta,
$$

$$
\hat{b}'(\tau) = -B^\top\hat{b}(\tau),
$$

$$
\hat{c}'(\tau) = i\hat{\varphi}q_t + \hat{b}^\top(\tau)\mu_t + \frac{1}{2}\hat{b}^\top(\tau)\Sigma\Sigma^\top\hat{b}(\tau) - \Big[\tilde{\kappa}_t - \lambda\hat{b}^\top(\tau)\Sigma\rho\Big]\hat{c}(\tau) + \frac{\lambda^2}{2}\hat{c}^2(\tau),
$$

where $\tau = S - t$. The boundary conditions follow from Equations (11) and (12) at $t = S$, and are given by

$$
\hat{a}(0) = 0,
$$

$$
\hat{b}(0) = i\hat{\xi},
$$

$$
\hat{c}(0) = i\hat{\omega}.
$$

As before, the above ODEs are Riccati equations, and there is an explicit solution for $\hat{b}(\tau)$ given by

$$
\hat{b}(\tau) = \exp(-B^\top\tau)i\hat{\xi}.
$$

The remaining two ODEs must be solved numerically, as before.

6 Data and Estimation

Having the characteristic function of the log swap rate, we can employ the techniques in [3] or [4] to price swaptions. The data consist of (i) constant maturity forward 6 months/12 months, (ii) swap rates at maturities 2, 3, 5, 10, 15, and 30 years, and (iii) at-the-money swaption premiums with maturities of 1, 2, 5, and 10 years. At each option maturity, we have three contracts with different underlying swap maturities: 2, 5, and 10 years. All interest rates and interest rate options are on U.S. dollars. The data are daily closing mid-quotes from March 7, 1997, through October 16, 2006 (2399 observations).

For parameter and state estimation of our two-factor ATSM, we employ a square-root unscented Kalman filter [7]. Figure 1 illustrates the estimated factors for both constant volatility and stochastic volatility. Figures 2 through 4 display market premiums versus model premiums for both constant volatility and stochastic volatility models for at-the-money straddles 2×5, 5×5, and 10×5. The figures clearly indicate that the stochastic volatility model provides a superior match to market premiums; however, the longer the maturity, the closer the constant volatility model comes to the stochastic volatility model.

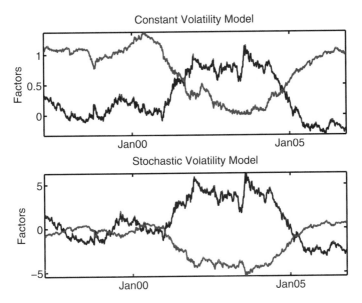

Fig. 1. Top: time series of the factors estimated for the constant volatility case. Bottom: time series of the factors estimated for the stochastic volatility case.

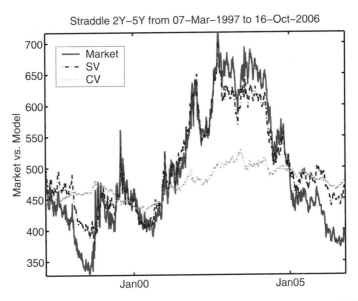

Fig. 2. Market premiums versus model premiums for both constant volatility and stochastic volatility; at-the-money straddle 2×5.

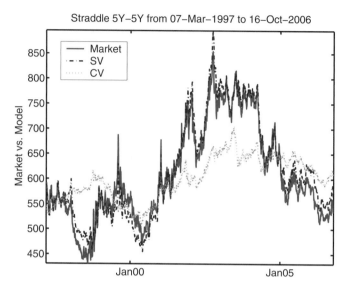

Fig. 3. Market premiums versus model premiums for both constant volatility and stochastic volatility: at-the-money straddle 5 × 5.

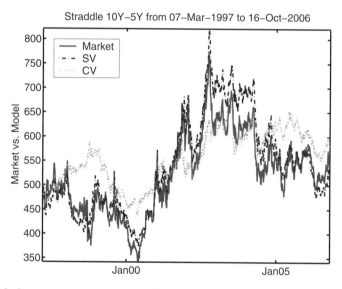

Fig. 4. Market premiums versus model premiums for both constant volatility and stochastic volatility; at-the-money straddle 10 × 5.

7 Findings and Observations

Having model premiums and time series of the factors, we regress the log of model premiums against the factors as follows,

$$\text{Regressed Premiums}_t = \exp(\alpha + \beta^\top x_t),$$

where α and β are obtained from regression, and x_t is the time series of the factors from estimation. For the same three at-the-money straddles—2×5, 5×5, and 10×5—we plot model premiums versus regressed premiums for both constant volatility and stochastic volatility models, as shown in Figures 5, 6, and 7, respectively. In the constant volatility model, model premiums are log-linear with respect to interest rate factors, but in stochastic volatility that is not the case.

We repeat the same procedure, but this time we regress the log of market premiums against the factors. The results are shown in Figures 8 through 10.

To establish a relationship between market premiums and the curve, we regress the difference of the ten-year swap rate and the five-year swap rate against the at-the-money straddle 5×5, and as before construct the regressed premiums. Figure 11 gives plots of regressed premiums versus market premiums and the z-score of the residual.

Our conclusions from the regressions are as follows. (a) Volatility is approximately log-linear in the factors, and (b) long-dated swaptions (straddles) are highly correlated to the slope of the yield curve. This makes logical sense, inasmuch swaption premiums are related to risk premiums, and the risk premium is approximated by the slope of the yield curve.

Acknowledgments

Ali Hirsa would like to thank Ashok Tikku for his comments and suggestions. We also thank participants of the Bachelier Finance Society Conference 2006 in Tokyo and participants of the Madan 60th Birthday Mathematical Finance Conference in College Park, Maryland for their comments and discussions. Errors are our own responsibility.

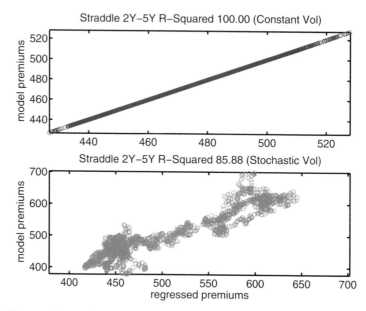

Fig. 5. Regressed premiums versus model premiums for at-the-money straddle 2×5: Constant volatility model at the top, and stochastic volatility model at the bottom.

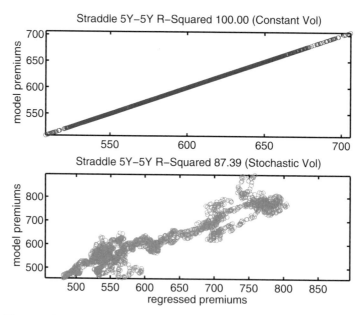

Fig. 6. Regressed premiums versus model premiums for at-the-money straddle 5×5: Constant volatility model at the top, and stochastic volatility model at the bottom.

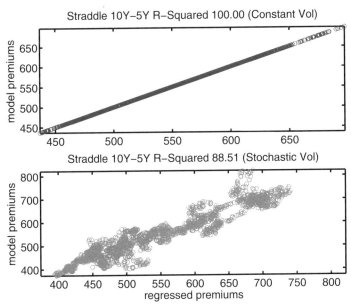

Fig. 7. Regressed premiums versus model premiums for at-the-money straddle 10 × 5: Constant volatility model at the top, and stochastic volatility model at the bottom.

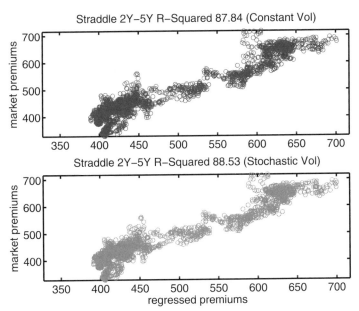

Fig. 8. Regressed premiums versus market premiums for at-the-money straddle 2 × 5: Constant volatility model at the top, and stochastic volatility model at the bottom.

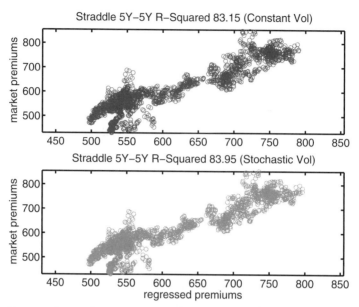

Fig. 9. Regressed premiums versus market premiums for at-the-money straddle 5 × 5: Constant volatility model at the top, and stochastic volatility model at the bottom.

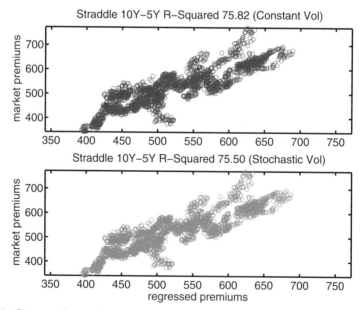

Fig. 10. Regressed premiums versus market premiums for at-the-money straddle 10 × 5: Constant volatility model at the top, and stochastic volatility model at the bottom.

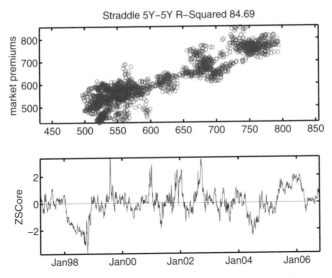

Fig. 11. Market premiums versus regressed premiums against the curve, and the z-score of the regression residuals.

References

1. A. Brace, T. Dunn, and G. Barton. Towards a central interest rate model. *Option Pricing, Interest Rates and Risk Management: Handbooks in Mathematical Finance*, eds. E. Jouini, J. Cvitanic, and M. Musiela, Cambridge University Press, 278–313, 2001.
2. A. Brace, D. Gatarek, and M. Musiela. The market model of interest rate dynamics. *Mathematical Finance*, 7:127–155, 1997.
3. P. Carr and D. B. Madan. Option valuation using the fast Fourier transform. *The Journal of Computational Finance*, 2(4):61–73, 1999.
4. K. Chourdakis. Option valuation using the fast Fourier transform. *Journal of Computational Finance*, 8(2):1–18, Winter 2004.
5. P. Collin-Dufresne and R. Goldstein. Pricing swaptions in affine models. *Journal of Derivatives*, 10(1):9–26, Fall 2002.
6. Q. Dai and K. Singleton. Specification analysis of affine term structure models. *Journal of Finance*, 55:1943–1978, 2000.
7. R. Van der Merwe and E.A. Wan. The square-root unscented Kalman filter for state and parameter estimation. *IEEE International Conference on Acoustics, Speech, and Signal Processing*, 6:3461–3464, 2001.
8. D. Heath, R. Jarrow, and A. Morton. Bond pricing and the term structure of interest rates: A discrete time approximation. *Journal of Financial and Quantitative Analysis*, 25:419–440, 1990.
9. C. Munk. Stochastic duration and fast coupon bond option pricing in multi-factor models. *Review of Derivative Research*, 3:157–181, 1999.
10. D.F. Schrager and A.J. Pelsser. Pricing swaptions and coupon bond options in affine term structure models. *Mathematical Finance*, 16(4):673–694, 2006.
11. K. Singleton and L. Umanstev. Pricing coupon bond and swaptions in affine term structure models. *Mathematical Finance*, 12:427–446, 2002.

Forward Evolution Equations for Knock-Out Options

Peter Carr[1] and Ali Hirsa[2]

[1] Bloomberg
731 Lexington Avenue
New York, NY 10022, USA
pcarr4@bloomberg.net

[2] Caspian Capital Management, LLC
745 Fifth Avenue
New York, NY 10151, USA
ali.hirsa@ixis-ccm.com

Summary. We derive forward partial integrodifferential equations (PIDEs) for pricing up-and-out and down-and-out call options when the underlying is a jump diffusion. We assume that the jump part of the returns process is an additive process. This framework includes the Variance-Gamma, finite moment logstable, Merton jump diffusion, Kou jump diffusion, Dupire, CEV, arcsinh normal, displaced diffusion, and Black–Scholes models as special cases.

Key words: Partial integrodifferential equation (PIDE); forward equations; knock-out options; jump diffusion; Lévy processes.

1 Introduction

Pricing and hedging derivatives consistent with the volatility smile has been a major research focus for over a decade. A breakthrough occurred in the mid-nineties with the recognition that in certain models, European option prices satisfied forward evolution equations in which the independent variables are the option's strike and maturity. More specifically, [12] showed that under deterministic carrying costs and a diffusion process for the underlying asset price, no-arbitrage implies that European option prices satisfy a certain partial differential equation (PDE), now called the Dupire equation. Assuming that one could observe European option prices of all strikes and maturities, then this forward PDE can be used to explicitly determine the underlier's instantaneous volatility as a function of the underlier's price and time. Once this volatility function is known, the value function for European, American, and many exotic options can be determined by a wide array of standard methods.

As this value function relates theoretical prices of these instruments to the underlier's price and time, it can also be used to determine many Greeks of interest as well.

Aside from their use in determining the volatility function, forward equations also serve a second useful purpose. Once one knows the volatility function either by an explicit specification or by a prior calibration, the forward PDE can be solved via finite differences to efficiently price a collection of European options of different strikes and maturities all written on the same underlying asset. Furthermore, as pointed out in [4], all the Greeks of interest satisfy the same forward PDE and hence can also be efficiently determined in the same way.

Since the original development of forward equations for European options in continuous models, several extensions have been proposed. For example, Esser and Schlag [14] developed forward equations for European options written on the forward price rather than the spot price. Forward equations for European options in jump diffusion models were developed in Andersen and Andreasen [1] and extended by Andreasen and Carr [3]. It is straightforward to develop the relevant forward equations for European binary options or for European power options by differentiating or integrating the forward equation for standard European options. Buraschi and Dumas [6] develop forward equations for compound options. In contrast to the PDEs determined by others, their evolution equation is an ordinary differential equation whose sole independent variable is the intermediate maturity date.

Given the close relationship between compound options and American options, it seems plausible that there might be a forward equation for American options. The development of such an equation has important practical implications, because all listed options on individual stocks are American-style. The Dupire equation cannot be used to infer the volatility function from market prices of American options, nor can it be used to efficiently value a collection of American options of differing strikes and maturities.

This problem is addressed for American calls on stocks paying discrete dividends in Buraschi and Dumas [6], and it is also considered in a lattice setting in Chriss [10]. In [8], we address the more difficult problem of pricing continuously exercisable American puts in continuous-time models. To do so, we depart from the diffusive models that characterize most of the previous research on forward equations in continuous time. To capture the smile, we assume that prices jump rather than assuming that the instantaneous volatility is a function of stock price and time. Dumas et al. [11] find little empirical support for the Dupire model, whereas there is a long history of empirical support for jump-diffusion models; three recent papers documenting support for such models are [2; 7; 9]. In particular, we assume that the returns on the underlying asset have stationary independent increments, or in other words that the log price is a Lévy process. Besides the [5] model, our framework includes as special cases the Variance-Gamma (VG) model of Madan et al. [18], the CGMY model of Carr et al. [7], the finite moment logstable model of

Carr and Wu [9], the Merton [19] and Kou [17] jump diffusion models, and the hyperbolic models of Eberlein et al. [13]. In all of these models except Black–Scholes, the existence of a jump component implies that the backward and forward equations contain an integral in addition to the usual partial derivatives. Despite the computational complications introduced by this term, we use finite differences to solve both of these fundamental partial integrodifferential equations (PIDEs). To illustrate that our forward PIDE is a viable alternative to the traditional backward approach, we calculate American option values in the diffusion-extended VG option pricing model and find very close agreement. For details on the application of finite differences to valuing American options in the VG model, see [16].

The approach to determining the forward equation for American options in [8] is to start with the well-known backward equation and then exploit the symmetries that essentially define Lévy processes. In the process of developing the forward equation, we also determine two hybrid equations of independent interest. The advantage of these hybrid equations over the forward equation is that they hold in greater generality. Depending on the problem at hand, these hybrid equations can also have large computational advantages over the backward or forward equations when the model has already been calibrated. In particular, the advantage of these hybrid equations over the backward equation is that they are more computationally efficient when one is interested in the variation of prices or Greeks across strike or maturity at a fixed time, for example, market close.

The first of these hybrid equations has the stock price and maturity as independent variables. The numerical solution of this hybrid equation is an alternative to the backward equation in producing a spot slide, which shows how American option prices vary with the initial spot price of the underlier. If one is interested in understanding how this spot slide varies with maturity, then our hybrid equation is much more efficient than the backward equation. This hybrid equation also has important implications for path-dependent options such as cliquets whose payoff directly depends on the particular level reached by an intermediate stock price.

Their second hybrid equation has the strike price and calendar time as independent variables. The numerical solution of this hybrid equation is an alternative to the forward equation in producing an implied volatility smile at a fixed maturity. If one is interested in understanding how the model predicts that this smile will change over time, then our hybrid equation is much more computationally efficient than the forward equation. This second hybrid equation also allows parameters to have a term structure, whereas our forward equation does not. Note, however, that implied volatility can have a term or strike structure in our Lévy setting. Hence, if one needs to efficiently value a collection of American options of different strikes in the time-dependent Black–Scholes model, then it is far more efficient to solve our hybrid equation than to use the standard backward equation.

In this paper, we focus on forward evolution equations for knock-out options. We derive forward partial integrodifferential equations for up-and-out and down-and-out call options when the model dynamics are jump diffusion, where jumps are additive in the log of the price. This framework includes VG, finite moment logstable, Merton and Kou jump diffusions, Dupire, CEV, arcsinh normal, displaced diffusion, and Black–Scholes models as special cases. The remainder of this paper is structured as follows. The next section introduces our assumptions for down-and-out calls and derives the forward PIDE for down-and-out calls. The following section introduces our setting and reviews the backward PIDE that governs up-and-out call values in this setting and then develops the forward equation for up-and-out call options. We present some numerical results afterwards, and the final section suggests further research.

2 Down-and-Out Calls

2.1 Assumptions and Notations

Throughout this paper, we assume the standard model of perfect capital markets, continuous trading, and no-arbitrage opportunities.

When a pure discount bond is used as numeraire, then it is well known that no-arbitrage implies that there exists a probability measure Q under which all nondividend-paying asset prices are martingales. Under this measure we assume that a stock price S_t obeys the following stochastic differential equation,

$$dS_t = [r(t)-q(t)]S_{t-}\,dt+a(S_{t-},t)dW_t+\int_{-\infty}^{\infty}S_{t-}(e^x-1)[\mu(dx,dt)-\nu(x,t)dxdt],$$

$$(1)$$

for all $t \in [0,\bar{T}]$, where the initial stock price $S_0 > 0$ is known, and $\bar{T}$ is some arbitrarily distant horizon. The process is Markov in itself because the coefficients of the stock price process at time t depend on the path only through S_{t-}, which is the prejump price at t. Thus, the dynamics are fully determined by the drift function $b(S,t) \equiv [r(t) - q(t)]S$, the (normal) volatility function $a(S,t)$, and the jump compensation function $\nu(x,t)$. The term dW_t denotes increments of a standard Brownian motion W_t defined on the time set $[0,\bar{T}]$ and on a complete probability space $(\Omega, \mathcal{F}, Q)$. The random measure $\mu(dx,dt)$ counts the number of jumps of size x in the log price at time t. The function $\{\nu(x,t), x \in \mathbb{R}, t \in [0,\bar{T}]\}$ is used to compensate the jump process $J_t \equiv \int_0^t \int_{-\infty}^{\infty} S_{t-}(e^x - 1)\mu(dx,ds)$, so that the last term in Equation (1) is the increment of a Q jump martingale. The function $\nu(x,t)$ must have the following properties.

1. $\nu(0,t) = 0$.
2. $\int_{-\infty}^{\infty}(x^2 \wedge 1)\nu(x,t)dx < \infty, \quad t \in [0,\bar{T}]$.

Thus, each price change is the sum of the increment in a general diffusion process with proportional drift and the increment in a pure jump martingale, where the latter is an additive process in the log price. We restrict the function $a(S, t)$ so that the spot price is always nonnegative and absorbing at the origin. A sufficient condition for keeping the stock price away from the origin is to bound the lognormal volatility. In particular, we set

$$a(0, t) = 0. \tag{2}$$

Hence, Equation (1) describes a continuous-time Markov model for the spot price dynamics, which is both arbitrage-free and consistent with limited liability. Aside from the Markov property, the main restrictions inherent in Equation (1) are the standard assumptions that interest rates, dividend yields, and the compensator do not depend on the spot price.

2.2 Analysis

Let time $t = 0$ denote the valuation date for a European down-and-out call option with strike price K, barrier $L \leq K$, initial spot $S_0 > L$, and maturity $T \geq 0$. Let $D_0^c(K, T)$ denote an initial price of the down-and-out call which is implied by the absence of arbitrage. Consider the product $e^{\int_t^T r(u)du}(S_t - K)^+$. By the Tanaka–Meyer formula,

$$
\begin{aligned}
(S_T - K)^+ = {}& e^{\int_0^T r(u)du}(S_0 - K)^+ + \int_0^T e^{\int_t^T r(u)du}\mathbf{1}_{\{S_{t-} > K\}}dS_t \\
&+ \int_0^T e^{\int_t^T r(u)du}\left\{\frac{a^2(S_{t-}, t)}{2}\delta(S_{t-} - K) - r(t)(S_{t-} - K)^+\right\}dt \\
&+ \int_0^T e^{\int_t^T r(u)du}\int_{-\infty}^{\infty}\left[(S_{t-}e^x - K)^+ - (S_{t-} - K)^+ \right. \\
&\qquad\qquad\left. - \mathbf{1}_{\{S_{t-} > K\}}S_{t-}(e^x - 1)\right]\mu(dx, dt),
\end{aligned}
$$

where $\mu(\cdot, \cdot)$ denotes the integer-valued counting measure and $\delta(\cdot)$ denotes the Dirac delta function, a generalized function characterized by two properties:

1. $\delta(x) = \begin{cases} 0 & \text{if } x \neq 0, \\ \infty & \text{if } x = 0; \end{cases}$
2. $\int_{-\infty}^{\infty}\delta(x)dx = 1$.

Multiplying by $e^{-\int_0^T r(u)du}\mathbf{1}_{\{\tau_L > T\}}$ and taking expectations on both sides under an equivalent martingale measure Q, we have

$$D_0^c(K,T) = (S_0 - K)^+ E_0^Q[\mathbf{1}_{\{\tau_L > T\}}]$$

$$+ \int_0^T e^{-\int_0^t r(u)du} E_0^Q\left[\mathbf{1}_{\{\tau_L > T\}}\mathbf{1}_{\{S_{t-} > K\}}[r(t) - q(t)]S_{t-}\right]dt$$

$$+ \int_0^T e^{-\int_0^t r(u)du}\left\{\frac{a^2(K,t)}{2} E_0^Q[\mathbf{1}_{\{\tau_L > T\}}\delta(S_{t-} - K)]\right.$$

$$\left. - r(t)E_0^Q[\mathbf{1}_{\{\tau_L > T\}}(S_{t-} - K)^+]\right\}dt$$

$$+ \int_0^T e^{-\int_0^t r(u)du} E_0^Q[\mathbf{1}_{\{\tau_L > T\}}] \int_{-\infty}^{\infty} \Big[(S_{t-}e^x - K)^+ - (S_{t-} - K)^+$$

$$- \mathbf{1}_{\{S_{t-} > K\}}S_{t-}(e^x - 1)\Big]\nu(x,t)dxdt.$$

Differentiating w.r.t. T implies

$$\frac{\partial}{\partial T}D_0^c(K,T) = -e^{-\int_0^T r(u)du} E_0^Q\left[\delta(\tau_L - T)(S_T - K)^+\right]$$

$$+ e^{-\int_0^T r(u)du} E_0^Q\left[\mathbf{1}_{\{\tau_L > T\}}\mathbf{1}_{\{S_{T-} > K\}}[r(T) - q(T)]S_{T-}\right]$$

$$+ \frac{a^2(K,T)}{2}e^{-\int_0^T r(u)du} E_0^Q[\mathbf{1}_{\{\tau_L > T\}}\delta(S_{T-} - K)]$$

$$- r(T)e^{-\int_0^T r(u)du} E_0^Q[\mathbf{1}_{\{\tau_L > T\}}(S_{T-} - K)^+]$$

$$+ e^{-\int_0^T r(u)du} E_0^Q\Big[\mathbf{1}_{\{\tau_L > T\}} \int_{-\infty}^{\infty} \Big((S_{T-}e^x - K)^+ - (S_{T-} - K)^+$$

$$- \mathbf{1}_{\{S_{T-} > K\}}S_{T-}(e^x - 1)\Big)\nu(x,T)dx\Big].$$

The payoff in the first term vanishes. Subtracting and adding

$$e^{-\int_0^T r(u)du} E_0^Q\left[\mathbf{1}_{\{\tau_L > T\}}[r(T) - q(T)]K\mathbf{1}_{\{S_{T-} > K\}}\right]$$

to the second term on the right-hand side (RHS) gives

$$\frac{\partial}{\partial T}D_0^c(K,T) = e^{-\int_0^T r(u)du} E_0^{QQ}\left[\mathbf{1}_{\{\tau_L > T\}}\mathbf{1}_{\{S_{T-} > K\}}[r(T) - q(T)](S_{T-} - K)\right]$$

$$+ e^{-\int_0^T r(u)du} E_0^Q\left[\mathbf{1}_{\{\tau_L > T\}}[r(T) - q(T)]K\mathbf{1}_{\{S_{T-} > K\}}\right]$$

$$+ \frac{a^2(K,T)}{2}\frac{\partial^2}{\partial K^2}D_0^c(K,T) - r(T)D_0^c(K,T) + e^{-\int_0^T r(u)du} E_0^Q\Big[\mathbf{1}_{\{\tau_L > T\}}$$

$$\int_{-\infty}^{\infty}\left(e^x(S_{T-} - Ke^{-x})^+ - \mathbf{1}_{\{S_{T-} > K\}}(S_{T-} - K + S_{T-}e^x - S_{T-})\right)\nu(x,T)dx\Big]$$

$$= [r(T) - q(T)]D_0^c(K,T) - [r(T) - q(T)]K\frac{\partial}{\partial K}D_0^c(K,T)$$

$$+ \frac{a^2(K,T)}{2}\frac{\partial^2}{\partial K^2}D_0^c(K,T) - r(T)D_0^c(K,T) + e^{-\int_0^T r(u)du}E_0^Q\left[\mathbf{1}_{\{\tau_L > T\}}\right.$$

$$\left.\int_{-\infty}^\infty e^x\left((S_{T-} - Ke^{-x})^+ - \mathbf{1}_{\{S_{T-} > K\}}(S_{T-} - Ke^{-x} + K - K)\right)\nu(x,T)dx\right]$$

$$= -q(T)D_0^c(K,T) - [r(T)-q(T)]K\frac{\partial}{\partial K}D_0^c(K,T) + \frac{a^2(K,T)}{2}\frac{\partial^2}{\partial K^2}D_0^c(K,T)$$

$$+ e^{-\int_0^T r(u)du}E_0^Q\left[\mathbf{1}_{\{\tau_L > T\}}\int_{-\infty}^\infty \left((S_{T-} - Ke^{-x})^+ - (S_{T-} - K)^+\right.\right.$$

$$\left.\left. - \frac{\partial}{\partial K}(S_{T-} - K)^+ K(e^{-x} - 1)\right)e^x\nu(x,T)dx\right]$$

$$= -q(T)D_0^c(K,T) - [r(T) - q(T)]K\frac{\partial}{\partial K}D_0^c(K,T) + \frac{a^2(K,T)}{2}\frac{\partial^2}{\partial K^2}D_0^c(K,T)$$

$$+ \int_{-\infty}^\infty \left[D_0^c(Ke^{-x},T) - D_0^c(K,T) - \frac{\partial}{\partial K}D_0^c(K,T)K(e^{-x} - 1)\right]e^x\nu(x,T)dx.$$

$$(3)$$

This PIDE holds on the domain $K \geq L, T \in [0,\bar{T}]$.

For a down-and-out call, the initial condition is

$$D_0^c(K,0) = (S_0 - K)^+, \qquad K \geq L.$$

Because a down-and-out call behaves as a standard call as its strike approaches infinity, we have

$$\lim_{K\uparrow\infty} D_0^c(K,T) = \lim_{K\uparrow\infty}\frac{\partial}{\partial K}D_0^c(K,T) = \lim_{K\uparrow\infty}\frac{\partial^2}{\partial K^2}D_0^c(K,T) = 0, \quad T \in [0,\bar{T}].$$

For a lower boundary condition, we note that a down-and-out call on a stock with the dynamics in Equation (1) has the same value prior to knocking out as a down-and-out call on a stock that absorbs at L. The second derivative of this latter call gives the r-discounted risk-neutral probability density for the event that the stock price has survived to at least T and is in the interval $(K, K + dK)$. Now it is well known that the appropriate boundary condition for an absorbing process is that this PDF vanishes on the boundary. Hence,

$$\lim_{K\downarrow L}\frac{\partial^2}{\partial K^2}D_0^c(K,T) = 0, \quad T \in [0,\bar{T}]. \tag{4}$$

Evaluating Equation (3) at $K = L$ and substituting in (4) implies

$$\frac{\partial D_0^c(L,T)}{\partial T} = \int_{-\infty}^\infty \left[D_0^c(Le^{-x},T) - D_0^c(L,T) - \frac{\partial}{\partial K}D_0^c(L,T)L(e^{-x} - 1)\right]$$

$$e^x\nu(x,T)dx - [r(T) - q(T)]L\frac{\partial}{\partial K}D_0^c(L,T) - q(T)D_0^c(K,T), \quad T \in [0,\bar{T}].$$

This is a Robin condition, as it involves the value and both its first partial derivatives along the boundary.

3 Up-and-Out Calls

3.1 The Backward Boundary Value Problem

Consider an up-and-out European call on a stock with a fixed maturity date $T \in [0, \bar{T}]$. At the first time that the stock price crosses an upper barrier H, the call knocks out. If the upper barrier is not touched prior to T, the call matures and pays $(S_T - K)^+$ at T, where $K \in [0, H)$ is the strike price.

We let τ_H denote the first passage time to H. We adopt the usual convention of setting this first passage time to infinity if the barrier is never hit. For $t \geq \tau$, the call has knocked out and is defined to be worthless at t. If $t < \tau$, then the stock price S_t must be in the continuation region $\mathcal{C} \equiv (S, t) \in (0, H) \times [0, T)$. While the call is alive, its value is given by a function, denoted $U(S, t)$, mapping $\mathcal{C}$ into the real line. In the interior of the continuation region, the partial derivatives, $\partial U / \partial t, \partial U / \partial S$, and $\partial^2 U / \partial S^2$ all exist as classical functions. No-arbitrage implies that the up-and-out call value function $U(S, t)$ satisfies the following deterministic PIDE in the continuation region $\mathcal{C}$; that is,

$$\int_{-\infty}^{\infty} \left[U(Se^x, t) - U(S, t) - \frac{\partial}{\partial S} U(S, t) S(e^x - 1) \right] \nu(x, t) dx + \frac{a^2(S, t)}{2} \frac{\partial^2 U(S, t)}{\partial S^2}$$

$$+ [r(t) - q(t)] S \frac{\partial U(S, t)}{\partial S} - r(t) U(S, t) + \frac{\partial U(S, t)}{\partial t} = 0, \quad \text{for } (S, t) \in \mathcal{C}. \quad (5)$$

A fortiori, the up-and-out call value function $U(S, t)$ solves a backward boundary value problem (BVP), consisting of the backward PIDE given by Equation (5) subject to the following boundary conditions.

$$U(S, T) = (S - K)^+, \quad S \in [0, H], \quad (6)$$

$$\lim_{S \downarrow 0} U(S, t) = 0, \quad t \in [0, T], \quad (7)$$

$$\lim_{S \uparrow H} U(S, t) = 0, \quad t \in [0, T]. \quad (8)$$

Equation (6) states that the up-and-out call is worth its intrinsic value at expiration. The value matching conditions given by Equations (7) and (8) show that at each $t \in [0, T)$, the up-and-out call's value tends to zero as the stock price approaches the origin or the barrier. For each t, the up-and-out call value is not in general differentiable in S at H.

Partial derivatives or integrals of the up-and-out call value with respect to K or T also satisfy a backward BVP. In particular, the second partial derivative of the up-and-out call value with respect to K, denoted $U_{kk}(S,t)$, solves the same PIDE as $U(S,t)$ in the continuation region $\mathcal{C}$; that is,

$$\int_{-\infty}^{\infty} \left[U_{kk}(Se^x, t) - U_{kk}(S,t) - \frac{\partial}{\partial S} U_{kk}(S,t) S(e^x - 1) \right] \nu(x,t) dx$$

$$+ \frac{a^2(S,t)}{2} \frac{\partial^2 U_{kk}(S,t)}{\partial S^2} + [r(t) - q(t)]S \frac{\partial U_{kk}(S,t)}{\partial S} - r(t)U_{kk}(S,t)$$

$$+ \frac{\partial U_{kk}(S,t)}{\partial t} = 0, \quad \text{for } (S,t) \in \mathcal{C}. \tag{9}$$

The partial derivative $U_{kk}(S,t)$ is subject to the following boundary conditions.

$$U_{kk}(S,T) = \delta(S - K), \quad S \in [0, H],$$
$$\lim_{S \downarrow 0} U_{kk}(S,t) = 0, \quad t \in [0, T],$$
$$\lim_{S \uparrow H} U_{kk}(S,t) = 0, \quad t \in [0, T].$$

3.2 Forward Propagation of Up-and-Out Call Values

Until now, we have been thinking of K and T as constants. In this section, we vary K and T, which will induce variation in the up-and-out call value. We also hold S and t constant at $S_0 \in (0, H)$ and 0, respectively. We let $u(K,T)$ denote the function relating the up-and-out call value to the strike price K and the maturity T when $(S,t) = (S_0, 0)$. Although we are interested in u for all $T \in [0, \bar{T}]$, we are only interested in u for all real $K \in [0, H]$.

Let $u_{kk}(K,T) \equiv U_{kk}(S_0, 0)$ be the function emphasizing the dependence of the second strike derivative of the up-and-out call value on K and T. The appendix shows that the adjoint of the backward PIDE given by Equation (9) governing u_{kk} is

$$\int_{-\infty}^{\infty} \left[e^{-x} u_{kk}(Ke^{-x}, T) + (e^x - 2)u_{kk}(K,T) + \frac{\partial}{\partial K} u_{kk}(K,T) K(e^x - 1) \right] \nu(x,T) dx$$

$$+ \frac{\partial^2}{\partial K^2} \left[\frac{a^2(K,T)}{2} u_{kk}(K,T) \right] - \frac{\partial}{\partial K} \{[r(T) - q(T)]K u_{kk}(K,T)\}$$

$$- r(T)u_{kk}(K,T) - \frac{\partial u_{kk}(K,T)}{\partial T} = 0, \tag{10}$$

for $K \in [0, H], T \in [0, \bar{T}]$. Note that

$$\frac{\partial^2}{\partial K^2} \{-[r(T) - q(T)]K v_k(K,T) - q(T)v(K,T)\}$$

$$= \frac{\partial}{\partial K} \{-[r(T) - q(T)]v_k(K,T) - [r(T) - q(T)]K v_{kk}(K,T) - q(T)v_k(K,T)\}$$

$$= -\frac{\partial}{\partial K} \{[r(T) - q(T)]K v_{kk}(K,T)\} - r(T)v_{kk}(K,T). \tag{11}$$

Also note that

$$e^x \frac{\partial^2}{\partial K^2} \{u(Ke^{-x}, T) - u(K, T) - u_k(K, T)K(e^{-x} - 1)\}$$

$$= e^x \frac{\partial}{\partial K} \{e^{-x} u_k(Ke^{-x}, T) - u_k(K, T) - u_{kk}(K, T)K(e^{-x} - 1)$$

$$- u_k(K, T)(e^{-x} - 1)\}$$

$$= \frac{\partial}{\partial K} \{u_k(Ke^{-x}, T) - u_k(K, T) - u_{kk}(K, T)K(1 - e^x)\}$$

$$= e^{-x} u_{kk}(Ke^{-x}, T) - u_{kk}(K, T) - u_{kkk}(K, T)K(1 - e^x) - u_{kk}(K, T)(1 - e^x)$$

$$= e^{-x} u_{kk}(Ke^{-x}, T) + (e^x - 2)u_{kk}(K, T) + u_{kkk}(K, T)K(e^x - 1). \tag{12}$$

Substituting Equations (11) and (12) into (10) implies

$$\frac{\partial^2}{\partial K^2} \left\{ \int_{-\infty}^{\infty} [u(Ke^{-x}, T) - u(K, T) - u_k(K, T)K(e^{-x} - 1)]e^x \nu(x, T)dx \right.$$

$$\left. + \frac{a^2(K, T)}{2} u_{kk}(K, T) - [r(T) - q(T)]Ku_k(K, T) - q(T)u(K, T) - \frac{\partial u(K, T)}{\partial T} \right\}$$

$$= 0, \quad K \in [0, H], T \in [0, \bar{T}].$$

Integrating on K twice implies

$$\int_{-\infty}^{\infty} [u(Ke^{-x}, T) - u(K, T) - u_k(K, T)K(e^{-x} - 1)]e^x \nu(x, T)dx$$

$$+ \frac{a^2(K, T)}{2} u_{kk}(K, T) - [r(T) - q(T)]Ku_k(K, T) - q(T)u(K, T) - \frac{\partial u(K, T)}{\partial T}$$

$$= A(T)K + B(T), \tag{13}$$

for $K \in [0, H], T \in [0, \bar{T}]$, and where $A(T)$ and $B(T)$ are independent of K. The forward PIDE is solved subject to an initial condition:

$$u(K, 0) = (S_0 - K)^+. \tag{14}$$

Suppose we regard the Lévy density $\nu(x, T)$ and volatility function $a(S, t)$ as given, and use Equation (13) subject to the initial condition given by Equation (14) to determine $u(K, T)$. Then the solution of the inhomogeneous PIDE given by Equation (13) subject to the initial condition given by (14) is not unique. As we don't know $A(T)$ and $B(T)$, we need two conditions just to determine the operator. For uniqueness, we may need two more independent boundary conditions as well. If both the origin and the upper barrier are accessible, then we will definitely need two additional independent boundary conditions to obtain uniqueness. However, if the origin is inaccessible (e.g., for geometric Brownian motion), then we can uniquely determine up-and-out call values without specifying a lower boundary condition.

We may alternatively suppose that all up-and-out call values are already known from the marketplace, and that the objective is to uniquely determine the Lévy density $\nu(x,t)$ and the volatility function $a(S,t)$. Under this perspective, one needs to supplement (13) subject to (14) with only two boundary conditions in order to determine $A(T)$ and $B(T)$. If one somehow knows the Lévy density, then it is straightforward to solve the algebraic equation (13) for $a(S,t)$.

We now assume that the objective is to determine $u(K,T)$ given $\nu(x,t)$ and $a(S,t)$. Differentiating Equation (13) with respect to K implies

$$
\begin{aligned}
A(T) &= \int_{-\infty}^{\infty} \left[e^{-x} u_k(Ke^{-x},T) - u_k(K,T) - \frac{\partial u_k}{\partial K}(K,T)K(e^{-x}-1) \right. \\
&\quad \left. - u_k(K,T)(e^{-x}-1) \right] e^x \nu(x,T)dx + \frac{\partial}{\partial K}\left\{ \frac{a^2(K,T)}{2} u_{kk}(K,T) \right. \\
&\quad \left. - [r(T)-q(T)]Ku_k(K,T) - q(T)u(K,T) - \frac{\partial u(K,T)}{\partial T} \right\} \\
&= \int_{-\infty}^{\infty} \left[u_k(Ke^{-x},T) - u_k(K,T) - \frac{\partial u_k}{\partial K}(K,T)K(1-e^{-x}) \right] \nu(x,T)dx \\
&\quad + \frac{a^2(K,T)}{2}\frac{\partial^2}{\partial K^2}u_k(K,T) \\
&\quad - \left\{ [r(T)-q(T)]K - a(K,T)\frac{\partial a}{\partial S}(K,T) \right\} \frac{\partial}{\partial K}u_k(K,T) \\
&\quad - r(T)u_k(K,T) - \frac{\partial u_k(K,T)}{\partial T}.
\end{aligned}
\tag{15}
$$

As the left-hand side (LHS) is invariant to K, the RHS is as well, and so we are free to determine $A(T)$ by evaluating the RHS at either $K=0$ or at $K=H$.

Once $A(T)$ is known, then from Equation (13),

$$
\begin{aligned}
B(T) &= \int_{-\infty}^{\infty} [u(Ke^{-x},T) - u(K,T) - u_k(K,T)K(e^{-x}-1)]e^x \nu(x,T)dx \\
&\quad + \frac{a^2(K,T)}{2}u_{kk}(K,T) - [r(T)-q(T)]Ku_k(K,T) - q(T)u(K,T) \\
&\quad - \frac{\partial u(K,T)}{\partial T} - A(T)K.
\end{aligned}
\tag{16}
$$

Once again, the LHS is invariant to K, and so we are free to determine $B(T)$ by evaluating the RHS at either $K=0$ or at $K=H$.

Appealing to the Feynman–Kac theorem, the solution to the backward BVP given by Equations (5) to (8) can be represented as

$$
u(K,T) = e^{-\int_0^T r(u)du} E^Q[\mathbf{1}_{\{\tau_H>T\}}(S_T-K)^+],
\tag{17}
$$

where the expectation is conditional on the known initial stock price S_0. The stock price process used to calculate these so-called risk-neutral expectations is the one solving the SDE given by Equation (1). At $K = H$, the function u gives the value of a call whose underlying must cross a knock-out barrier to finish above the strike:

$$u(H, T) = 0. \tag{18}$$

Differentiating Equation (17) w.r.t. K implies that

$$u_k(K, T) = -e^{-\int_0^T r(u)du} E^Q[\mathbf{1}_{\{\tau_H > T\}} \mathbf{1}_{\{S_T > K\}}]. \tag{19}$$

At $K = H$, the function u_k gives the value of a short binary call whose underlying must cross a knock-out barrier to finish above the strike:

$$u_k(H, T) = 0. \tag{20}$$

Differentiating Equation (19) w.r.t. K implies that

$$u_{kk}(K, T) = e^{-\int_0^T r(u)du} E^Q[\mathbf{1}_{\{\tau_H > T\}} \delta(S_T - K)],$$

where $\delta(\cdot)$ denotes the Dirac delta function. Hence, the function $u_{kk}(K, T)$ gives the r-discounted risk-neutral probability density for the event that the stock price process survives to T and that the time T stock price is in $(K, K + dK)$:

$$u_{kk}(K, T) = e^{-\int_0^T r(u)du} \frac{Q(\tau_H > T, S_T \in (K, K + dK))}{dK}. \tag{21}$$

Hence, as the strike price K approaches the knock-out barrier H, the probability of surviving beyond T goes to zero:

$$\lim_{K \uparrow H} u_{kk}(K, T) = 0. \tag{22}$$

Differentiating Equation (21) w.r.t. K implies that the slope of the discounted survival probability in K is given by

$$u_{kkk}(K, T) = e^{-\int_0^T r(u)du} E^Q[\mathbf{1}_{\{\tau_H > T\}} \delta^{(1)}(S_T - K)],$$

where $\delta^{(1)}(\cdot)$ denotes the first derivative of a delta function. When evaluated at $K = H$, u_{kkk} does not appear to simplify. We later show how the information at $K = 0$ allows us to relate u_{kkk} to the PDF for the first passage time τ_H.

Differentiating Equation (17) w.r.t. T implies that

$$\frac{\partial}{\partial T} u(K, T) = -r(T)u(K, T) + e^{-\int_0^T r(u)du} E^Q[\delta(\tau_H - T)(S_T - K)^+]. \tag{23}$$

One may interpret the second term as the value of a call whose notional is contingent on the first passage time to H being the call's maturity T. For any $H \leq K$, this term vanishes. Hence, evaluating Equation (23) at $K = H$ implies

$$\frac{\partial}{\partial T}u(H,T) = 0, \tag{24}$$

applying Equation (18). Differentiating Equation (19) w.r.t. T implies that

$$\frac{\partial}{\partial T}u_k(K,T) = -r(T)u_k(K,T) + e^{-\int_0^T r(u)du}E^Q[\delta(\tau_H - T)\mathbf{1}_{\{S_T > K\}}]. \tag{25}$$

The only difference between Equations (23) and (25) is that the second term now represents the value of a binary call with contingent notional. Hence,

$$\frac{\partial^2}{\partial T \partial K}u(H,T) = 0, \tag{26}$$

applying Equation (20).

We have less in the way of additional boundary conditions as $K \downarrow 0$. Because an up-and-out call has less value than a standard call, we have an upper bound on the value at $K = 0$:

$$u(0,T) \leq S_0 e^{-\int_0^T q(u)du}.$$

Similarly, because an up-and-out binary call has less value than a standard binary, we have a lower bound on the absolute slope in K at $K = 0$:

$$u_k(0,T) \geq -e^{-\int_0^T r(u)du}.$$

As an up-and-out butterfly spread has the same or lower value than a standard butterfly spread, we have

$$u_{kk}(0,T) = 0. \tag{27}$$

Similarly, as an up-and-out vertical spread of butterfly spreads has the same or lower value than a vertical spread of butterfly spreads, we have

$$u_{kkk}(0,T) = 0. \tag{28}$$

Evaluating Equation (23) at $K = 0$ implies

$$\frac{\partial}{\partial T}u(0,T) = -r(T)u(0,T) + e^{-\int_0^T r(u)du}E^Q[\delta(\tau_H - T)S_T]. \tag{29}$$

As the stock price must be worth H in order for the call holder to receive anything, Equation (29) yields a simple expression for the discounted first passage time density:

$$\phi(H,T) \equiv e^{-\int_0^T r(u)du}\frac{Q(\tau_H \in (T, T+dT))}{dT} = \frac{1}{H}\left[r(T)u(0,T) + \frac{\partial}{\partial T}u(0,T)\right].$$

When $\nu(x,t)$ and $a(S,t)$ are known and $u(K,T)$ is to be determined, the RHS is not known exante. However, when $\nu(x,t)$ and $a(S,t)$ are not known and when one can observe market prices of up-and-out calls with zero strikes (which are up-and-out shares with no dividends received prior to T), then the RHS is observable. Evaluating Equation (25) at $K = 0$ implies

$$\frac{\partial}{\partial T}u_k(0,T) = -r(T)u_k(0,T) + e^{-\int_0^T r(u)du}E^Q[\delta(\tau_H - T)].$$

Hence, we have another simple expression for the discounted first passage time density:

$$\phi(H,T) = r(T)u_k(0,T) + \frac{\partial}{\partial T}u_k(0,T). \tag{30}$$

Evaluating Equation (15) at $K = H$ and substituting in Equations (20), (22), and (26) implies that for an up-and-out call,

$$A(T) = \int_0^\infty u_k(He^{-x},T)\nu(x,T)dx + \frac{a^2(H,T)}{2}\frac{\partial^2}{\partial K^2}u_k(H,T). \tag{31}$$

Evaluating Equation (16) at $K = H$ and substituting in Equations (18), (20), (22), (24), and (31) implies that for an up-and-out call,

$$B(T) = \int_0^\infty u(He^{-x},T)e^x\nu(x,T)dx$$
$$- \left[\int_0^\infty u_k(He^{-x},T)\nu(x,T)dx + \frac{a^2(H,T)}{2}\frac{\partial^2}{\partial K^2}u_k(H,T)\right]H. \tag{32}$$

We can alternatively try to determine $A(T)$ and $B(T)$ using boundary conditions for $K = 0$. Evaluating Equation (15) at $K = 0$ and substituting in Equations (2), (27), and (28) implies that for an up-and-out call,

$$A(T) = -r(T)u_k(0,T) - \frac{\partial u_k(0,T)}{\partial T} = -\phi(H,T), \tag{33}$$

applying Equation (30). Evaluating Equation (16) at $K = 0$ and substituting in Equations (2), (27), (28), and (33) implies that for an up-and-out call,

$$B(T) = -q(T)u(0,T) - \frac{\partial u(0,T)}{\partial T}. \tag{34}$$

Using $A(T)$ and $B(T)$ determined by Equations (31) and (32), respectively, (13) becomes

$$\int_{-\infty}^\infty [u(Ke^{-x},T) - u(K,T) - u_k(K,T)K(e^{-x}-1)]e^x\nu(x,T)dx$$
$$+ \frac{a^2(K,T)}{2}u_{kk}(K,T) - [r(T)-q(T)]Ku_k(K,T) - q(T)u(K,T) - \frac{\partial u(K,T)}{\partial T}$$
$$= \left[\int_0^\infty u_k(He^{-x},T)\nu(x,T)dx + \frac{a^2(H,T)}{2}u_{kkk}(H,T)\right](K - H)$$
$$+ \int_0^\infty u(He^{-x},T)e^x\nu(x,T)dx, \tag{35}$$

for $K \in (0, H), T \in [0, \bar{T}]$ and for $H > S_0$. Recall that for an up-and-out call, the initial condition is

$$u(K, 0) = (S_0 - K)^+, \quad K \in [0, H), \tag{36}$$

and for $H > S_0$. For boundary conditions, Equations (18), (20), (22), (27), and (28) are all available. As a result, we have more than enough independent boundary conditions to uniquely determine $u(K, T)$. Note that the forward operator is not local as it acts on the function $u(K, T)$ and its derivatives at both $K < H$ and at $K = H$.

As Equations (31) and (33) both yield expressions for $A(T)$, it follows that

$$\int_0^\infty u_k(He^{-x}, T)\nu(x, T)dx + \frac{a^2(H, T)}{2}\frac{\partial^2}{\partial K^2}u_k(H, T) = -\phi(H, T). \tag{37}$$

Substituting Equation (37) in (35) implies that the forward PIDE for an up-and-out call can also be written as

$$\frac{\partial u(K, T)}{\partial T} = \int_{-\infty}^\infty [u(Ke^{-x}, T) - u(K, T) - u_k(K, T)K(e^{-x} - 1)]e^x\nu(x, T)dx$$

$$+ \frac{a^2(K, T)}{2}u_{kk}(K, T) - [r(T) - q(T)]Ku_k(K, T) - q(T)u(K, T)$$

$$- \phi(H, T)(H - K) - \int_0^\infty u(He^{-x}, T)e^x\nu(x, T)dx. \tag{38}$$

To interpret this PIDE financially, first note that if an investor buys a calendar spread of up-and-out calls, then the initial cost is given by the LHS. The first term on the RHS arises only from paths that survive to T and cross K then. It can be shown that this first term is the initial value of a path-dependent claim that pays the overshoots of the strike at T. The second term on the RHS arises only from paths that survive to T and finish at K. Consider the infinite position in the later maturing call at time $t = T$ if the option survives until then. This position will have infinite time value when $S_T = K$ and zero value otherwise. The greater is the local variance rate at $S_T = K$, the greater is this conditional time value and the more valuable is this position initially. The next two terms arise only from paths that survive to T and finish above K. They capture the additional carrying costs of stock and bond that are embedded in the time value of the later maturing call. The operator given by the first four terms on the RHS also represents the present value of benefits obtained at T when an investor buys a calendar spread of standard or down-and-out calls. In contrast, the last two terms in Equation (38) have no counterpart for calendar spreads in standard or down-and-out calls. To interpret them financially, note that

$$\phi(H,T)(H-K) + \int_0^\infty u(He^{-x},T)e^x\nu(x,T)dx$$

$$= e^{-\int_0^T r(u)du} E_0^Q[\delta(\tau_H - T)(H - K)]$$

$$+ e^{-\int_0^T r(u)du} E_0^Q\left[\int_0^\infty (S_{T-} - He^{-x})^+ e^x \nu(x,T)dx\right]$$

$$= e^{-\int_0^T r(u)du} E_0^Q[\delta(\tau_H - T)\mathbf{1}_{\{S_T \geq H\}}(H - K)]$$

$$+ e^{-\int_0^T r(u)du} E_0^Q\left[\int_0^\infty (S_{T-}e^x - H)^+ \mathbf{1}_{\{\tau_H \geq T\}} e^x \nu(x,T)dx\right]$$

$$= e^{-\int_0^T r(u)du} E_0^Q\left[\delta(\tau_H - T)(S_T - H)^+\right].$$

Thus, the last two terms in Equation (38) represent the discounted expected value of the payoff from a call struck at H if the first passage time to H is T. Note that the possibility of this loss can cause the calendar spread value to be negative.

If we take the up-and-out call prices as given by the market, and even supposing that we know the Lévy density $\nu(x,t)$, then solving Equation (13) or (38) for the local volatility function $a(S,t)$ is problematic, unless we somehow know $a(H,T)$ or $\phi(H,T)$ ex ante. Fortunately, this problem is solved by using $A(T)$ and $B(T)$ determined by Equations (33) and (34), respectively, instead. In this case, Equation (13) becomes:

$$\int_{-\infty}^\infty [u(Ke^{-x},T) - u(K,T) - u_k(K,T)K(e^{-x} - 1)]e^x\nu(x,T)dx$$

$$+ \frac{a^2(K,T)}{2} u_{kk}(K,T) - [r(T) - q(T)]Ku_k(K,T) - q(T)u(K,T) - \frac{\partial u(K,T)}{\partial T}$$

$$= \left[-r(T)u_k(0,T) - \frac{\partial u_k(0,T)}{\partial T}\right]K - q(T)u(0,T) - \frac{\partial u(0,T)}{\partial T}. \tag{39}$$

When the up-and-out call prices are given by the market and the Lévy density $\nu(x,t)$ is known, then solving Equation (39) for the local volatility function $a(S,t)$ is straightforward. Note that our assumption that the origin is inaccessible was crucial for achieving this result.

If the Lévy density $\nu(x,t)$ and volatility function $a(S,t)$ are instead given, then one must try to solve Equation (39) for the up-and-out call value function $u(K,T)$ on the domain $K \in (0,H), T \in [0,\bar{T}]$. Once again, this forward operator is not local. The PIDE (39) is also solved subject to the initial condition given by Equation (36). Once again, we have the five boundary conditions (18), (20), (22), (27), and (28) available. As usual, we need at least one boundary condition to uniquely determine $u(K,T)$ once the operator is determined. In this specification, we also need to solve for the four functions $u(0,T), u_k(0,T), (\partial u_k(0,T))/\partial T$, and $(\partial u(0,T))/\partial T$ to determine the operator. Hence, it appears that $u(K,T)$ is not determined uniquely by the forward BVP involving (39). Fortunately, it is determined by the forward

BVP involving (35), and so we are able to solve for either $a(S,t)$ or $u(K,T)$ through use of the appropriate forward BVP.

4 Numerical Examples

We employ the same methodology used in [8] and [16] to numerically solve the backward and forward PIDEs for both down-and-out and up-and-out calls. For our numerical examples, we take $\nu(x)dx$ to be the Lévy density for the VG process in the following form.

$$\nu(x) = \frac{e^{-\lambda_p x}}{\nu x} \quad \text{for } x > 0 \quad \text{and} \quad \nu(x) = \frac{e^{-\lambda_n |x|}}{\nu |x|} \quad \text{for } x < 0$$

and

$$\lambda_p = \left(\frac{\theta^2}{\sigma^4} + \frac{2}{\sigma^2 \nu} \right)^{1/2} - \frac{\theta}{\sigma^2} \quad \lambda_n = \left(\frac{\theta^2}{\sigma^4} + \frac{2}{\sigma^2 \nu} \right)^{1/2} + \frac{\theta}{\sigma^2},$$

where σ, ν, and θ are VG parameters.

We consider the following local volatility surface,

$$\sigma(K,T) = 0.3e^{-T} (100/K)^{0.2},$$

which is plotted in Figure 1. Other parameter values for our numerical experiments are as follows: spot $S_0 = 100$, risk-free rate $r = 6\%$, dividend rate $q = 2$, and VG parameters $\sigma = 0.3$, $\nu = 0.25$, $\theta = -0.3$.

Numerical Results on Up-and-Out Calls

In the case of up-and-out calls, the up-barrier is taken to be $H = 140$. In the backward case, for each maturity and each strike, we solve the backward PIDE and extract the value for time 0 and spot 100 as shown in Figures 2 through 4. In Figure 2, the left graph illustrates the value for up-and-out calls for 3-month maturity and strike 90, and the right graph displays the value for up-and-out calls for 3-month maturity and strike 110. In Figure 3, the left graph illustrates the value for up-and-out calls for 6-month maturity and strike 90, and the right graph displays the value for up-and-out calls for 6-month maturity and strike 110. In Figure 4, the left graph illustrates the value for up-and-out calls for 12-month maturity and strike 90, and the right graph displays the value for up-and-out calls for 12-month maturity and strike 110. In the forward case, however, we just solve the forward PIDE once and extract the values at these maturities and strikes as shown in Figure 5. Table 1 summarizes the results for up-and-out calls from both backward and forward PIDEs, where as expected, the two prices match pretty closely.

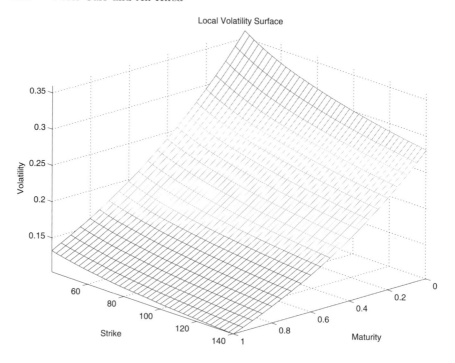

Fig. 1. Local volatility surface.

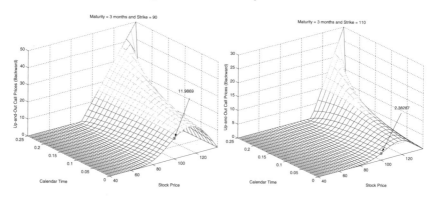

Fig. 2. On the left, up-and-out call prices for 3-month maturity and strike 90; on the right, up-and-out call prices for 3-month maturity and strike 110.

Numerical Results on Down-and-Out Calls

Table 2 illustrates the numerical results for down-and-out calls for both backward and forward PIDEs for maturities: 3, 6, and 12 months, strikes: 90 and 110. For down-and-out calls, the down-barrier is assumed to be $L = 60$. As before, in the backward case, for each maturity and each strike, we solve the

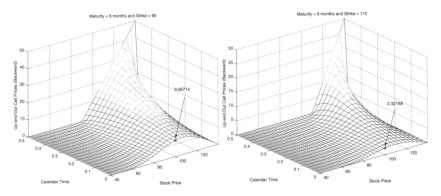

Fig. 3. On the left, up-and-out call prices for 6-month maturity and strike 90; on the right, up-and-out call prices for 6-month maturity and strike 110.

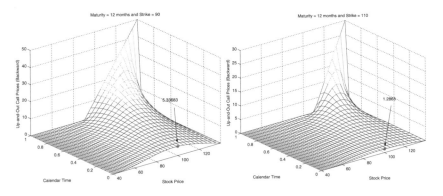

Fig. 4. On the left, up-and-out call prices for 12-month maturity and strike 90; on the right, up-and-out call prices for 12-month maturity and strike 110.

Table 1. Up-and-out call prices for three maturities and two strikes (up-barrier $H = 140$).

Maturity		$T_1 = 0.25$		$T_2 = 0.5$		$T_3 = 1.0$	
Barrier	Strike	Bwd	Fwd	Bwd	Fwd	Bwd	Fwd
140	90	11.9869	11.98901	9.56714	9.56918	5.33683	5.33786
	110	2.38287	2.38951	2.32168	2.32895	1.28613	1.28012

backward PIDE and extract the value for time 0 and spot at 100. In the forward case, we solve it once and extract the values at (K_i, T_j). As we expected, the prices are nearly identical.

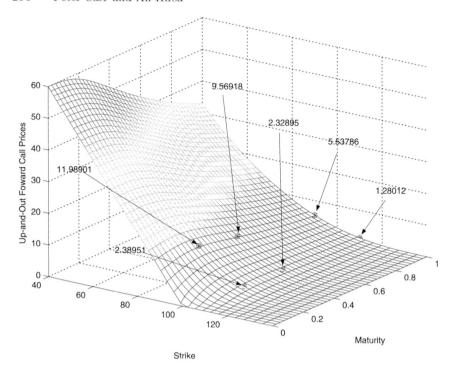

Fig. 5. Up-and-out call prices for maturities 3, 6, and 12 months and strikes 90 and 110.

Table 2. Down-and-out call prices for three maturities and and two strikes (down-barrier $L = 60$).

Maturity		$T_1 = 0.25$		$T_2 = 0.5$		$T_3 = 1.0$	
Barrier	Strike	Bwd	Fwd	Bwd	Fwd	Bwd	Fwd
60	90	13.8837	13.8798	16.8732	16.8801	21.3305	21.3345
	110	3.57784	3.57891	6.99384	6.98564	12.1307	12.1295

5 Future Research

We would like to extend this work to other kind of barrier options such as no-touches and double-barrier options. One can also attempt to extend the dynamics assumption to stochastic volatility and stock jump arrival rates.

Acknowledgments

We would like to thank participants of the ICBI Global Derivatives Conference 2003 in Barcelona, participants of Risk Europe 2003 in Paris, participants of the Control and Dynamical Systems Lecture Series at the Institute for Systems Research at the University of Maryland at College Park, and students at the Mathematics of Finance Programs at Courant Institute and Columbia University for their comments and discussions. We would also like to thank Robert Kohn and Pedro Judice of Courant Institute for their comments and suggestions. Errors are our own responsibility.

References

1. L. Andersen and J. Andreasen. Jumping smiles. *Risk*, 12:65–68, November 1999.
2. T. Andersen, L Benzoni, and J. Lund. An empirical investigation of continuoustime equity return models. *Journal of Finance*, 57:1239–1284, June 2002.
3. J. Andreasen and P. Carr. Put-call reversal. Manuscript, University of Aarhus.
4. J. Anreasen. Implied modelling, stable implementation, hedging, and duality. Manuscript, University of Aarhus, 1998.
5. F. Black and M. Scholes. The pricing of options and corporate liabilities. *Journal of Political Economy*, 81:637–654, 1973.
6. A. Buraschi and B. Dumas. The forward valuation of compound options. *Journal of Derivatives*, 9:8–17, Fall 2001.
7. P. Carr, H. Geman, D.B. Madan, and M. Yor. The fine structure of asset returns: An empirical investigation. *Journal of Business*, 75(2):305–332, April 2002.
8. P. Carr and A. Hirsa. Why be backward? *Risk*, 16(1):103–107, January 2003.
9. P. Carr and L. Wu. The finite moment logstable process and option pricing. *Journal of Finance*, 58(2):753–770, April 2003.
10. N. Chriss. Transatlantic trees. *Risk*, 9(7):45–48, July 1996.
11. B. Dumas, J. Fleming, and R. Whaley. Implied volatilities: Empirical tests. *Journal of Finance*, (53):2059–2106, 1998.
12. B. Dupire. Pricing with a smile. *Risk*, 7(1):18–20, January 1994.
13. E. Eberlein, U. Keller, and K. Prause. New insights into smile, mispricing, and value at risk: The hyperbolic model. *Journal of Business*, 71:371–406, 1998.
14. A. Esser and C. Schlag. A note on forward and backward partial differential equations for derivative contracts with forwards as underlyings in foreign exchange risk. Edited by Jurgen Hakala and Uwe Wystup, Chapter 12, *Risk Books*, 2002.
15. I.I. Gihman and A.V. Skorohod. *Stochastic Differential Equations*. Springer-Verlag, 1972.
16. A. Hirsa and D.B. Madan. Pricing American options under variance gamma. *Journal of Computational Finance*, 7(2):63–80, Winter 2003.
17. S.G. Kou. A jump-diffusion model for option pricing. *Management Science*, 48:1086–1101, 2002.
18. D.B. Madan, P. Carr, and E.C. Chang. The variance gamma process and option pricing. *European Finance Review*, 2:79–105, 1998.
19. R. Merton. Option pricing when underlying stock. *Journal of Financial Economics*, 3:125–144, 1976.

Appendix: Adjoint of Backward PIDE

From Gihman and Skorohod [15, p. 297],

$$\frac{\partial}{\partial T}e^{-\int_t^T r(u)du}E_t f(S_T) =$$

$$-r(T)E_t e^{-\int_t^T r(u)du}f(S_T) + e^{-\int_t^T r(u)du}E_t[(\mathcal{L}_T f)(S_T)], \qquad (40)$$

where

$$\mathcal{L}_T f(S) \equiv \int_{-\infty}^{\infty}[f(Se^x) - f(S) - f'(S)S(e^x - 1)]\,\nu(x,T)dx$$

$$+ \frac{a^2(S,T)}{2}f''(S) + [r(T) - q(T)]Sf'(S).$$

Taking $f(S) = \delta(S - K)$, Equation (40) becomes

$$\frac{\partial}{\partial T}e^{-\int_t^T r(u)du}E_t\delta(S_T - K) + r(T)e^{-\int_t^T r(u)du}E_t\delta(S_T - K)$$

$$= e^{-\int_t^T r(u)du}E_t[(\mathcal{L}_T\delta(S - K))(S_T)]. \qquad (41)$$

The LHS of Equation (41) equals

$$\frac{\partial}{\partial T}u_{kk}(K,T) + r(T)u_{kk}(K,T),$$

whereas the RHS of Equation (41) equals

$$\int_0^{\infty}u_{kk}(L,T)\Big\{\int_{-\infty}^{\infty}\Big[\delta(Le^x - K) - \delta(L - K) - \delta^{(1)}(L - K)L(e^x - 1)\Big]\nu(x,T)dx$$

$$+ \frac{a^2(L,T)}{2}\delta^{(2)}(L - K) + [r(T) - q(T)]L\delta^{(1)}(L - K)\Big\}dL$$

$$= \int_{-\infty}^{\infty}\Big[e^{-x}u_{kk}(Ke^{-x},T) - u_{kk}(K,T) +$$

$$\Big[\frac{\partial u_{kk}}{\partial K}(K,T)K + u_{kk}(K,T)\Big](e^x - 1)\Big]\nu(x,T)dx$$

$$+ \frac{\partial^2}{\partial K^2}\Big[\frac{a^2(K,T)}{2}u_{kk}(K,T)\Big] - \frac{\partial}{\partial K}\{[r(T) - q(T)]Ku_{kk}(K,T)\}$$

$$= \int_{-\infty}^{\infty}\Big[e^{-x}u_{kk}(Ke^{-x},T) + (e^x - 2)u_{kk}(K,T) + \frac{\partial u_{kk}}{\partial K}(K,T)K(e^x - 1)\Big]\nu(x,T)dx$$

$$+ \frac{\partial^2}{\partial K^2}\Big[\frac{a^2(K,T)}{2}u_{kk}(K,T)\Big] - \frac{\partial}{\partial K}\{[r(T) - q(T)]Ku_{kk}(K,T)\}.$$

Thus, the adjoint is

$$\frac{\partial}{\partial T} u_{kk}(K,T) \tag{42}$$

$$= \int_{-\infty}^{\infty} \left[e^{-x} u_{kk}(Ke^{-x},T) + (e^x - 2)u_{kk}(K,T) + \frac{\partial u_{kk}}{\partial K}(K,T)K(e^x - 1) \right] \nu(x,T)dx$$

$$+ \frac{\partial^2}{\partial K^2} \left[\frac{a^2(K,T)}{2} u_{kk}(K,T) \right]$$

$$- \frac{\partial}{\partial K} \{ [r(T) - q(T)]K u_{kk}(K,T) \} - r(T)u_{kk}(K,T).$$

Mean Reversion Versus Random Walk in Oil and Natural Gas Prices

Hélyette Geman

Birkbeck, University of London, United Kingdom
& ESSEC Business School, Cergy-Pontoise, France
hgeman@ems.bbk.ac.uk

Summary. The goals of the paper are as follows: (i) review some qualitative properties of oil and gas prices in the last 15 years; (ii) propose some mathematical elements towards a definition of mean reversion that would not be reduced to the form of the drift in a stochastic differential equation; (iii) conduct econometric tests in order to conclude whether mean reversion still exists in the energy commodity price behavior. Regarding the third point, a clear "break" in the properties of oil and natural gas prices and volatility can be exhibited in the period 2000–2001.

Key words: Oil and gas markets; mean reversion; invariant measure.

1 Introduction

Energy commodity prices have been rising at an unprecedented pace over the last five years. As depicted in Figure 1, an investment of $100 made in January 2002 in the global Dow Jones–AIG Commodity Index had more than doubled by July 2006, whereas Figure 2 indicates that an investment of $100 in the Dow Jones–AIG Energy subindex had turned into $500 in July 2005. Among the numerous explanations for this phenomenon, we may identify the severe tensions on oil and their implications for other fossil fuels that may be substitutes. The increase of oil prices is driven by demand growth, particularly in Asia where Chinese consumption rose by 900,000 barrels per day, mostly accounted for by imports.

At the world level, the issue of "peak oil"—the date at which half of the reserves existing at the beginning of time are (will be) consumed—is the subject of intense debates. Matthew Simmons asks in his book, *Twilight in the Desert*, whether there is a significant amount of oil left in the soil of Saudi Arabia. The concern of depleting reserves in the context of an exhaustive commodity such as oil is certainly present on market participants' minds, and in turn, on the trajectory depicted in Figure 2.

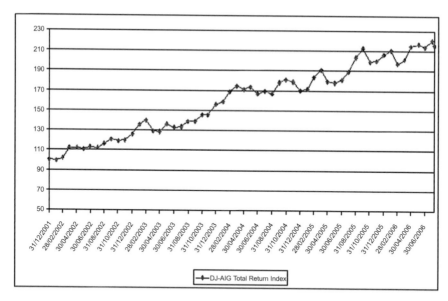

Fig. 1. Dow Jones–AIG Total Return Index over the period January 2002–July 2006.

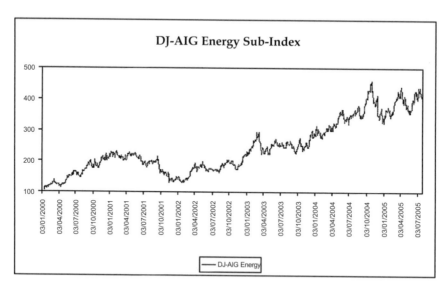

Fig. 2. Dow Jones–AIG Energy Sub-Index over the period January 2000–July 2005.

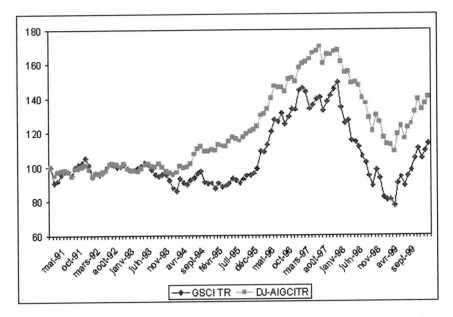

Fig. 3. Goldman Sachs Commodity Index Total Return and Dow Jones–AIG Commodity Index total return over the period February 1991–December 1999.

The financial literature on commodity price modeling started with the pioneering paper by Gibson and Schwartz [7]. In the spirit of the Black–Scholes–Merton [2] formula, they use a geometric Brownian motion for oil spot prices. Given the behavior of commodity prices during the 1990s depicted in Figure 3 by the two major commodity indexes, [10] introduces a mean-reverting drift in the stochastic differential equation driving oil price dynamics; [14; 4; 12] keep this mean-reversion representation for oil, electricity, and bituminous coal.

The goal of this paper is to revisit the modeling of oil and natural gas prices in the light of the trajectories observed in the recent past (see Figure 2), as well as the definition of mean reversion from a general mathematical perspective. Note that this issue matters also for key quantities in finance such as stochastic volatility. Fouque et al. [5] are interested in the property of clustering exhibited by the volatility of asset prices. Their view is that volatility is "bursty" in nature, and burstiness is closely related to mean reversion, because a bursty process is returning to its mean (at a speed that depends on the length of the burst period).

2 Some Elements on Mean Reversion in Diffusion from a Mathematical Perspective

The long-term behavior of continuous-time Markov processes has been the subject of much attention, starting with the work of Has'minskii [8]. Accordingly, the long-term evolution of the price of an exhaustible commodity, such as oil or copper, is a topic of major concern in finance, given the world geopolitical and economic consequences of this issue. In what follows, (X_t) essentially has the economic interpretation of a log-price. We start with a process (X_t) defined as the solution of a stochastic differential equation

$$dX_t = b(X_t)\,dt + \sigma(X_t)\,dW_t,$$

where (W_t) is a standard Brownian motion on a probability space $(\Omega, \mathcal{F}, \mathcal{P})$ describing the randomness of the economy. We know from Itô that if b and σ are Lipschitz, there exists a unique solution to the equation. If only b is Lipschitz and σ Holder of coefficient $\frac{1}{2}$, we still have existence and uniqueness of the solution. In both cases, the process (X_t) will be Markov and the drift $b(X_t)$ will contain the representation of the trend perceived at date t for future spot prices.

Given a process (X_t), we are in finance particularly interested in the possible existence of a distribution for X_0 such that, for any positive t, X_t has the same distribution. This distribution, if it exists, may be viewed as an equilibrium state for the process. We now recall the definition of an invariant probability measure for a Markov process $(X_t)_{t \geq 0}$, whose semi-group is denoted $(P_t)_{t \geq 0}$ and satisfies the property that, for any bounded measurable function, $P_t f(x) = E[f(X_t)]$.

Definition 1. *(i) A measure μ is said to be* invariant *for the process (X_t) if and only if*

$$\int \mu(dx)\,P_t f(x) = \int \mu(dx)\,f(x),$$

for any bounded function f. (ii) μ is invariant for (X_t) if and only if $\mu P_t = \mu$. Equivalently, the law of $(X_{t+u})_{u \geq 0}$ is independent of t if we start at date 0 with the measure μ.

Proposition 1. *The existence of an invariant measure implies that the process (X_t) is stationary. If (X_t) admits a limit law independent of its initial state, then this limit law is an invariant measure.*

Proof. The first part of the proposition is nothing but one of the two forms of the definition above. Now suppose $E[f(X_t)] \to_{t \to \infty} \int \mu(dy)\,f(y)$, for any bounded function f. Then

$$E_X[f(X_{t+s})] = E_X[P_s f(X_t)] \to_{t \to \infty} \int \mu(dy)\,f(y).$$

Hence, $\mu P_s = \mu$, and μ is invariant.   □

Proposition 2.   *1. The Ornstein–Uhlenbeck process admits a finite invariant measure, and this measure is Gaussian.*

2. The Cox–Ingersoll–Ross (or square-root) process also has a finite invariant measure.

3. The arithmetic Brownian motion (as do all Lévy processes) admits the Lebesgue measure as an invariant measure, hence, not finite.

4. A (squared) Bessel process exhibits the same property, namely an infinite invariant measure.

Proof.   1. For $\phi : \mathbb{R} \to \mathbb{R}$ in C^2, consider the Smoluchowski equation

$$dX_t = \phi'(X_t)\, dt + dW_t,$$

where W_t is a standard Brownian motion. Then the measure $\mu(dx) = e^{2\phi(x)}dx$ is invariant for the process (X_t). If we consider now an Ornstein–Uhlenbeck process (with a standard deviation equal to 1 for implicity), then $dX_t = (a - bX_t)\,dt + dW_t$, $a, b > 0$, $\phi'(x) = a - bx$, $\phi(x) = c + ax - bx^2/2$, and $\mu(dx) = e^{-bx^2 + 2ax + c}dx$ is an invariant measure (normalized to 1 through c), and we recognize the Gaussian density. In the general case of an Ornstein–Uhlenbeck process reverting to the mean m, and with a standard deviation equal to σ, the invariant measure will be $\mathcal{N}(m, \sigma^2)$.

2. We recall the definition of the squared-root (or CIR) process introduced in finance by Cox et al. [3]:

$$dX_t = (\delta - bX_t)\,dt + \sigma\sqrt{X_t}dW_t.$$

We remember that a CIR process is the square of the norm of a δ-dimensional Ornstein–Uhlenbeck process, where δ is the drift of the CIR process at 0. The semi-group of a CIR process is the radial projection of the semi-group of an Ornstein–Uhlenbeck. If we note v, the image of μ, by the radial projection

$$\int v(dr)\,\phi(r) = \int \mu(dx)\, P_t(\phi|\cdot|)(x) = \int \mu(dx)\, P_t\phi(|x|) = \int v(dr)\, P_t\phi(r).$$

Hence v, the image of μ by the norm application, is invariant for P_t.

3. Consider a Lévy process $(\mathcal{L}_t)$ and P_t its semi-group. Then

$$P_t(x, f) = E_0[f(x + \mathcal{L}_t)],$$

$$\int dx P_t(x, f) =_{Fubini} E_0\left[\int f(x + \mathcal{L}_t)\,dx\right] = \int f(y)\,dy.$$

Consequently, the Lebesgue measure is invariant for the process $(\mathcal{L}_t)$.

4. We use the well-known relationship between a Bessel process and the norm of Brownian motion and also obtain an infinite invariant measure for Bessel processes.   □

Having covered the fundamental types of Markovian diffusions used in finance, we are led to propose the following definition.

Definition 2. *Given a Markov diffusion (X_t), we say that the process (X_t) exhibits* mean reversion *if and only if it admits a finite invariant measure.*

Remarks.

1. The definition does not necessarily involve the drift of a stochastic differential equation satisfied by (X_t), as also suggested in [11].
2. It allows inclusion of high-dimensional non-Markovian processes driving energy commodity prices or volatility levels.
3. Following [13], we can define the set

$$T_P = \{\text{probability measure } \mu \text{ such that } \mu P_t = \mu \; \forall t \geq 0\}.$$

Then the set T_P is convex and closed for the tight convergence topology (through Feller's property) and possibly empty. If T_P is not the empty set and compact (for the tight convergence topology), there is at least one extremal probability μ^* in T_P. Then the process (X_t) is ergodic for this measure μ^*: for any set $A \in \mathcal{F}_\infty^X$ that is invariant by the time translators $(\theta_t)_{t \geq 0}$, then $P_{\mu^*}(A) = 0$ or 1, where we classically denote $\mathcal{F}_\infty^X$ the natural filtration of the process (X_t). The time-translation operator θ_t is defined on the space Ω by $(\theta_t(X))_s = X_{t+s}$. It follows by Birkhoff's theorem that, for any function $F \in L^1(P_{\mu^*})$,

$$\frac{1}{t} \int_0^t F(X_s)\, ds \to E_{P_{\mu^*}}[F(X)] \quad P_{\mu^*} - \text{a.s.}$$

when $t \to \infty$. The interpretation of this result is the following: the long-run time average of a bounded function of the ergodic process (X_t) is close to its statistical average with respect to its invariant distribution. This property is crucial in finance, as the former quantity is the only one we can hope to compute using an historical database of the process (X_t).

3 An Econometric Approach to Mean Reversion in Energy Commodity Prices

We recall the classical steps in testing mean reversion in a series of prices (X_t). The objective is to check whether in the representation

$$X_{t+1} = \rho X_t + \varepsilon_t,$$

the coefficient ρ is significantly different from 1. The $\mathcal{H}_0$ hypothesis is the existence of a unit root (i.e., $\rho = 1$). A p-value smaller than 0.05 allows one to reject the $\mathcal{H}_0$ hypothesis with a confidence level higher than 0.95, in which case the process is of the mean-reverting type. Otherwise, a unit root is uncovered, and the process is of the "random walk" type. The higher the p-value, the more the random walk model is validated.

3.1 Mean-Reversion Tests

They are fundamentally of two types:

(i) The Augmented Dickey–Fuller (ADF) consists in estimating the regression coefficient of $p(t)$ on $p(t-1)$. If this coefficient is significantly below 1, it means that the process is mean reverting; if it is close to 1, the process is a random walk.

(ii) The Phillips–Perron test consists in searching for a unit root in the equation linking X_t and X_{t+1}. Again, a high p-value reinforces the hypothesis of a unit root.

3.2 Statistical Properties Observed on Oil and Natural Gas Prices

For crude oil,

- A mean-reversion pattern prevails over the period 1994–2000,
- It changes into a random walk (arithmetic Brownian motion) as of 2000.

Whereas for natural gas,

- There is a mean-reversion pattern until 1999,
- Since 2000, a change into a random walk occurs, but with a lag compared to oil prices,
- During both periods, seasonality of gas prices tends to blur the signals.

For U.S. natural gas prices over the period January 1994–October 2004, spot prices are proxied by the New York Mercantile Exchange (NYMEX) one-month futures contract. Over the entire period January 1994–October 2004, the ADF p-value is 0.712 and the Phillips–Perron p-value is 0.1402, whereas over the period January 1999–October 2004, the ADF p-value is 0.3567 and the Phillips–Perron p-value is 0.3899. Taking instead log-prices, the numbers become

Jan 94–Oct 04	Jan 99–Oct 04
ADF p-value = 0.0863	ADF p-value = 0.4452
Phillips–Perron p-value = 0.0888	Phillips–Perron p-value = 0.4498

Over the last five years of the period, the arithmetic Brownian motion assumption clearly prevails and mean reversion seems to have receded.

For West Texas Intermediate (WTI) oil spot prices over the same period January 1994–October 2004, again spot prices are proxied by NYMEX one-month futures prices, and the tests are conducted for log-prices.

1994–2004	Jan 1999–Oct 2004
ADF p-value 0.651	ADF p-value 0.7196
Phillips–Perron 0.5048	Phillips–Perron 0.5641

The mean-reversion assumption is strongly rejected over the whole period and even more so over the recent one. Because of absence of seasonality, the behavior of a random walk is more pronounced in the case of oil log-prices than in the case of natural gas.

4 The Economic Literature on Mean Reversion in Commodity Prices

In [1], the term structure of futures prices is tested over the period January 1982 to December 1991, for which mean reversion is found in the 11 markets examined, and it is also concluded that the magnitude of mean reversion is large for agricultural commodities and crude oil, and substantially less for metals. Rather than examining evidence of ex-post reversion using time series of asset prices, [1] uses price data from futures contracts with various horizons to test whether investors expect prices to revert. The "price discovery" element in forward and futures prices is related to the famous "rational expectations hypothesis" long tested by economics, for interest rates in particular (cf. [9]), and stating that forward rates are unbiased predictors of future spot rates. The paper [1] analyzes the relation between price levels and the slope of the futures term structure defined by the difference between a long maturity future contract and the first nearby. Assuming that futures prices are unbiased expectations (under the real probability measure) of future spot prices, an inverse relation between prices and this slope constitutes evidence that investors expect mean reversion in spot prices, as it implies a lower expected future spot when prices rise. The authors conclude the existence of mean reversion for oil prices over the period 1982–1999; however, the same computations conducted over the period 2000–2005 lead to inconclusive results.

Pindyck [12] analyzes 127 years (1870–1996) of data on crude oil and bituminous coal, obtained from the U.S. Department of Commerce. Using a unit root test, he shows that prices mean revert to stochastically fluctuating trend lines that represent long-run total marginal costs but are themselves unobservable. He also finds that during the time period of analysis, the random walk distribution for log-prices (i.e., the geometric Brownian motion for spot prices) is a much better approximation for coal and gas than oil. As suggested by Figure 4, the recent period (2000–2006) has been quite different.

One way to reconcile the findings in [12] and the properties we observed in the recent period described in this paper, is to "mix" mean reversion for the spot price towards a long-term value of oil prices driven by a geometric Brownian motion. The following three-state variable model also incorporates stochastic volatility.

$$dS_t = a\left(L_t - S_t\right) S_t \, dt + \sigma_t \, S_t \, dW_t^1,$$
$$dy_t = \alpha\left(b - y_t\right) \, dt + \eta\sqrt{y_t} \, d\,W_t^2, \qquad \text{where } y_t = \sigma_t^2,$$
$$dL_t = \mu L_t \, dt + \xi L_t \, dW_t^3,$$

where the Brownian motions are positively correlated. The positive correlation between W^1 and W^2 accounts for the "inverse leverage" effect that prevails for commodity prices (in contrast to the "leverage effect" observed in the equity markets), whereas the positive correlation between W^1 and W^3 translates the fact that news of depleted reserves will generate a rise in both spot and long-term oil prices.

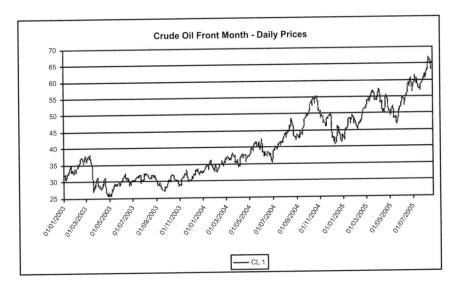

Fig. 4. NYMEX Crude Oil Front: month daily prices over the period January 2003–August 2005.

5 Conclusion

From the methodological standpoint, more work remains to be done in order to analyze in a unified setting (i) the mathematical properties of existence of an invariant measure and ergodicity for a stochastic process, and (ii) the mean-reversion behavior as it is intuitively perceived in the field of finance. From an economic standpoint, the modeling of oil and natural gas prices should incorporate the recent perception by market participants of the importance of reserves uncertainty and exhaustion of these reserves.

References

1. H. Bessembinder, J. Coughenour, P. Seguin, and M. Smoller. Mean reversion in equilibrium asset prices: Evidence from the futures term structure. *Journal of Finance*, 50:361–375, 1995.
2. F. Black and M. Scholes. The pricing of options and corporate liabilities. *Journal of Political Economy*, 81:637–659, 1973.
3. J.C. Cox, J.E. Ingersoll, and S.A. Ross. A theory of the term structure of interest rates. *Econometrica*, 53:385–408, 1985.
4. A. Eydeland and H. Geman. Pricing power derivatives. *Risk*, 71–73, September 1998.
5. J.P. Fouque, G. Papanicoleou, and K.R. Sircar. *Derivatives in Financial Markets with Stochastic Volatility*. Cambridge University Press, 2000.

6. H. Geman. *Commodities and Commodity Derivatives: Modeling and Pricing for Agriculturals, Metals and Energy*. Wiley, 2005.
7. R. Gibson and E.S. Schwartz. Stochastic convenience yield and the pricing of oil contingent claims. *Journal of Finance*, 45:959–976, 1990.
8. R.Z. Has'minskii. *Stochastic Stability of Differential Equations*. Sitjthoff & Noordhoff, 1980.
9. J.M. Keynes. *The Applied Theory of Money*. Macmillan, 1930.
10. R.H. Litzenberger and N. Rabinowitz. Backwardation in oil futures markets: Theory and empirical evidence. *Journal of Finance*, 50(5):1517–1545, 1995.
11. M. Musiela. Some issues on asset price modeling. Isaac Newton Institute Workshop on Mathematical Finance, 2005.
12. R. Pindyck. The dynamics of commodity spot and futures markets: A primer. *Energy Journal*, 22(3):1–29, 2001.
13. G. Pages. Sur quelques algorithmes recursifs pour les probabilites numériques. *ESAIM, Probability and Statistics*, 5:141–170, 2001.
14. E.S. Schwartz. The stochastic behavior of commodity prices: Implications for valuation and hedging. *Journal of Finance*, 52(3):923–973, 1997.
15. Simmons, M.R. *Twilight in the Desert: The Coming Saudi Shock and the World Economy*. John Wiley & Sons, 1995.

Part III

Credit Risk and Investments

Beyond Hazard Rates: A New Framework for Credit-Risk Modelling

Dorje C. Brody,[1] Lane P. Hughston,[2] and Andrea Macrina[2]

[1] Department of Mathematics
Imperial College
London SW7 2BZ, United Kingdom
`dorje@imperial.ac.uk`

[2] Department of Mathematics
King's College London
The Strand, London WC2R 2LS, United Kingdom
`lane.hughston@kcl.ac.uk`, `andrea.macrina@kcl.ac.uk`

Summary. A new approach to credit risk modelling is introduced that avoids the use of inaccessible stopping times. Default events are associated directly with the failure of obligors to make contractually agreed payments. Noisy information about impending cash flows is available to market participants. In this framework, the market filtration is modelled explicitly, and is assumed to be generated by one or more independent market information processes. Each such information process carries partial information about the values of the market factors that determine future cash flows. For each market factor, the rate at which true information is provided to market participants concerning the eventual value of the factor is a parameter of the model. Analytical expressions that can be readily used for simulation are presented for the price processes of defaultable bonds with stochastic recovery. Similar expressions can be formulated for other debt instruments, including multi-name products. An explicit formula is derived for the value of an option on a defaultable discount bond. It is shown that the value of such an option is an increasing function of the rate at which true information is provided about the terminal payoff of the bond. One notable feature of the framework is that it satisfies an overall dynamic consistency condition that makes it suitable as a basis for practical modelling situations where frequent recalibration may be necessary.

Key words: Credit risk; credit derivatives; incomplete information; information-based asset pricing; market filtration; Bayesian inference; Brownian bridge process.

1 Introduction and Summary

Models for credit risk management and, in particular, for the pricing of credit derivatives are usually classified into two types: structural and reduced-form.

For general overviews of these approaches, see, for example, Jeanblanc and Rutkowski [25], Hughston and Turnbull [18], Bielecki and Rutkowski [4], Duffie and Singleton [12], Schönbucher [35], Bielecki et al. [3], Lando [29], and Elizalde [13].

There is a divergence of opinion in the literature as to the relative merits of the structural and reduced-form methodologies. Both approaches have strengths, but there are also shortcomings. Structural models attempt to account at some level of detail for the events leading to default (see, e.g., Merton [33], Black and Cox [5], Leland and Toft [30], Hilberink and Rogers [17], Jarrow et al. [23], and Hull and White [19]). One general problem of the structural approach is that it is difficult in a structural model to deal systematically with the multiplicity of situations that can lead to default. For this reason, structural models are sometimes viewed as unsatisfactory as a basis for a practical modelling framework, particularly when multi-name products such as nth-to-default swaps and collateralised debt obligations (CDOs) are involved.

Reduced-form models are more commonly used in practice, on account of their tractability and because fewer assumptions are required about the nature of the debt obligations involved and the circumstances that might lead to default (see, e.g., Flesaker et al. [14], Jarrow and Turnbull [22], Duffie et al. [10], Jarrow et al. [20], Lando [28], Madan and Unal [31], [32], Duffie and Singleton [11], and Jarrow and Yu [24]). Most reduced-form models are based on the introduction of a random time of default, modelled as the time at which the integral of a random intensity process first hits a certain critical level, this level itself being an independent random variable. An unsatisfactory feature of such intensity models is that they do not adequately take into account the fact that defaults are typically associated directly with a failure in the delivery of a contractually agreed cash flow—for example, a missed coupon payment. Another drawback of the intensity approach is that it is not well adapted to the situation where one wants to model the rise and fall of credit spreads, which can in reality be due in part to changes in the level of investor confidence.

The purpose of this paper is to introduce a new class of reduced-form credit models in which these problems are addressed. The modelling framework that we develop is broadly in the spirit of the incomplete-information approaches of Kusuoka [27], Duffie and Lando [9], Cetin et al. [6], Gieseke [15], Gieseke and Goldberg [16], Jarrow and Protter [21], and others. In our approach, no attempt is made as such to bridge the gap between the structural and the intensity-based models; rather, by abandoning the need for the intensity-based approach, we are able to formulate a class of reduced-form models that exhibit a high degree of intuitively natural behaviour and analytic tractability. Our approach is to build an economic model for the information that market participants have about future cash flows.

For simplicity, we assume that the underlying default-free interest rate system is deterministic. The cash flows of the debt obligation—in the case of a coupon bond, the coupon payments and the principal repayment—are modelled by a collection of random variables, and default is identified as the event of the first such payment that fails to achieve the terms specified in the contract. We assume that partial information about each such cash flow is available at earlier times to market participants; however, the information of the actual value of the payout is obscured by a Gaussian noise process that is conditioned to vanish once the time of the required cash flow is reached. We proceed under these assumptions to derive an exact expression for the bond price process.

In the case of a defaultable discount bond admitting two possible payouts (either the full principal, or some partial recovery payment), we derive an exact expression for the value of an option on the bond. Remarkably, this turns out to be a formula of the Black–Scholes–Merton type. In particular, the parameter σ that governs the rate at which the true value of the impending cash flow is revealed to market participants, against the background of the obscuring noise process, turns out to play the role of a volatility parameter in the associated option pricing formula; this interpretation is reinforced with the observation that the option price is an increasing function of this parameter.

The structure of the paper is as follows. In Section 2, we introduce the notion of an 'information process' as a mechanism for modelling the market filtration. In the case of a credit-risky discount bond, the information process carries partial information about the terminal payoff, and it is assumed that the market filtration is generated by this process. The price process of the bond is obtained by taking the conditional expectation of the payout, and the properties of the resulting formula are analysed in the case of a binary payout. In Section 3, these results are extended to the situation of a defaultable discount bond with stochastic recovery, and a proof is given for the Markov property of the information process. In Section 4, we work out the dynamics of the price of a defaultable bond, and we show that the bond price process satisfies a diffusion equation. An explicit construction is presented for the Brownian motion that drives this process. This is given in Equation (20). We also work out an expression for the volatility of the bond price, given in Equation (28). In Section 5, we simulate the resulting price processes and show that for suitable values of the information flow-rate parameter introduced in Equation (4) the default of a credit-risky discount bond occurs in effect as a 'surprise.' In Section 6, we establish a decomposition formula similar to that obtained by Lando [28] in the case of a bond with partial recovery. In Section 7, we show that the framework has a general 'dynamical consistency' property. This has important implications for applications of the resulting models. In deriving these results we make use of special orthogonality properties of the Brownian bridge process (see, e.g., Yor [36; 37]). In Sections 8 and 9, we consider options on defaultable bonds and work out explicit formulae for the price processes of such options. In doing so, we introduce a novel change-of-measure technique that enables us to calculate explicitly various expectations involving Brownian bridge processes. This technique is of some interest in its own right. One of the consequences of the fact that the bond price process is a diffusion in this framework is that explicit formulae can be worked out for the delta-hedging of option positions. In Section 10, we extend the general theory to the situation where there are multiple cash flows. We look in particular at the case of a coupon bond and derive an explicit formula for the price process of such a bond. In Sections 11, 12 and 13, we consider complex debt instruments, and construct appropriate information-based valuation formulae for credit default swaps (CDSs) and portfolios of credit-risky instruments.

2 The Information-Based Approach

The object of this paper is to build an elementary modelling framework in which matters related to credit are brought to the forefront. Accordingly, we assume that the background default-free interest-rate system is deterministic: this assumption allows us to focus attention on credit-related issues; it also permits us to derive explicit expressions for certain types of credit derivative prices. The general philosophy is that we should try to sort out credit-related matters first, before attempting to incorporate a stochastic default-free system of interest rates.

   As a further simplifying feature we take the view that default events are directly linked with anomalous cash flows. Thus default is not something that happens 'in the abstract,' but rather is associated with the failure of an agreed cash flow to materialise at the required time. In this way we improve the theory by eradicating the superfluous use of inaccessible stopping times.

   Our theory is based on modelling the flow of partial information to market participants about impending debt payments. As usual, we model the financial markets with the specification of a probability space $(\Omega, \mathcal{F}, Q)$ with filtration $\{\mathcal{F}_t\}_{0 \leq t < \infty}$. The probability measure Q is understood to be the risk-neutral measure, and $\{\mathcal{F}_t\}$ is understood to be the market filtration. All asset-price processes and other information-providing processes accessible to market participants are adapted to $\{\mathcal{F}_t\}$. Contrary to the usual practice, we shall model $\{\mathcal{F}_t\}$ explicitly, rather than simply regarding it as being given.

   The real probability measure does not enter into the present investigation. We assume the absence of arbitrage and the existence of a pricing kernel (cf. Cochrane [8], and references cited therein). With these conditions the existence of a unique risk-neutral measure is ensured, even though the market may be incomplete. We assume further that the default-free discount-bond system, denoted $\{P_{tT}\}_{0 \leq t \leq T < \infty}$, can be written in the form $P_{tT} = P_{0T}/P_{0t}$, where the function $\{P_{0t}\}_{0 \leq t < \infty}$ is differentiable and strictly decreasing, and satisfies $0 < P_{0t} \leq 1$ and $\lim_{t \to \infty} P_{0t} = 0$. Under these assumptions it follows that if the integrable random variable H_T represents a cash flow occurring at T, then its value H_t at any earlier time t is given by

$$H_t = P_{tT} E[H_T | \mathcal{F}_t]. \tag{1}$$

   Now let us consider, as a first example, the case of a simple credit-risky discount bond that matures at time T to pay a principal of h_1 dollars, if there is no default. In the event of default, the bond pays h_0 dollars, where $h_0 < h_1$. When two such payoffs are possible we call the resulting structure a 'binary' discount bond. In the case given by $h_1 = 1$ and $h_0 = 0$ we call the resulting defaultable debt obligation a 'digital' bond.

Let us write p_0 for the probability that the bond will pay h_0, and p_1 for the probability that the bond will pay h_1. The probabilities are risk-neutral, and hence build in any risk adjustments required in expectations needed in order to deduce prices. Thus if we write B_{0T} for the price at time 0 of the credit-risky discount bond we have

$$B_{0T} = P_{0T}(p_0 h_0 + p_1 h_1).\tag{2}$$

It follows that, providing we know the market data B_{0T} and P_{0T}, we can infer the *a priori* probabilities p_0 and p_1:

$$p_0 = \frac{1}{h_1 - h_0}\left(h_1 - \frac{B_{0T}}{P_{0T}}\right), \qquad p_1 = \frac{1}{h_1 - h_0}\left(\frac{B_{0T}}{P_{0T}} - h_0\right).\tag{3}$$

Given this setup we proceed to model the bond-price process $\{B_{tT}\}_{0 \le t \le T}$. We suppose that the true value of H_T is not fully accessible until time T; that is, we assume H_T is $\mathcal{F}_T$-measurable, but not $\mathcal{F}_t$-measurable for $t < T$. We assume, nevertheless, that partial information regarding the value of the principal repayment H_T is available at earlier times. This information will in general be imperfect; one is looking into a crystal ball, so to speak, but the image is cloudy. Our model for such cloudy information is of a simple type that allows for analytic tractability. More precisely, we assume that the following $\{\mathcal{F}_t\}$-adapted process is accessible to market participants:

$$\xi_t = \sigma H_T t + \beta_{tT}.\tag{4}$$

We call $\{\xi_t\}$ a *market information process*. The process $\{\beta_{tT}\}_{0 \le t \le T}$ appearing here is a standard Brownian bridge on the time interval $[0, T]$. Thus $\{\beta_{tT}\}$ is a Gaussian process satisfying $\beta_{0T} = 0$ and $\beta_{TT} = 0$, such that $E[\beta_{tT}] = 0$ and $E[\beta_{sT}\beta_{tT}] = s(T-t)/T$ for all s, t satisfying $0 \le s \le t \le T$. We assume that $\{\beta_{tT}\}$ is independent of H_T, and thus represents pure noise. Market participants do not have direct access to $\{\beta_{tT}\}$; that is to say, $\{\beta_{tT}\}$ is not assumed to be adapted to $\{\mathcal{F}_t\}$. We can think of $\{\beta_{tT}\}$ as representing the rumour, speculation, misrepresentation, overreaction, and general disinformation often occurring, in one form or another, in connection with financial activity.

Clearly the choice (4) can be generalised to include a wider class of models enjoying similar qualitative features. The present analysis sticks with (4) for the sake of definiteness and tractability. Indeed, the choice of $\{\xi_t\}$ defined by (4) has many attractive features, and can be regarded as a convenient 'standard' model for an information process. The motivation for the use of a bridge process to represent the noise is intuitively as follows. We assume that initially all available market information is taken into account in the determination of the price; in the case of a credit-risky discount bond, the relevant information is embodied in the *a priori* probabilities. After the passage of time, new rumours and stories start circulating, and we model this by taking into account that the variance of the Brownian bridge steadily increases for the first half of its trajectory. Eventually, however, the variance falls to zero at the maturity of the bond, when the 'moment of truth' arrives.

The parameter σ in this model represents the rate at which the true value of H_T is 'revealed' as time progresses. Thus, if σ is low, then the value of H_T is effectively hidden until very near the maturity date of the bond; on the other hand, if σ is high, then we can think of H_T as being revealed quickly. The parameter σ has the units $\sigma \sim [\text{time}]^{-1/2}[\text{price}]^{-1}$. A rough measure for the timescale τ_D over which information is revealed is $\tau_D = 1/\sigma^2(h_1 - h_0)^2$. In particular, if $\tau_D \ll T$, then the value of H_T will be revealed rather early in the history of the bond, for example, after the passage of a few multiples of τ_D. In this situation, if default is 'destined' to occur, even though the initial value of the bond is high, then this will be signalled by a rapid decline in the bond price. On the other hand, if $\tau_D \gg T$, then H_T will only be revealed at the last minute, so to speak, and the default will come as a surprise. It is by virtue of this feature of the framework that the use of inaccessible stopping times can be avoided.

To make a closer inspection of the default timescale, we proceed as follows. For simplicity, we assume that the only market information available about H_T at times earlier than T comes from observations of $\{\xi_t\}$. Specifically, if we denote $\mathcal{F}_t^\xi$ the subalgebra of $\mathcal{F}_t$ generated by $\{\xi_s\}_{0 \le s \le t}$, then our simplifying assumption is that $\{\mathcal{F}_t\} = \{\mathcal{F}_t^\xi\}$. That is to say, we assume that the market filtration is generated by the information process.

Now we are in a position to determine the price-process $\{B_{tT}\}_{0 \le t \le T}$ for a credit-risky bond with payout H_T. In particular, the bond price is given by

$$B_{tT} = P_{tT} H_{tT}, \tag{5}$$

where H_{tT} denotes the conditional expectation of the bond payout:

$$H_{tT} = E\left[H_T \,\middle|\, \mathcal{F}_t^\xi\right]. \tag{6}$$

It turns out that this conditional expectation can be worked out explicitly. The result is given by the following expression:

$$H_{tT} = \frac{p_0 h_0 \exp\left[\frac{T}{T-t}\left(\sigma h_0 \xi_t - \frac{1}{2}\sigma^2 h_0^2 t\right)\right] + p_1 h_1 \exp\left[\frac{T}{T-t}\left(\sigma h_1 \xi_t - \frac{1}{2}\sigma^2 h_1^2 t\right)\right]}{p_0 \exp\left[\frac{T}{T-t}\left(\sigma h_0 \xi_t - \frac{1}{2}\sigma^2 h_0^2 t\right)\right] + p_1 \exp\left[\frac{T}{T-t}\left(\sigma h_1 \xi_t - \frac{1}{2}\sigma^2 h_1^2 t\right)\right]}. \tag{7}$$

Thus we see that there exists a function $H(x, y)$ of two variables such that $H_{tT} = H(\xi_t, t)$, and as a consequence that the bond price can be expressed as a function of the market information.

Because $\{\xi_t\}$ is given by a combination of the random bond payout and an independent Brownian bridge, it is straightforward to simulate trajectories for $\{B_{tT}\}$. The bond price trajectories rise and fall randomly in line with the fluctuating information about the likely final payoff: in this respect, the resulting model is more satisfactory than an intensity-based model, where the only basis for shifts in credit spreads is through a random change in the default intensity.

The details of the derivation of the formula presented above are given in the next section. First, however, let us verify that the expression in (7) converges, as t approaches T, to the 'actual' value of H_T. The proof is as follows. Suppose that the actual value of the payout is $H_T = h_0$. In that case the information process takes the form

$$\xi_t = \sigma h_0 t + \beta_{tT} \tag{8}$$

Inserting this expression into (7), we divide the numerator and the denominator by the coefficient of $p_0 h_0$. As a consequence we obtain

$$H_{tT} = \frac{p_0 h_0 + p_1 h_1 \exp\left[-\frac{T}{T-t}\left(\frac{1}{2}\sigma^2(h_1 - h_0)^2 t - \sigma(h_1 - h_0)\beta_{tT}\right)\right]}{p_0 + p_1 \exp\left[-\frac{T}{T-t}\left(\frac{1}{2}\sigma^2(h_1 - h_0)^2 t - \sigma(h_1 - h_0)\beta_{tT}\right)\right]}. \tag{9}$$

We observe that as t approaches T the bond price converges to h_0, as required. A similar argument shows that if $H_T = h_1$, then the bond price converges to h_1. We note, in line with our heuristic arguments concerning the characteristic timescale τ_D, that the parameter $\sigma^2(h_1 - h_0)^2$ governs the speed at which H_{tT} converges to its terminal value. In particular, if the *a priori* probability of no default is high (say, $p_1 \approx 1$), and if σ is very small, and if in fact $H_T = h_0$, then it will only be when t is near T that serious decay in the bond price will set in.

3 Defaultable Discount Bond Price Processes

Let us now consider the more general situation where the discount bond pays the value $H_T = h_i$ $(i = 0, 1, \ldots, n)$ with *a priori* probability $Q(H_T = h_i) = p_i$. For convenience we assume $h_n > h_{n-1} > \cdots > h_1 > h_0$. The case $n = 1$ corresponds to the binary bond we have just considered. In the general situation we think of $H_T = h_n$ as the case of no default, and the other cases as various possible degrees of recovery.

Although for simplicity we work with a discrete payout spectrum for H_T, the continuous case can be formulated analogously. In that situation we assign a fixed *a priori* probability p_1 to the case of no default, and a continuous probability distribution function $p_0(x) = Q(H_T < x)$ for values of x less than or equal to h, satisfying $p_1 + p_0(h) = 1$.

Now defining the information process $\{\xi_t\}$ as before, we want to find the conditional expectation (6). It follows from the Markovian property of $\{\xi_t\}$, which is established below, that to compute (6), it suffices to take the conditional expectation of H_T with respect to the subalgebra generated by the random variable ξ_t alone. Therefore, writing $H_{tT} = E[H_T|\xi_t]$ we have

$$H_{tT} = \sum_i h_i \pi_{it}, \tag{10}$$

where $\pi_{it} = Q(H_T = h_i|\xi_t)$ is the conditional probability that the credit-risky bond pays out h_i. That is to say, $\pi_{it} = E[\mathbf{1}_{\{H_T=h_i\}}|\xi_t]$.

To show that the information process satisfies the Markov property, we need to verify that

$$Q(\xi_t \le x|\mathcal{F}_s^\xi) = Q(\xi_t \le x|\xi_s) \tag{11}$$

for all $x \in \mathbb{R}$ and all s, t such that $0 \le s \le t \le T$. It suffices to show that

$$Q\left(\xi_t \le x|\xi_s, \xi_{s_1}, \xi_{s_2}, \ldots, \xi_{s_k}\right) = Q\left(\xi_t \le x|\xi_s\right) \tag{12}$$

for any collection of times $t, s, s_1, s_2, \ldots, s_k$ such that $T \ge t > s > s_1 > s_2 > \cdots > s_k > 0$. First, we remark that for any times t, s, s_1 satisfying $t > s > s_1$ the Gaussian random variables β_{tT} and $\beta_{sT}/s - \beta_{s_1T}/s_1$ have vanishing covariance, and thus are independent. More generally, for $s > s_1 > s_2 > s_3$ the random variables $\beta_{sT}/s - \beta_{s_1T}/s_1$ and $\beta_{s_2T}/s_2 - \beta_{s_3T}/s_3$ are independent. Next, we note that $\xi_s/s - \xi_{s_1}/s_1 = \beta_{sT}/s - \beta_{s_1T}/s_1$. It follows that

$$
\begin{aligned}
&Q\left(\xi_t \le x|\xi_s, \xi_{s_1}, \xi_{s_2}, \ldots, \xi_{s_k}\right) \\
&= Q\left(\xi_t \le x\Big|\xi_s, \frac{\xi_s}{s} - \frac{\xi_{s_1}}{s_1}, \frac{\xi_{s_1}}{s_1} - \frac{\xi_{s_2}}{s_2}, \ldots, \frac{\xi_{s_{k-1}}}{s_{k-1}} - \frac{\xi_{s_k}}{s_k}\right) \\
&= Q\left(\xi_t \le x\Big|\xi_s, \frac{\beta_{sT}}{s} - \frac{\beta_{s_1T}}{s_1}, \frac{\beta_{s_1T}}{s_1} - \frac{\beta_{s_2T}}{s_2}, \ldots, \frac{\beta_{s_{k-1}T}}{s_{k-1}} - \frac{\beta_{s_kT}}{s_k}\right). \tag{13}
\end{aligned}
$$

However, because ξ_s and ξ_t are independent of the remaining random variables $\beta_{sT}/s - \beta_{s_1T}/s_1$, $\beta_{s_1T}/s_1 - \beta_{s_2T}/s_2$, $\ldots$, $\beta_{s_{k-1}T}/s_{k-1} - \beta_{s_kT}/s_k$, the desired Markov property follows immediately.

Next we observe that the *a priori* probability p_i and the *a posteriori* probability π_{it} are related by the following version of the Bayes formula:

$$Q(H_T = h_i|\xi_t) = \frac{p_i\rho(\xi_t|H_T = h_i)}{\sum_i p_i\rho(\xi_t|H_T = h_i)}. \tag{14}$$

Here the conditional density function $\rho(\xi|H_T = h_i)$, $\xi \in \mathbb{R}$, for the random variable ξ_t is defined by the relation

$$Q\left(\xi_t \le x|H_T = h_i\right) = \int_{-\infty}^x \rho(\xi|H_T = h_i)\,d\xi, \tag{15}$$

and is given explicitly by

$$\rho(\xi|H_T = h_i) = \frac{1}{\sqrt{2\pi t(T-t)/T}} \exp\left(-\frac{1}{2}\frac{(\xi - \sigma h_i t)^2}{t(T-t)/T}\right). \tag{16}$$

This expression can be deduced from the fact that conditional on $H_T = h_i$ the random variable ξ_t is normally distributed with mean $\sigma h_i t$ and variance $t(T-t)/T$. As a consequence of (14) and (16), we see that

$$\pi_{it} = \frac{p_i \exp\left[\frac{T}{T-t}(\sigma h_i \xi_t - \frac{1}{2}\sigma^2 h_i^2 t)\right]}{\sum_i p_i \exp\left[\frac{T}{T-t}(\sigma h_i \xi_t - \frac{1}{2}\sigma^2 h_i^2 t)\right]}. \tag{17}$$

It follows then, on account of (10), that

$$H_{tT} = \frac{\sum_i p_i h_i \exp\left[\frac{T}{T-t}\left(\sigma h_i \xi_t - \frac{1}{2}\sigma^2 h_i^2 t\right)\right]}{\sum_i p_i \exp\left[\frac{T}{T-t}\left(\sigma h_i \xi_t - \frac{1}{2}\sigma^2 h_i^2 t\right)\right]}. \tag{18}$$

This is the desired expression for the conditional expectation of the bond payoff. In particular, for the binary case $i = 0, 1$ we recover formula (7). The discount-bond price B_{tT} is then given by (5), with H_{tT} defined as in (18).

4 Defaultable Discount Bond Dynamics

In this section we analyse the dynamics of the defaultable bond price process $\{B_{tT}\}$ determined in Section 3. The key relation we need for working out the dynamics of the bond price is that the conditional probability $\{\pi_{it}\}$ satisfies a diffusion equation of the form

$$d\pi_{it} = \frac{\sigma T}{T-t}(h_i - H_{tT})\pi_{it}\,dW_t. \tag{19}$$

In particular, we can show that the process $\{W_t\}_{0 \le t < T}$ arising here, defined by the expression

$$W_t = \xi_t + \int_0^t \frac{1}{T-s}\xi_s\,ds - \sigma T \int_0^t \frac{1}{T-s}H_{sT}\,ds, \tag{20}$$

is an $\{\mathcal{F}_t\}$-Brownian motion. The fact that $\{\pi_{it}\}$ satisfies (19) with $\{W_t\}$ defined as in (20) can be obtained directly from (17) by an application of Itô's lemma. We need to use the relation $(d\xi_t)^2 = dt$, which follows from the relation $(d\beta_{tT})^2 = dt$. The fact that $\{W_t\}$ is an $\{\mathcal{F}_t\}$-Brownian motion can then be verified by showing that $\{W_t\}$ is an $\{\mathcal{F}_t\}$-martingale and that $(dW_t)^2 = dt$. We proceed as follows.

To prove that $\{W_t\}$ is an $\{\mathcal{F}_t\}$-martingale, we need to show that $E[W_u|\mathcal{F}_t] = W_t$ for $0 \le t \le u < T$. First we note that it follows from (20) and the Markov property of $\{\xi_t\}$ that

$$E[W_u|\mathcal{F}_t] = W_t + E\left[(\xi_u - \xi_t)|\xi_t\right] + E\left[\int_t^u \frac{1}{T-s}\xi_s\,ds\bigg|\xi_t\right]$$

$$- \sigma T E\left[\int_t^u \frac{1}{T-s}H_{sT}\,ds\bigg|\xi_t\right]. \tag{21}$$

Formula (21) can be simplified if we recall that $H_{sT} = E[H_T|\xi_s]$ and use the tower property in the last term on the right. Inserting the definition (4) into the second and third terms on the right in (21), we then have:

$$E[W_u|\mathcal{F}_t] = W_t + E[\sigma H_T u + \beta_{uT}|\xi_t] - E[\sigma H_T t + \beta_{tT}|\xi_t]$$

$$+ \sigma E[H_T|\xi_t]\int_t^u \frac{s}{T-s}\,ds + E\left[\int_t^u \frac{1}{T-s}\beta_{sT}\,ds\bigg|\xi_t\right]$$

$$- \sigma E[H_T|\xi_t]\int_t^u \frac{T}{T-s}\,ds. \tag{22}$$

It follows immediately that all of the terms involving the random variable H_T cancel each other in (22). This leads us to the following relation:

$$E[W_u|\mathcal{F}_t] = W_t + E[\beta_{uT}|\xi_t] - E[\beta_{tT}|\xi_t] + \int_t^u \frac{1}{T-s}E[\beta_{sT}|\xi_t]\,ds. \tag{23}$$

Next we use the tower property and the independence of $\{\beta_{tT}\}$ and H_T to deduce that

$$E[\beta_{uT}|\xi_t] = E[E[\beta_{uT}|H_T,\beta_{tT}]|\xi_t] = E[E\,[\beta_{uT}|\beta_{tT}]|\xi_t]. \tag{24}$$

To calculate the inner expectation $E[\beta_{uT}|\beta_{tT}]$, we use the fact that the random variable $\beta_{uT}/(T-u) - \beta_{tT}/(T-t)$ is independent of the random variable β_{tT}. This can be established by calculating their covariance, and using the relation $E[\beta_{uT}\beta_{tT}] = t(T-u)/T$. We conclude after a short calculation that

$$E[\beta_{uT}|\beta_{tT}] = \frac{T-u}{T-t}\beta_{tT}. \tag{25}$$

Inserting this result into (24), we obtain the following formula:

$$E[\beta_{uT}|\xi_t] = \frac{T-u}{T-t}E[\beta_{tT}|\xi_t]. \tag{26}$$

Applying this formula to the second and fourth terms on the right side of (23), we immediately deduce that $E[W_u|\mathcal{F}_t] = W_t$. That establishes that $\{W_t\}$ is an $\{\mathcal{F}_t\}$-martingale. Now we need to show that $(dW_t)^2 = dt$. This follows if we insert (4) into the definition of $\{W_t\}$ above and use again the fact that $(d\beta_{tT})^2 = dt$. Hence, by virtue of Lévy's criterion (see, e.g., Karatzas and Shreve [26]), we conclude that $\{W_t\}$ is an $\{\mathcal{F}_t\}$-Brownian motion.

The Brownian motion $\{W_t\}$, the existence of which we have just established, can be regarded as part of the information accessible to market participants. Unlike β_{tT}, the value of W_t contains 'real' information relevant to the bond payoff. It follows from (10) and (19) that for the discount bond dynamics, we have

$$dB_{tT} = r_t B_{tT}\, dt + \Sigma_{tT}\, dW_t. \tag{27}$$

Here $r_t = -\partial \ln P_{0t}/\partial t$ is the short rate, and the absolute bond volatility Σ_{tT} is given by

$$\Sigma_{tT} = \frac{\sigma T}{T-t} P_{tT} V_{tT}, \tag{28}$$

where V_{tT} is the conditional variance of the terminal payoff H_T, defined by

$$V_{tT} = \sum_i (h_i - H_{tT})^2 \pi_{it}. \tag{29}$$

We observe that as the maturity date is approached the absolute discount bond volatility will be high unless the conditional probability has most of its mass concentrated around the 'true' outcome; this ensures that the correct level can be eventually reached.

It follows as a consequence of (20) that $\{\xi_t\}$ satisfies the following stochastic differential equation:

$$d\xi_t = \frac{1}{T-t}\left(\sigma T H(\xi_t, t) - \xi_t\right) dt + dW_t. \tag{30}$$

We see that $\{\xi_t\}$ is a diffusion process; and because $H(\xi_t, t)$ is monotonic in its dependence on ξ_t, we deduce that $\{B_{tT}\}$ is also a diffusion process. To establish that B_{tT} is increasing in ξ_t, we note that $P_{tT} H'(\xi_t, t) = \Sigma_{tT}$, where $H'(\xi, t) = \partial H(\xi, t)/\partial \xi$. It is interesting to observe that, in principle, instead of 'deducing' the dynamics of $\{B_{tT}\}$ from the information-based arguments of the previous sections, we might have simply 'proposed' on an *ad hoc* basis the one-factor diffusion process (30), noting that it leads to the correct default dynamics. This reasoning shows that our framework can be viewed, if desired, as leading to purely 'classical' financial models, based on observable price processes.

5 Simulation of Bond Price Processes

The present framework allows for a highly efficient simulation methodology for the dynamics of defaultable bonds. In the case of a defaultable discount bond, all we need to do is to simulate the dynamics of $\{\xi_t\}$. For each run of the simulation, we choose at random a value for H_T (in accordance with the *a priori* probabilities), and a sample path for the Brownian bridge. That is to say, each simulation corresponds to a choice of $\omega \in \Omega$, and for each such choice we simulate the path $\xi_t(\omega) = \sigma t H_T(\omega) + \beta_{tT}(\omega)$ for $t \in [0, T]$. One way to simulate a Brownian bridge is to write $\beta_{tT} = B_t - (t/T)B_T$, where $\{B_t\}$ is a standard Brownian motion. It is straightforward to verify that if $\{\beta_{tT}\}$ is defined in this way then it has the correct mean and covariance. Because the bond price at t is expressed directly as a function of the random variable ξ_t, this means that pathwise simulation of the bond price trajectory is feasible for any number of recovery levels.

In Figure 1 we present some sample trajectories of the defaultable bond price process for various values of σ. These illustrations are fascinating inasmuch as they show that for small values of σ the default comes almost as a 'surprise' near the maturity date of the bond; whereas for large values of σ the default, if it occurs, effectively takes place early in the life of bond.

6 Digital Bonds and Binary Bonds with Partial Recovery

It is interesting to ask whether in the case of a binary bond with partial recovery, with possible payoffs $\{h_0, h_1\}$, the bond-price process admits the representation

$$B_{tT} = P_{tT}h_0 + D_{tT}(h_1 - h_0). \qquad (31)$$

Here D_{tT} denotes the value of a digital credit-risky bond that at maturity pays a unit of currency with probability p_1 and zero with probability $p_0 = 1 - p_1$. Thus h_0 is the amount that is guaranteed, whereas $h_1 - h_0$ is the part that is 'at risk.' It is well known that such a relation can be deduced in intensity-based models. The problem now is to find a process $\{D_{tT}\}$ consistent with our scheme such that (31) holds. It turns out that this can be achieved.

Suppose we consider a digital payoff structure $D_T \in \{0, 1\}$ for which the information-flow parameter σ is replaced by $\bar{\sigma} = \sigma(h_1 - h_0)$. In other words, in establishing the appropriate dynamics for $\{D_{tT}\}$ we 'renormalise' σ by replacing it with $\bar{\sigma}$. The information available to market participants in the case of the digital bond is represented by the process $\{\bar{\xi}_t\}$ defined by

$$\bar{\xi}_t = \bar{\sigma} D_T t + \beta_{tT}. \qquad (32)$$

It follows from (18) that the corresponding digital bond price is given by

$$D_{tT} = P_{tT} \frac{p_1 \exp\left[\frac{T}{T-t}\left(\bar{\sigma}\bar{\xi}_t - \frac{1}{2}\bar{\sigma}^2 t\right)\right]}{p_0 + p_1 \exp\left[\frac{T}{T-t}\left(\bar{\sigma}\bar{\xi}_t - \frac{1}{2}\bar{\sigma}^2 t\right)\right]}. \qquad (33)$$

A calculation making use of (7) then allows us to confirm that (31) holds, where D_{tT} is given by (33). Thus even though *prima facie* the general binary bond process (7) does not appear to admit a decomposition of the form (31), we see that it does, once a suitably renormalised value for the market information flow-rate parameter has been introduced.

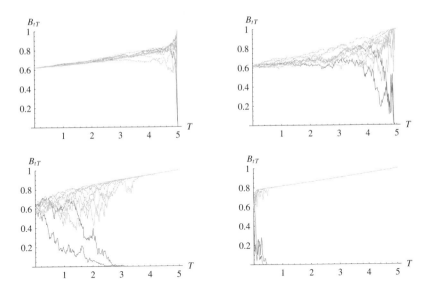

Fig. 1. Bond price processes for various information flow rates. The parameter σ governs the rate at which information is released to market participants concerning the payout of a defaultable discount bond. Four values of σ are illustrated, given by .04, .2, 1, and 5. The bond has a maturity of five years, and the default-free interest-rate system has a constant short rate given by $r = 5\%$. The *a priori* probability of default is taken to be 20%. For low values of σ, collapse of the bond price, if it occurs, takes place only 'at the last minute.'

More generally, if the bond has a number of recovery levels, so the random variable H_T can take the values $\{h_0, h_1, \ldots, h_n\}$, then h_0 can be regarded as the 'risk-free' component of the bond. The bond price process admits an additive decomposition into two parts, namely, a default-free discount bond that pays h_0, and a credit-risky discount bond that pays $\bar{h}_i = h_i - h_0$ with *a priori* probability p_i. This decomposition is given by

$$B_{tT} = P_{tT}h_0 + P_{tT} \frac{\sum\limits_{i=1}^{n} p_i \bar{h}_i \exp\left[\frac{T}{T-t}\left(\sigma \bar{h}_i \bar{\xi}_t - \frac{1}{2}\sigma^2 \bar{h}_i^2 t\right)\right]}{p_0 + \sum\limits_{i=1}^{n} p_i \exp\left[\frac{T}{T-t}\left(\sigma \bar{h}_i \bar{\xi}_t - \frac{1}{2}\sigma^2 \bar{h}_i^2 t\right)\right]}, \tag{34}$$

where $\bar{\xi}_t = \sigma(H_T - h_0)t + \beta_{tT}$ is the resulting price-shifted information process. Note that $\bar{h}_0 = 0$, which makes the summation in the numerator begin from $i = 1$; thus the second term in the right-hand side of (34) is the multiple-recovery-level analogue of formula (33) above.

7 Dynamic Consistency and Model Calibration

The technique of renormalising the information flow rate has another useful application. It turns out that the information-based framework exhibits a property that might appropriately be called 'dynamic consistency.' Loosely speaking, the question is: if the information process is given as described, but then we reinitialise the model at some specified intermediate time, is it still the case that the dynamics of the model moving forward from that intermediate time can be consistently represented by an information process?

To answer this question we proceed as follows. First we define a standard Brownian bridge over the interval $[t, T]$ to be a Gaussian process $\{\gamma_{uT}\}_{t \leq u \leq T}$ satisfying $\gamma_{tT} = 0$, $\gamma_{TT} = 0$, $E[\gamma_{uT}] = 0$ for all $u \in [t, T]$, and $E[\gamma_{uT}\gamma_{vT}] = (u - t)(T - v)/(T - t)$ for all u, v such that $t \leq u \leq v \leq T$. We make note of the following observation.

Let $\{\beta_{tT}\}_{0 \leq t \leq T}$ be a standard Brownian bridge over the interval $[0, T]$, and define the process $\{\gamma_{uT}\}_{t \leq u \leq T}$ by

$$\gamma_{uT} = \beta_{uT} - \frac{T - u}{T - t} \beta_{tT}. \qquad (35)$$

Then $\{\gamma_{uT}\}_{t \leq u \leq T}$ is a standard Brownian bridge over the interval $[t, T]$, and is independent of $\{\beta_{sT}\}_{0 \leq s \leq t}$.

The result is easily proved by use of the covariance relation $E[\beta_{tT}\beta_{uT}] = t(T - u)/T$. We need to recall also that a necessary and sufficient condition for a pair of Gaussian random variables to be independent is that their covariance should vanish. Now let the information process $\{\xi_s\}_{0 \leq s \leq T}$ be given, and fix an intermediate time $t \in (0, T)$. Then for all $u \in [t, T]$, let us define a process $\{\eta_u\}$ by

$$\eta_u = \xi_u - \frac{T - u}{T - t} \xi_t. \qquad (36)$$

We claim that $\{\eta_u\}$ is an information process over the time interval $[t, T]$. In fact, a short calculation establishes that

$$\eta_u = \tilde{\sigma} H_T(u - t) + \gamma_{uT}, \qquad (37)$$

where $\{\gamma_{uT}\}_{t \leq u \leq T}$ is a standard Brownian bridge over the interval $[t, T]$, independent of H_T, and the new information flow rate is given by $\tilde{\sigma} = \sigma T/(T - t)$. The interpretation of these results is as follows. The 'original' information process proceeds from time 0 up to time t. At that time we can recalibrate the model by taking note of the value of the random variable ξ_t, and introducing the reinitialised information process $\{\eta_u\}$. The new information process depends on H_T; but because the value of ξ_t is supplied, the *a priori* probability distribution for H_T now changes to the appropriate *a posteriori* distribution consistent with the information gained from the knowledge of ξ_t at time t.

These interpretive remarks can be put into a more precise form as follows. Let $0 \leq t \leq u < T$. What we want is a formula for the conditional probability π_{iu} expressed in terms of the information η_u and the 'new' a priori probability π_{it}. Such a formula exists, and is given as follows:

$$\pi_{iu} = \frac{\pi_{it}\exp\left[\frac{T-t}{T-u}\left(\tilde{\sigma}h_i\eta_u - \frac{1}{2}\tilde{\sigma}^2 h_i^2(u-t)\right)\right]}{\sum_i \pi_{it}\exp\left[\frac{T-t}{T-u}\left(\tilde{\sigma}h_i\eta_u - \frac{1}{2}\tilde{\sigma}^2 h_i^2(u-t)\right)\right]}. \tag{38}$$

This remarkable relation can be verified by substituting the given expressions for π_{it}, η_u, and $\tilde{\sigma}$ into the right-hand side of (38). But (38) has the same structure as the original formula (17) for π_{it}, and thus we see that the model exhibits manifest dynamic consistency.

8 Options on Credit-Risky Bonds

We now turn to consider the pricing of options on credit-risky bonds. As we demonstrate shortly, in the case of a binary bond there is an exact solution for the valuation of European-style vanilla options. The resulting expression for the option price exhibits a structure that is strikingly analogous to that of the Black–Scholes–Merton option pricing formula.

We consider the value at time 0 of an option that is exercisable at a fixed time $t > 0$ on a credit-risky discount bond that matures at time $T > t$. The value C_0 of a call option is

$$C_0 = P_{0t}E\left[(B_{tT} - K)^+\right], \tag{39}$$

where K is the strike price. Inserting formula (5) for B_{tT} into the valuation formula (39) for the option, we obtain

$$
\begin{aligned}
C_0 &= P_{0t}\, E\left[(P_{tT}H_{tT} - K)^+\right] \\
&= P_{0t}\, E\left[\left(\sum_{i=0}^{n} P_{tT}\pi_{it}h_i - K\right)^+\right] \\
&= P_{0t}\, E\left[\left(\frac{1}{\Phi_t}\sum_{i=0}^{n} P_{tT}p_{it}h_i - K\right)^+\right] \\
&= P_{0t}\, E\left[\frac{1}{\Phi_t}\left(\sum_{i=0}^{n}\left(P_{tT}h_i - K\right)p_{it}\right)^+\right].
\end{aligned}
\tag{40}
$$

Here the random variables p_{it}, $i = 0, 1, \ldots, n$, are the 'unnormalised' conditional probabilities, defined by

$$p_{it} = p_i\exp\left[\frac{T}{T-t}\left(\sigma h_i\xi_t - \frac{1}{2}\sigma^2 h_i^2 t\right)\right]. \tag{41}$$

Then $\pi_{it} = p_{it}/\Phi_t$ where $\Phi_t = \sum_i p_{it}$, or, more explicitly,

$$\Phi_t = \sum_{i=0}^{n} p_i \exp\left[\frac{T}{T-t}\left(\sigma h_i \xi_t - \tfrac{1}{2}\sigma^2 h_i^2 t\right)\right]. \tag{42}$$

Our plan now is to use the factor $1/\Phi_t$ appearing in (40) to make a change of probability measure on $(\Omega, \mathcal{F}_t)$. To this end, we fix a time horizon u at or beyond the option expiration but before the bond maturity, so $t \leq u < T$. We define a process $\{\Phi_t\}_{0 \leq t \leq u}$ by use of the expression (42), where now we let t vary in the range $[0, u]$. It is a straightforward exercise in Itô calculus, making use of (30), to verify that

$$d\Phi_t = \sigma^2 \left(\frac{T}{T-t}\right)^2 H_{tT}^2 \Phi_t \, dt + \sigma \frac{T}{T-t} H_{tT} \Phi_t \, dW_t. \tag{43}$$

It follows then that

$$d\Phi_t^{-1} = -\sigma \frac{T}{T-t} H_{tT} \Phi_t^{-1} dW_t, \tag{44}$$

and hence that

$$\Phi_t^{-1} = \exp\left(-\sigma \int_0^t \frac{T}{T-s} H_{sT} \, dW_s - \tfrac{1}{2}\sigma^2 \int_0^t \frac{T^2}{(T-s)^2} H_{sT}^2 \, ds\right). \tag{45}$$

Because $\{H_{sT}\}$ is bounded, and $s \leq u < T$, we see that the process $\{\Phi_s^{-1}\}_{0 \leq s \leq u}$ is a martingale. In particular, because $\Phi_0 = 1$, we deduce that $E^Q[\Phi_t^{-1}] = 1$, where t is the option maturity date, and hence that the factor $1/\Phi_t$ in (40) can be used to effect a change of measure on $(\Omega, \mathcal{F}_t)$. Writing B for the new probability measure thus defined, we have

$$C_0 = P_{0t} \, E^B\left[\left(\sum_{i=0}^{n}\left(P_{tT} h_i - K\right) p_{it}\right)^+\right]. \tag{46}$$

We call B the 'bridge' measure because it has the special property that it makes $\{\xi_s\}_{0 \leq s \leq t}$ a B-Gaussian process with mean zero and covariance $E^B[\xi_r \xi_s] = r(T-s)/T$ for $0 \leq r \leq s \leq t$. In other words, with respect to the measure B, and over the interval $[0, t]$, the information process has the law of a standard Brownian bridge over the interval $[0, T]$. Armed with this fact, we proceed to calculate the expectation in (46).

The proof that $\{\xi_s\}_{0 \leq s \leq t}$ has the claimed properties under the measure B is as follows. For convenience we introduce a process $\{W_t^*\}_{0 \leq t \leq u}$ which we define as the following Brownian motion with drift in the Q-measure:

$$W_t^* = W_t + \sigma \int_0^t \frac{T}{T-s} H_{sT} \, ds. \tag{47}$$

It is straightforward to check that on $(\Omega, \mathcal{F}_t)$ the process $\{W_t^*\}_{0 \leq t \leq u}$ is a Brownian motion with respect to the measure defined by use of the density martingale $\{\Phi_t^{-1}\}_{0 \leq t \leq u}$ given by (45). It then follows from the definition of $\{W_t\}$, given in (20), that

$$W_t^* = \xi_t + \int_0^t \frac{1}{T-s} \xi_s \, ds. \tag{48}$$

Taking the stochastic differential of each side of this relation, we deduce that

$$d\xi_t = -\frac{1}{T-t} \xi_t \, dt + dW_t^*. \tag{49}$$

We note, however, that (49) is the stochastic differential equation satisfied by a Brownian bridge (see, e.g., Karatzas and Shreve [26], Yor [37], and Protter [34]) over the interval $[0, T]$. Thus we see that in the measure B defined on $(\Omega, \mathcal{F}_t)$, the process $\{\xi_s\}_{0 \leq s \leq t}$ has the properties of a standard Brownian bridge over $[0, T]$, albeit restricted to the interval $[0, t]$.

For the transformation back from B to Q on $(\Omega, \mathcal{F}_t)$, the appropriate density martingale $\{\Phi_t\}_{0 \leq t \leq u}$ with respect to B is given by

$$\Phi_t = \exp\left(\sigma \int_0^t \frac{T}{T-s} H_{sT} \, dW_s^* - \tfrac{1}{2}\sigma^2 \int_0^t \frac{T^2}{(T-s)^2} H_{sT}^2 \, ds\right). \tag{50}$$

The crucial point that follows from this analysis is that the random variable ξ_t is B-Gaussian. In the case of a binary discount bond, therefore, the relevant expectation for determining the option price can be carried out by standard techniques, and we are led to a formula of the Black–Scholes–Merton type. In particular, for a binary bond, (46) reads

$$C_0 = P_{0t} E^B \left[\left((P_{tT} h_1 - K) p_{1t} + (P_{tT} h_0 - K) p_{0t} \right)^+ \right], \tag{51}$$

where p_{0t} and p_{1t} are given by

$$p_{0t} = p_0 \exp\left[\tfrac{T}{T-t} \left(\sigma h_0 \xi_t - \tfrac{1}{2}\sigma^2 h_0^2 t\right)\right],$$
$$p_{1t} = p_1 \exp\left[\tfrac{T}{T-t} \left(\sigma h_1 \xi_t - \tfrac{1}{2}\sigma^2 h_1^2 t\right)\right]. \tag{52}$$

To compute the value of (51), there are essentially three different cases that have to be considered: (i) $P_{tT} h_1 > P_{tT} h_0 > K$, (ii) $K > P_{tT} h_1 > P_{tT} h_0$, and (iii) $P_{tT} h_1 > K > P_{tT} h_0$. In case (i) the option is certain to expire in the money. Thus, making use of the fact that ξ_t is B-Gaussian with mean zero and variance $t(T-t)/T$, we see that $E^B[p_{it}] = p_i$; hence in case (i) we have $C_0 = B_{0T} - P_{0t} K$. In case (ii) the option expires out of the money, and thus $C_0 = 0$. In case (iii) the option can expire in or out of the money, and there is a 'critical' value of ξ_t above which the argument of (51) is positive. This is obtained by setting the argument of (51) to zero and solving for ξ_t. Writing $\bar{\xi}_t$ for the critical value, we find that $\bar{\xi}_t$ is determined by the relation

$$\frac{T}{T-t}\sigma(h_1 - h_0)\bar{\xi}_t = \ln\left[\frac{p_0(P_{tT}h_0 - K)}{p_1(K - P_{tT}h_1)}\right] + \tfrac{1}{2}\sigma^2(h_1^2 - h_0^2)\tau, \qquad (53)$$

where $\tau = tT/(T - t)$. Next we note that inasmuch as ξ_t is B-Gaussian with mean zero and variance $t(T - t)/T$, for the purpose of computing the expectation in (51) we can set $\xi_t = Z\sqrt{t(T - t)/T}$, where Z is B-Gaussian with zero mean and unit variance. Then writing $\bar{Z}$ for the corresponding critical value of Z, we obtain

$$\bar{Z} = \frac{\ln\left[\frac{p_0(P_{tT}h_0 - K)}{p_1(K - P_{tT}h_1)}\right] + \tfrac{1}{2}\sigma^2(h_1^2 - h_0^2)\tau}{\sigma\sqrt{\tau}(h_1 - h_0)}. \qquad (54)$$

With this expression at hand, we can work out the expectation in (51). We are thus led to the following option pricing formula:

$$C_0 = P_{0t}\left[p_1(P_{tT}h_1 - K)N(d^+) - p_0(K - P_{tT}h_0)N(d^-)\right]. \qquad (55)$$

Here d^+ and d^- are defined by

$$d^{\pm} = \frac{\ln\left[\frac{p_1(P_{tT}h_1 - K)}{p_0(K - P_{tT}h_0)}\right] \pm \tfrac{1}{2}\sigma^2(h_1 - h_0)^2\tau}{\sigma\sqrt{\tau}(h_1 - h_0)}. \qquad (56)$$

A short calculation shows that the corresponding option delta, defined by $\Delta = \partial C_0/\partial B_{0T}$, is given by

$$\Delta = \frac{(P_{tT}h_1 - K)N(d^+) + (K - P_{tT}h_0)N(d^-)}{P_{tT}(h_1 - h_0)}. \qquad (57)$$

This can be verified by using (3) to determine the dependency of the option price C_0 on the initial bond price B_{0T}.

It is interesting to note that the parameter σ plays a role like that of the volatility parameter in the Black–Scholes–Merton model. The more rapidly information is 'leaked out' about the true value of the bond repayment, the higher the volatility. It is straightforward to verify that the option price has a positive vega, that is, C_0 is an increasing function of σ. This means that we can use bond option prices (or, equivalently, caps and floors) to back out an implied value for σ, and hence to calibrate the model. Writing $\mathcal{V} = \partial C_0/\partial\sigma$ for the corresponding option vega, we obtain the following positive expression:

$$\mathcal{V} = \frac{1}{\sqrt{2\pi}}\,\mathrm{e}^{-rt-(1/2)A}(h_1 - h_0)\sqrt{\tau p_0 p_1(P_{tT}h_1 - K)(K - P_{tT}h_0)}, \qquad (58)$$

where

$$A = \frac{1}{\sigma^2 \tau (h_1 - h_0)^2} \left(\ln \left[\frac{p_1 (P_{tT} h_1 - K)}{p_0 (K - P_{tT} h_0)} \right] \right)^2 + \tfrac{1}{4} \sigma^2 \tau (h_1 - h_0)^2. \quad (59)$$

We remark that in the more general case of a stochastic recovery, a semi-analytic option pricing formula can be obtained that, for practical purposes, can be regarded as fully tractable. In particular, starting from (46), we consider the case where the strike price K lies in the range $P_{tT} h_{k+1} > K > P_{tT} h_k$ for some value of $k \in \{0, 1, \ldots, n\}$. It is an exercise to verify that there exists a unique critical value of ξ_t such that the summation appearing in the argument of the $\max(x, 0)$ function in (46) vanishes. Writing $\bar{\xi}_t$ for the critical value, which can be obtained by numerical methods, we define the scaled critical value $\bar{Z}$ as before, by setting $\bar{\xi}_t = \bar{Z} \sqrt{t(T-t)/T}$. A calculation then shows that the option price is given by the following expression:

$$C_0 = P_{0t} \sum_{i=0}^{n} p_i \left(P_{tT} h_i - K \right) N(\sigma h_i \sqrt{\tau} - \bar{Z}). \quad (60)$$

9 Bond Option Price Processes

In the previous section we obtained the initial value C_0 of an option on a binary credit-risky bond. In the present section we determine the price process $\{C_s\}_{0 \le s \le t}$ of such an option. We fix the bond maturity T and the option maturity t. Then the price C_s of a call option at time $s \le t$ is given by

$$\begin{aligned} C_s &= P_{st} E \left[(B_{tT} - K)^+ | \mathcal{F}_s \right] \\ &= \frac{P_{st}}{\Phi_s} E^B \left[\Phi_t (B_{tT} - K)^+ | \mathcal{F}_s \right] \\ &= \frac{P_{st}}{\Phi_s} E^B \left[\left(\sum_{i=0}^{n} \left(P_{tT} h_i - K \right) p_{it} \right)^+ \Bigg| \mathcal{F}_s \right]. \end{aligned} \quad (61)$$

We recall that p_{it}, defined in (41), is a function of ξ_t. The calculation can thus be simplified by use of the fact that $\{\xi_t\}$ is a B-Brownian bridge. To determine the conditional expectation (61) we note that the B-Gaussian random variable Z_{st} defined by

$$Z_{st} = \frac{\xi_t}{T-t} - \frac{\xi_s}{T-s} \quad (62)$$

is independent of $\{\xi_u\}_{0 \le u \le s}$. We can then express $\{p_{it}\}$ in terms of ξ_s and Z_{st} by writing

$$p_{it} = p_i \exp \left[\frac{T}{T-s} \sigma h_i T \xi_s - \frac{1}{2} \frac{T}{T-t} \sigma^2 h_i^2 t + \sigma h_i Z_{st} T \right]. \quad (63)$$

Substituting (63) into (61), we find that C_s can be calculated by taking an expectation involving the random variable Z_{st}, which has mean zero and variance v_{st}^2, given by

$$v_{st}^2 = \frac{t-s}{(T-t)(T-s)}. \tag{64}$$

In the case of a call option on a binary discount bond that pays h_0 or h_1, we can obtain a closed-form expression for (61). In that case the option price is given as follows:

$$C_s = \frac{P_{st}}{\Phi_s} E^B\left[\left((P_{tT}h_0 - K)p_{0t} + (P_{tT}h_1 - K)p_{1t}\right)^+ \Big| \mathcal{F}_t\right]. \tag{65}$$

Substituting (63) in (65) we find that the expression in the expectation is positive only if the inequality $Z_{st} > \bar{Z}$ is satisfied, where

$$\bar{Z} = \frac{\ln\left[\frac{\pi_{0s}(K - P_{tT}h_0)}{\pi_{1s}(P_{tT}h_1 - K)}\right] + \frac{1}{2}\sigma^2(h_1^2 - h_0^2)v_{st}^2 T}{\sigma v_{st} T(h_1 - h_0)}. \tag{66}$$

It is convenient to set $Z_{st} = v_{st}Z$, where Z is a B-Gaussian random variable with zero mean and unit variance. The computation of the expectation in (65) then reduces to a pair of Gaussian integrals, and we obtain

$$C_s = P_{st}\left[\pi_{1s}(P_{tT}h_1 - K)N(d_s^+) - \pi_{0s}(K - P_{tT}h_0)N(d_s^-)\right], \tag{67}$$

where the conditional probabilities $\{\pi_{is}\}$ are as defined in (17), and

$$d_s^\pm = \frac{\ln\left[\frac{\pi_{1s}(P_{tT}h_1 - K)}{\pi_{0s}(K - P_{tT}h_0)}\right] \pm \frac{1}{2}\sigma^2 v_{st}^2 T^2(h_1 - h_0)^2}{\sigma v_{st}T(h_1 - h_0)}. \tag{68}$$

We note that $d_s^+ = d_s^- + \sigma v_{st}T(h_1 - h_0)$, and that $d_0^\pm = d^\pm$.

One important feature of the model worth pointing out in the present context is that a position in a bond option can be hedged with a position in the underlying bond. This is because the option price process and the underlying bond price process are one-dimensional diffusions driven by the same Brownian motion. Because C_t and B_{tT} are both monotonic in their dependence on ξ_t, it follows that C_t can be expressed as a function of B_{tT}; the delta of the option can then be obtained in the conventional way as the derivative of the option price with respect to the underlying. In the case of a binary bond, the resulting hedge ratio process $\{\Delta_s\}_{0 \le s \le t}$ is given by

$$\Delta_s = \frac{(P_{tT}h_1 - K)N(d_s^+) + (K - P_{tT}h_0)N(d_s^-)}{P_{tT}(h_1 - h_0)}. \tag{69}$$

This brings us to another interesting point. For certain types of instruments, it may be desirable to model the occurrence of credit events (e.g., credit-rating downgrades) taking place at some time preceding a cash-flow date. In particular, we may wish to consider contingent claims based on such events. In the present framework we can regard such contingent claims as derivative structures for which the payoff is triggered by the level of $\{\xi_t\}$. For example, it may be that a credit event is established if B_{tT} drops below some specific level, or if the credit spread widens beyond some threshold. As a consequence, a number of different types of contingent claims can be valued by use of barrier option methods in this framework (cf. Albanese et al. [1], Chen and Filipović [7], and Albanese and Chen [2]).

10 Coupon Bonds: The X-Factor Approach

The discussion so far has focused on simple structures, such as discount bonds and options on discount bonds. One of the advantages of the present approach, however, is that its tractability extends to situations of a more complex nature. In this section we consider the case of a credit-risky coupon bond. One should regard a coupon bond as being a rather complicated instrument from the point of view of credit risk management, because default can occur at any of the coupon dates. The market will in general possess partial information concerning all of the future coupon payments, as well as the principal payment.

As an illustration, we consider a bond with two payments remaining: a coupon H_{T_1} at time T_1, and a coupon plus the principal totalling H_{T_2} at time T_2. We assume that if default occurs at T_1, then no further payment is made at T_2. On the other hand, if the T_1-coupon is paid, default may still occur at T_2. We model this by setting

$$H_{T_1} = \mathbf{c} X_{T_1} \quad \text{and} \quad H_{T_2} = (\mathbf{c} + \mathbf{p}) X_{T_1} X_{T_2}, \tag{70}$$

where X_{T_1} and X_{T_2} are independent random variables each taking the values $\{0, 1\}$, and the constants $\mathbf{c}$ and $\mathbf{p}$ denote the coupon and principal. Let us write $\{p_0^{(1)}, p_1^{(1)}\}$ for the a priori probabilities that $X_{T_1} = \{0, 1\}$, and $\{p_0^{(2)}, p_1^{(2)}\}$ for the a priori probabilities that $X_{T_2} = \{0, 1\}$. We introduce a pair of information processes

$$\xi_t^{(1)} = \sigma_1 X_{T_1} t + \beta_{tT_1}^{(1)} \quad \text{and} \quad \xi_t^{(2)} = \sigma_2 X_{T_2} t + \beta_{tT_2}^{(2)}, \tag{71}$$

where $\{\beta_{tT_1}^{(1)}\}$ and $\{\beta_{tT_2}^{(2)}\}$ are independent Brownian bridges, and σ_1 and σ_2 are parameters. Then for the credit-risky coupon-bond price process we have

$$S_t = \mathbf{c} P_{tT_1} E\left[X_{T_1} | \xi_t^{(1)}\right] + (\mathbf{c} + \mathbf{p}) P_{tT_2} E\left[X_{T_1} | \xi_t^{(1)}\right] E\left[X_{T_2} | \xi_t^{(2)}\right]. \tag{72}$$

The two conditional expectations appearing in this formula can be worked out explicitly using the techniques already described. The result is

$$E\left[X_{T_i}|\xi_t^{(i)}\right] = \frac{p_1^{(i)}\exp\left[\frac{T_i}{T_i-t}\left(\sigma_i\xi_t^{(i)} - \frac{1}{2}\sigma_i^2 t\right)\right]}{p_0^{(i)} + p_1^{(i)}\exp\left[\frac{T_i}{T_i-t}\left(\sigma_i\xi_t^{(i)} - \frac{1}{2}\sigma_i^2 t\right)\right]}, \tag{73}$$

for $i = 1, 2$. It should be evident that in the case of a bond with two payments remaining we obtain a natural 'two-factor' model, the factors being the two independent Brownian motions arising in connection with the information processes $\{\xi_t^{(i)}\}_{i=1,2}$. Similarly, if there are n outstanding coupons, we model the payments by $H_{T_k} = \mathbf{c}X_{T_1}\cdots X_{T_k}$ for $k \leq n-1$ and $H_{T_n} = (\mathbf{c}+\mathbf{p})X_{T_1}\cdots X_{T_n}$, and introduce the information processes

$$\xi_t^{(i)} = \sigma_i X_{T_i}t + \beta_{tT_i}^{(i)} \qquad (i = 1, 2, \ldots, n). \tag{74}$$

The case of n outstanding payments leads to an n-factor model. The independence of the random variables $\{X_{T_i}\}_{i=1,2,\ldots,n}$ implies that the price of a credit-risky coupon bond admits a closed-form expression analogous to (72).

With a slight modification of these expressions, we can consider the case when there is partial recovery in the event of default. In the two-coupon example discussed above, for instance, we can extend the model by saying that in the event of default on the first coupon the effective recovery rate (as a percentage of coupon plus principal) is R_1, whereas in the case of default on the final payment, the recovery rate is R_2. Then $H_{T_1} = \mathbf{c}X_{T_1} + R_1(\mathbf{c}+\mathbf{p})(1 - X_{T_1})$ and $H_{T_2} = \mathbf{c} + \mathbf{p}X_{T_1}X_{T_2} + R_2(\mathbf{c}+\mathbf{p})X_{T_1}(1 - X_{T_2})$. A further extension of this reasoning allows for the introduction of a stochastic recovery rate.

11 Credit Default Swaps

Swap-like structures can also be readily treated. For example, in the case of a basic credit default swap (CDS), we have a series of premium payments, each of the amount $\mathbf{g}$, made to the seller of protection. The payments continue until the failure of a coupon payment in the reference bond, at which point a payment $\mathbf{n}$ is made to the buyer of protection.

As an illustration, we consider two reference coupons, letting X_{T_1} and X_{T_2} be the associated independent market factors, following the pattern of the previous example. We assume for simplicity that the default-swap premium payments are made immediately after the coupon dates. Then the value of the default swap, from the point of view of the seller of protection, is

$$V_t = \mathbf{g}P_{tT_1}E\left[X_{T_1}|\xi_t^{(1)}\right] - \mathbf{n}P_{tT_1}E\left[1 - X_{T_1}|\xi_t^{(1)}\right]$$
$$+ \mathbf{g}P_{tT_2}E\left[X_{T_1}|\xi_t^{(1)}\right]E\left[X_{T_2}|\xi_t^{(2)}\right]$$
$$- \mathbf{n}P_{tT_2}E\left[X_{T_1}|\xi_t^{(1)}\right]E\left[1 - X_{T_2}|\xi_t^{(2)}\right]. \tag{75}$$

After some rearrangement of terms, this can be expressed more compactly as follows:

$$V_t = -\mathbf{n}P_{tT_1} + [(\mathbf{g}+\mathbf{n})P_{tT_1} - \mathbf{n}P_{tT_2}] E\left[X_{T_1}|\xi_t^{(1)}\right]$$
$$+ (\mathbf{g}+\mathbf{n})P_{tT_2} E\left[X_{T_1}|\xi_t^{(1)}\right] E\left[X_{T_2}|\xi_t^{(2)}\right], \tag{76}$$

which can then be written explicitly in terms of the expressions given in (73). A similar approach can be adapted in the multi-name credit situation. The point that we would like to emphasise is that there is a good deal of flexibility available in the manner in which the various cash flows can be modelled to depend on one another, and in many situations tractable expressions emerge that can be used as the basis for the modelling of complex credit instruments.

12 Baskets of Credit-Risky Bonds

We consider now the valuation problem for a basket of bonds where there are correlations in the payoffs. We shall obtain a closed-form expression for the value of a basket of defaultable bonds with various maturities.

For definiteness we consider a set of digital bonds. It is convenient to label the bonds in chronological order with respect to maturity. We let H_{T_1} denote the payoff of the bond that expires first; we let H_{T_2} $(T_2 \geq T_1)$ denote the payoff of the first bond that matures after T_1; and so on. In general, the various bond payouts will not be independent. We propose to model this set of dependent random variables in terms of an underlying set of independent market factors. To achieve this, we let X denote the random variable associated with the payoff of the first bond: $H_{T_1} = X$. The random variable X takes on the values $\{1,0\}$ with *a priori* probabilities $\{p, 1-p\}$. The payoff of the second bond H_{T_2} can then be represented in terms of three independent random variables: $H_{T_2} = XX_1 + (1-X)X_0$. Here X_0 takes the values $\{1,0\}$ with the probabilities $\{p_0, 1-p_0\}$, and X_1 takes the values $\{1,0\}$ with the probabilities $\{p_1, 1-p_1\}$. Clearly, the payoff of the second bond is unity if and only if the random variables (X, X_0, X_1) take the values $(0,1,0)$, $(0,1,1)$, $(1,0,1)$, or $(1,1,1)$. Because these random variables are independent, the *a priori* probability that the second bond does not default is $p_0 + p(p_1 - p_0)$, where p is the *a priori* probability that the first bond does not default. To represent the payoff of the third bond, we introduce four additional independent random variables:

$$H_{T_3} = XX_1X_{11} + X(1-X_1)X_{10} + (1-X)X_0X_{01}$$
$$+ (1-X)(1-X_0)X_{00}. \tag{77}$$

The market factors $\{X_{ij}\}_{i,j=0,1}$ here take the values $\{1,0\}$ with probabilities $\{p_{ij}, 1-p_{ij}\}$. It is a matter of elementary combinatorics to determine the *a priori* probability that $H_{T_3} = 1$ in terms of p, $\{p_i\}$, and $\{p_{ij}\}$.

The scheme above can be extended to represent the payoff of a generic bond in the basket, with an expression of the following form:

$$H_{T_{n+1}} = \sum_{\{k_j\}=1,0} X^{\omega(k_1)} X_{k_1}^{\omega(k_2)} X_{k_1 k_2}^{\omega(k_3)} \cdots X_{k_1 k_2 \cdots k_{n-1}}^{\omega(k_n)} X_{k_1 k_2 \cdots k_{n-1} k_n}. \quad (78)$$

Here, for any random variable X we define $X^{\omega(0)} = 1 - X$ and $X^{\omega(1)} = X$. The point to observe is that if we have a basket of N digital bonds with arbitrary *a priori* default probabilities and correlations, then we can introduce $2^N - 1$ independent digital random variables to represent the N correlated random variables associated with the bond payoffs.

One advantage of the decomposition into independent market factors is that we retain analytical tractability for the pricing of the basket. In particular, because the random variables $\{X_{k_1 k_2 \cdots k_n}\}$ are independent, it is natural to introduce $2^N - 1$ independent Brownian bridges to represent the noise that veils the values of the independent market factors:

$$\xi_t^{k_1 k_2 \cdots k_n} = \sigma_{k_1 k_2 \cdots k_n} X_{k_1 k_2 \cdots k_n} t + \beta_{tT_{n+1}}^{k_1 k_2 \cdots k_n}. \quad (79)$$

The number of independent factors grows rapidly with the number of bonds in the portfolio. As a consequence, a market that consists of correlated bonds is in general highly incomplete. This fact provides an economic justification for the creation of products such as CDSs and CDOs that enhance the 'hedgeability' of such portfolios.

13 Homogeneous Baskets

In the case of a 'homogeneous' basket, the number of independent factors determining the payoff of the portfolio can be reduced. We assume now that the basket contains n defaultable discount bonds, each maturing at time T, and each paying 0 or 1, with the same *a priori* probability of default. This situation is of interest as a first step in the analysis of the more general setup. Our goal is to model default correlations in the portfolio, and in particular to model the flow of market information concerning default correlation. Let us write H_T for the payoff at time T of the homogeneous portfolio, and set

$$H_T = n - X_1 - X_1 X_2 - X_1 X_2 X_3 - \cdots - X_1 X_2 \ldots X_n, \quad (80)$$

where the random variables $\{X_j\}_{j=1,2,\ldots,n}$, each taking the values $\{0,1\}$, are assumed to be independent. Thus if $X_1 = 0$, then $H_T = n$; if $X_1 = 1$ and $X_2 = 0$, then $H_T = n - 1$; if $X_1 = 1$, $X_2 = 1$, and $X_3 = 0$, then $H_T = n - 2$; and so on. Now suppose we write $p_j = Q(X_j = 1)$ and $q_j = Q(X_j = 0)$ for $j = 1, 2, \ldots, n$. Then $Q(H_T = n) = q_1$, $Q(H_T = n - 1) = p_1 q_2$, $Q(H_T = n - 2) = p_1 p_2 q_3$, and so on. More generally, we have $Q(H_T = n - k) = p_1 p_2 \cdots p_k q_{k+1}$. Thus, for example, if $p_1 \ll 1$ but $p_2, p_3, \ldots, p_k$ are large, then we are in a situation of low default probability and high default correlation; that is to say, the probability of a default occurring in the portfolio is small, but conditional on at least one default occurring, the probability of several defaults is high.

The market will take a view on the likelihood of various numbers of defaults occurring in the portfolio. We model this by introducing a set of independent information processes $\{\eta_t^j\}$ defined by

$$\eta_t^j = \sigma_j X_j t + \beta_{tT}^j, \tag{81}$$

where $\{\sigma_j\}_{j=1,2,\ldots,n}$ are parameters, and $\{\beta_{tT}^j\}_{j=1,2,\ldots,n}$ are independent Brownian bridges. The market filtration $\{\mathcal{F}_t\}$ is generated collectively by $\{\eta_t^j\}_{j=1,2,\ldots,n}$, and for the portfolio value $H_t = P_{tT} E\,[H_T|\mathcal{F}_t]$, we have

$$H_t = P_{tT}\Big[n - E_t[X_1] - E_t[X_1]E_t[X_2] - \cdots - E_t[X_1]E_t[X_2]\ldots E_t[X_n]\Big]. \tag{82}$$

The conditional expectations appearing here can be calculated by means of formulae established earlier in the paper. The resulting dynamics for $\{H_t\}$ can then be used to describe the evolution of correlations in the portfolio. For example, if $E_t[X_1]$ is low and $E_t[X_2]$ is high, then the conditional probability at time t of a default at time T is small, whereas if $E_t[X_1]$ were to increase suddenly, then the conditional probability of two or more defaults at T would rise as a consequence. Thus, the model is sufficiently rich to admit a detailed account of the correlation dynamics of the portfolio. The losses associated with individual tranches can be identified, and derivative structures associated with such tranches can be defined. For example, a digital option that pays out in the event that there are three or more defaults has the payoff structure $H_T^{(3)} = X_1 X_2 X_3$. The homogeneous portfolio model has the property that the dynamics of equity-level and mezzanine-level tranches involve a relatively small number of factors. The market prices of tranches can be used to determine the *a priori* probabilities, and the market prices of options on tranches can be used to fix the information-flow parameters.

Acknowledgments

The authors are grateful to T. Bielecki, I. Buckley, T. Dean, G. Di Graziano, A. Elizalde, B. Flesaker, T. Hurd, M. Jeanblanc, B. Meister, C. Rogers, M. Rutkowski, and M. Zervos, for helpful comments.

We are grateful to seminar participants at the Isaac Newton Institute, Cambridge, 3 May, 19 June, and 8 July 2005; at the Mathematics in Finance Conference, Kruger Park, RSA, 9 August 2005; at the Imperial College Workshop on Mathematical Finance and Stochastic Analysis, 23 August 2005; at CEMFI (Centro de Estudios Monetarios y Financieros), Madrid, 3 October 2005; at the Winter School in Financial Mathematics, Lunteren, the Netherlands, 24 January 2006; at the First Conference on Advanced Mathematical Methods for Finance, Side, Antalya, Turkey, 26–29 April 2006; and at the

Bachelier Finance Society Fourth World Congress, Tokyo, Japan, 17–19 August 2006, where drafts of this paper have been presented, for their comments.

This work was carried out in part while Lane Hughston and Andrea Macrina were participants in the Isaac Newton Institute's Developments in Quantitative Finance programme (April–July 2005). We thank I. Buckley for assistance with the figures. Dorje Brody and Lane Hughston acknowledge support from The Royal Society; Lane Hughston and Andrea Macrina acknowledge support from EPSRC grant number GR/S22998/01; Andrea Macrina thanks the Public Education Authority of the Canton of Bern, Switzerland, and the UK Universities ORS scheme.

Finally, it is our pleasure to dedicate this paper to Dilip Madan on the occasion of his sixtieth birthday, in acknowledgment of his numerous significant contributions to mathematical finance and in recognition of the inspiration and encouragement he has so often and so generously given to others working in the subject.

References

1. C. Albanese, J. Campolieti, O. Chen, and A. Zavidonov. Credit barrier models. *Risk*, 16:109–113, 2003.
2. C. Albanese and O. Chen. Implied migration rates from credit barrier models. *J. Banking and Finance*, 30:607–626, 2006.
3. T.R. Bielecki, M. Jeanblanc, and M. Rutkowski. Modelling and valuation of credit risk. In *Stochastic Methods in Finance*, Bressanone Lectures 2003, eds. M. Fritelli and W. Runggaldier, LNM 1856, Springer, 2004.
4. T.R. Bielecki and M. Rutkowski. *Credit Risk: Modelling, Valuation and Hedging*. Springer, 2002.
5. F. Black and J.C. Cox. Valuing corporate securities: Some effects of bond indenture provisions. *J. Finance*, 31:351–367, 1976.
6. U. Cetin, R. Jarrow, P. Protter, and Y. Yildrim. Modelling credit risk with partial information. *Ann. Appl. Prob.*, 14:1167–1172, 2004.
7. L. Chen and D. Filipović. A simple model for credit migration and spread curves. *Finance and Stochastics*, 9:211–231, 2005.
8. J.H. Cochrane. *Asset Pricing*. Princeton University Press, 2005.
9. D. Duffie and D. Lando. Term structure of credit spreads with incomplete accounting information. *Econometrica*, 69:633–664, 2001.
10. D. Duffie, M. Schroder, and C. Skiadas. Recursive valuation of defaultable securities and the timing of resolution of uncertainty. *Ann. Appl. Prob.*, 6:1075–1090, 1996.
11. D. Duffie and K.J. Singleton. Modelling term structures of defaultable bonds. *Rev. Financial Studies*, 12:687–720, 1999.
12. D. Duffie and K.J. Singleton. *Credit Risk: Pricing, Measurement and Management*. Princeton University Press, 2003.
13. A. Elizalde. Credit risk models: I Default correlation in intensity models, II Structural models, III Reconciliation structural-reduced models, IV Understanding and pricing CDOs. www.abelelizalde.com, 2005.

14. B. Flesaker, L.P. Hughston, L. Schreiber, and L. Sprung. Taking all the credit. *Risk*, 7:104–108, 1994.
15. K. Giesecke. Correlated default with incomplete information. *J. Banking and Finance*, 28:1521–1545, 2004.
16. K. Giesecke and L.R. Goldberg. Sequential default and incomplete information. *J. Risk*, 7:1–26, 2004.
17. B. Hilberink and L.C.G. Rogers. Optimal capital structure and endogenous default. *Finance and Stochastics*, 6:237–263, 2002.
18. L.P. Hughston and S. Turnbull. Credit risk: Constructing the basic building blocks. *Economic Notes*, 30:281–292, 2001.
19. J. Hull and A. White. Merton's model, credit risk and volatility skews. *J. Credit Risk*, 1:1–27, 2004.
20. R.A. Jarrow, D. Lando, and S.M. Turnbull. A Markov model for the term structure of credit risk spreads. *Rev. Financial Stud.*, 10:481–528, 1997.
21. R.A. Jarrow and P. Protter. Structural versus reduced form models: A new information based perspective. *J. Invest. Management*, 2:1–10, 2004.
22. R.A. Jarrow and S.M. Turnbull. Pricing derivatives on financial securities subject to credit risk. *J. Finance*, 50:53–85, 1995.
23. R.A. Jarrow, D. van Deventer, and X. Wang. A robust test of Merton's structural model for credit risk. *J. Risk*, 6:39–58, 2003.
24. R.A. Jarrow and F. Yu. Counterparty risk and the pricing of defaultable securities. *J. Finance*, 56:1765–1799, 2001.
25. M. Jeanblanc and M. Rutkowski. Modelling of default risk: An overview. In *Mathematical Finance: Theory and Practice*, Higher Education Press, 2000.
26. I. Karatzas and S. Shreve. *Brownian Motion and Stochastic Calculus*, 2nd edition. Springer, 1991.
27. S. Kusuoka. A remark on default risk models. *Advances Math. Economics*, 1:69-82, 1999.
28. D. Lando. On Cox processes and credit risky securities. *Rev. Derivatives Research*, 2:99-120, 1998.
29. D. Lando. *Credit Risk Modelling*. Princeton University Press, 2004.
30. H.E. Leland and K.B. Toft. Optimal capital structure, endogenous bankruptcy and the term structure of credit spreads. *J. Finance*, 50:789–819, 1996.
31. D. Madan and H. Unal. Pricing the risks of default. *Rev. Derivatives Research*, 2:121–160, 1998.
32. D. Madan and H. Unal. A two-factor hazard-rate model for pricing risky debt and the term structure of credit spreads. *J. Financial and Quantitative Analysis*, 35:43–65, 2000.
33. R. Merton. On the pricing of corporate debt: The risk structure of interest rates. *J. Finance*, 29:449–470, 1974.
34. P.E. Protter. *Stochastic Integration and Differential Equations*. Springer, 2004.
35. P.J. Schönbucher. *Credit Derivatives Pricing Models*. John Wiley & Sons, 2003.
36. M. Yor. *Some Aspects of Brownian Motion, Part I: Some Special Functionals*. Birkhäuser Verlag, 1992.
37. M. Yor. *Some Aspects of Brownian Motion, Part II: Some Recent Martingale Problems*. Birkhäuser Verlag, 1996.

A Generic One-Factor Lévy Model for Pricing Synthetic CDOs

Hansjörg Albrecher,[1,2] Sophie A. Ladoucette,[3] and Wim Schoutens[3]

[1] Johann Radon Institute
Austrian Academy of Sciences
Altenbergerstrasse 69
A-4040 Linz, Austria

[2] Graz University of Technology
Department of Mathematics
Steyrergasse 30
A-8010 Graz, Austria
albrecher@tugraz.at

[3] Katholieke Universiteit Leuven
Department of Mathematics
W. de Croylaan 54
B-3001 Leuven, Belgium
sophie.ladoucette@wis.kuleuven.be, wim@schoutens.be

Summary. The one-factor Gaussian model is well known not to fit the prices of the different tranches of a collateralized debt obligation (CDO) simultaneously, leading to the implied correlation smile. Recently, other one-factor models based on different distributions have been proposed. Moosbrucker [12] used a one-factor Variance-Gamma (VG) model, Kalemanova et al. [7] and Guégan and Houdain [6] worked with a normal inverse Gaussian (NIG) factor model, and Baxter [3] introduced the Brownian Variance-Gamma (BVG) model. These models bring more flexibility into the dependence structure and allow tail dependence. We unify these approaches, describe a generic one-factor Lévy model, and work out the large homogeneous portfolio (LHP) approximation. Then we discuss several examples and calibrate a battery of models to market data.

Key words: Lévy processes; collateralized debt obligation (CDO); credit risk; credit default; large homogeneous portfolio approximation.

1 Introduction

A collateralized debt obligation (CDO) can be defined as a transaction that transfers the credit risk on a reference portfolio of assets. A standard feature of a CDO structure is the *tranching* of credit risk. Credit tranching refers to creating multiple tranches of securities that have varying degrees of seniority and risk exposure. The risk of loss on the reference portfolio is then divided into tranches of increasing seniority in the following way. The *equity* tranche is the first to be affected by losses in the event of one or more defaults in the portfolio. If losses exceed the value of this tranche, they are absorbed by the *mezzanine* tranche(s). Losses that have not been absorbed by the other tranches are sustained by the *senior* tranche and finally by the *super-senior* tranche. Each tranche then protects the ones senior to it from the risk of loss on the underlying portfolio. When tranches are issued, they usually receive a rating by rating agencies. The CDO issuer typically determines the size of the senior tranche so that it is AAA-rated. Likewise, the CDO issuer generally designs the other tranches so that they achieve successively lower ratings. The CDO investors take on exposure to a particular tranche, effectively selling credit protection to the CDO issuer, and in turn collecting premiums (spreads).

We are interested in pricing tranches of *synthetic* CDOs. A synthetic CDO is a CDO backed by *credit default swaps* (CDSs) rather than bonds or loans; that is, the reference portfolio is composed of CDSs. Recall that a CDS offers protection against default of an underlying entity over some time horizon. The term synthetic is used because CDSs permit synthetic exposure to credit risk. By contrast, a CDO backed by ordinary bonds or loans is called a *cash* CDO. Synthetic CDOs recently have become very popular.

It turns out that the pricing of synthetic CDO tranches only involves loss distributions over different time horizons (see Section 4). Then, we may think of using the large homogeneous portfolio (LHP) approximation to compute the premiums of these tranches. This convenient method is well known in the credit portfolio field and permits us to approximate the loss distribution which is computationally intensive.

Assuming a one-factor model approach for modeling correlated defaults of the different names in the reference portfolio together with the conditional independence of these defaults on a common market factor leads to a simplification of the calculation of the loss distribution (see Section 2). The one-factor Gaussian model is well known not to fit the prices of the different tranches of a CDO simultaneously, leading to the implied correlation smile. Recently, other one-factor models based on different distributions have been proposed. Moosbrucker [12] used a one-factor Variance-Gamma (VG) model, Kalemanova et al. [7] and Guégan and Houdain [6] worked with a normal inverse Gaussian (NIG) factor model, and Baxter [3] introduced the Brownian Variance-Gamma (BVG) model. These models bring more flexibility into the dependence structure and allow tail dependence. We unify these approaches in Section 2, describe a generic one-factor Lévy model, and work out the LHP approximation. With such a model, the distribution function of any name's asset value is analytically known, bringing significant improvement with respect to computation times. In Section 3, we discuss several examples, including the

Table 1. Standard synthetic CDO structure on the DJ iTraxx Europe index.

Reference Portfolio	Tranche Name	$K_1(\%)$	$K_2(\%)$
	Equity	0	3
125	Junior mezzanine	3	6
CDS	Senior mezzanine	6	9
names	Senior	9	12
	Super-senior	12	22

Gaussian, shifted gamma, shifted inverse Gaussian, VG, NIG, and Meixner cases. Finally, we calibrate a battery of one-factor LHP Lévy models to market quotes of a tranched iTraxx and give the prices generated by these models.

In 2004, the main traded CDS indices were merged into a single family under the names DJ iTraxx (Europe and Asia) and DJ CDX (North America and emerging markets). These indices provide established portfolios upon which standardized tranches can be structured, allowing a more transparent and liquid market for CDO tranches.

Take the example of the DJ iTraxx Europe index. It consists of a portfolio composed of 125 actively traded names in terms of CDS volume, with an equal weighting given to each, and remains static over its lifetime of six months, except for entities defaulting which are then eliminated from the index. It is possible to invest in standardized tranches of the index via the tranched iTraxx which is nothing else but a synthetic CDO on a static portfolio. In other words, a tranched CDS index is a synthetic CDO based on a CDS index, where each tranche references a different segment of the loss distribution of the underlying CDS index. The main advantage of such a synthetic CDO relative to other CDOs is that it is standardized. In Table 1, we give the standard synthetic CDO structure on the DJ iTraxx Europe index.

We end with some notation. By $f^{[-1]}$, we mean the inverse function of f. The indicator function of any set or event A is denoted $\mathbf{1}_A$. We use the abbreviation a.s. for almost surely, and the symbol $\overset{a.s.}{\longrightarrow}$ stands for almost sure convergence. The gamma function is denoted $\Gamma(x) := \int_0^{+\infty} t^{x-1}e^{-t}\,dt$, $x > 0$. The modified Bessel function of the third kind with real index ζ is denoted $K_\zeta(x) := \frac{1}{2}\int_0^{+\infty} t^{\zeta-1}\exp\bigl(-x(t+t^{-1})/2\bigr)\,dt$, $x > 0$. The notation $X \sim F$ means that the random variable X follows the distribution F.

2 Generic One-Factor Lévy Model

2.1 Lévy Process

Suppose ϕ is the characteristic function of a distribution. If for every positive integer n, ϕ is also the nth power of a characteristic function, we say that the distribution is *infinitely divisible*. One can define for any infinitely divisible distribution a stochastic process, $X = \{X_t, t \geq 0\}$, called a *Lévy process*, which starts at zero, has stationary independent increments, and such that the distribution of an increment over $[s, s + t]$, $s, t \geq 0$ (i.e., $X_{t+s} - X_s$) has ϕ^t as characteristic function.

The function $\psi := \log \phi$ is called the *characteristic exponent*, and it satisfies the following Lévy–Khintchine formula (see Bertoin [4]),

$$\psi(z) = i\gamma z - \frac{\varsigma^2}{2} z^2 + \int_{-\infty}^{+\infty} \left(\exp(izx) - 1 - izx \mathbf{1}_{\{|x|<1\}} \right) \nu(dx), \quad z \in \mathbb{R},$$

where $\gamma \in \mathbb{R}$, $\varsigma^2 \geq 0$, and ν is a measure on $\mathbb{R} \backslash \{0\}$ with $\int_{-\infty}^{+\infty} \min(1, x^2) \, \nu(dx) < \infty$. From the Lévy–Khintchine formula, one sees that, in general, a Lévy process consists of three independent parts: a linear deterministic part, a Brownian part, and a pure jump part. We say that our infinitely divisible distribution has a triplet of Lévy characteristics $[\gamma, \varsigma^2, \nu(dx)]$. The measure $\nu(dx)$ is called the *Lévy measure* of X, and it dictates how the jumps occur. Jumps of sizes in the set A occur according to a Poisson process with parameter $\int_A \nu(dx)$. If $\varsigma^2 = 0$ and $\int_{-1}^{+1} |x| \, \nu(dx) < \infty$, it follows from standard Lévy process theory (e.g., Bertoin [4] and Sato [16]) that the process is of finite variation. For more details about the applications of Lévy processes in finance, we refer to Schoutens [20].

2.2 Generic One-Factor Lévy Model

Next, we model a portfolio of n obligors; each obligor has the same weight in the portfolio. Later on we focus on a homogeneous portfolio, but we start with the general situation where each obligor has some recovery value R_i in case of default and some individual default probability term structure $p_i(t)$, $t \geq 0$, which is the probability that obligor i will default before time t.

Fix a time horizon T. For the modeling, let us start with a mother infinitely divisible distribution L. Let $X = \{X_t, t \in [0, 1]\}$ be a Lévy process based on that infinitely divisible distribution, such that X_1 follows the law L. Note that we only work with Lévy processes with time running over the unit interval. Denote the distribution function of X_t by H_t, $t \in [0, 1]$, and assume it is continuous. Assume further that the distribution is standardized in the sense that $E[X_1] = 0$ and $\text{Var}[X_1] = 1$. In terms of ψ, this means that $\psi'(0) = 0$ and $\psi''(0) = -1$. Then, it is not that hard to prove that $\text{Var}[X_t] = t$.

Let $X = \{X_t, t \in [0, 1]\}$ and $X^{(i)} = \{X_t^{(i)}, t \in [0, 1]\}$, $i = 1, \ldots, n$ be independent and identically distributed Lévy processes (so all processes are independent of each other and are based on the same mother infinitely divisible distribution L).

Next, we propose the generic one-factor Lévy model. Let $\rho \in (0, 1)$. We assume that the asset value of obligor $i = 1, \ldots, n$ at time T is of the form:

$$A_i(T) = X_\rho + X_{1-\rho}^{(i)}. \tag{1}$$

Each $A_i(T)$ has by the stationary and independent increments property the same distribution as the mother distribution L with distribution function H_1. Indeed, the sum of an increment of the process over a time interval of length ρ and an independent increment over a time interval of length $1 - \rho$ follows the distribution of an increment over an interval of unit length, that is, follows the law L. As a consequence, $E[A_i(T)] = 0$ and $\text{Var}[A_i(T)] = 1$. Furthermore, the asset values of any two obligors i and j $(i \neq j)$ are correlated with linear correlation coefficient ρ. Indeed, one readily computes

$$\text{Corr}[A_i(T), A_j(T)] = \frac{E[A_i(T)A_j(T)] - E[A_i(T)]\,E[A_j(T)]}{\sqrt{\text{Var}[A_i(T)]}\sqrt{\text{Var}[A_j(T)]}}$$
$$= E[A_i(T)A_j(T)] = E[X_\rho^2] = \rho.$$

So, starting from any mother standardized infinitely divisible law, we can set up a one-factor model with the required correlation.

We say that the ith obligor *defaults at time T* if its firm value $A_i(T)$ falls below some preset barrier $K_i(T)$, that is, if $A_i(T) \leq K_i(T)$. In order to match default probabilities under this model with default probabilities $p_i(T)$ observed in the market, we have to set $K_i(T) = H_1^{[-1]}(p_i(T))$. Indeed, it follows that $P(A_i(T) \leq K_i(T)) = P(A_i(T) \leq H_1^{[-1]}(p_i(T))) = H_1(H_1^{[-1]}(p_i(T))) = p_i(T)$.

Notice that conditional on the common factor X_ρ, the firm values and the defaults are independent.

From now on, we assume that the portfolio is *homogeneous*; that is,

- All obligors have the same default barrier $(K_i(T) = K(T), i = 1, \ldots, n)$, and hence the same marginal default distribution $(p_i(T) = p(T))$.
- All obligors have the same recovery rate $(R_i = R, i = 1, \ldots, n)$.
- All obligors have the same notional amount: denoting the total portfolio notional by N, we then set $N_i = N/n$ for all $i = 1, \ldots, n$.

Let us denote the number of defaults in the portfolio until time T by $D_{T,n}$. The probability of having exactly k defaults equals

$$P(D_{T,n} = k) = \int_{-\infty}^{+\infty} P(D_{T,n} = k|X_\rho = y)\,dH_\rho(y), \quad k = 0, \ldots, n.$$

Conditional on $X_\rho = y$, the probability of having k defaults is (because of independence)

$$P(D_{T,n} = k|X_\rho = y) = \binom{n}{k} p(y;T)^k (1 - p(y;T))^{n-k},$$

where $p(y;T)$ denotes the probability that the firm's value $A_i(T)$ is below the barrier $K(T)$, given that the systematic factor X_ρ takes the value y; that is,

$$p(y;T) := P(A_i(T) \leq K(T)|X_\rho = y)$$
$$= P(X_\rho + X_{1-\rho}^{(i)} \leq K(T)|X_\rho = y)$$
$$= P(X_{1-\rho}^{(i)} \leq K(T) - y)$$
$$= H_{1-\rho}(K(T) - y).$$

Substituting then yields

$$P(D_{T,n} = k) = \binom{n}{k} \int_{-\infty}^{+\infty} (H_{1-\rho}(K(T) - y))^k$$

$$(1 - H_{1-\rho}(K(T) - y))^{n-k} \, dH_\rho(y).$$

The loss fraction on the portfolio notional at time T given by

$$L_{T,n}^{\mathrm{HP}} := \frac{1-R}{n} \sum_{i=1}^{n} \mathbf{1}_{\{A_i(T) \leq K(T)\}} \tag{2}$$

is clearly one-to-one related with the number of defaults. We then obtain the following distribution function for the portfolio loss fraction,

$$P\left(L_{T,n}^{\mathrm{HP}} \leq \frac{k(1-R)}{n}\right) = \sum_{i=0}^{k} P(D_{T,n} = i), \quad k = 0, \ldots, n.$$

In the next subsection, we consider a method to approximate the latter distribution that turns out to be of prime interest in the CDO pricing (see Section 4).

Now, denote for small x by $\lambda_{ij}(x) := P(A_j(T) \leq x|A_i(T) \leq x)$, $i \neq j$, a measure for the dependence in the tail. Using (conditional) independence arguments yields

$$\lambda_{ij}(x) = \frac{P(X_\rho + X_{1-\rho}^{(j)} \leq x, X_\rho + X_{1-\rho}^{(i)} \leq x)}{P(X_\rho + X_{1-\rho}^{(i)} \leq x)}$$

$$= \frac{E\left[P\left(X_\rho + X_{1-\rho}^{(j)} \leq x, X_\rho + X_{1-\rho}^{(i)} \leq x \middle| X_\rho\right)\right]}{H_1(x)}$$

$$= \int_{-\infty}^{+\infty} \frac{P(X_{1-\rho}^{(j)} \leq x - y, X_{1-\rho}^{(i)} \leq x - y)}{H_1(x)} \, dH_\rho(y)$$

$$= \int_{-\infty}^{+\infty} \frac{H_{1-\rho}^2(x-y)}{H_1(x)} \, dH_\rho(y).$$

The limit $\lambda_{ij} := \lim_{x \to -\infty} \lambda_{ij}(x)$ is then the well-known *lower tail dependence coefficient* of $A_i(T)$ and $A_j(T)$. Lower tail dependence exactly captures the probability of concordant down movements of the underlying asset values. The formula being stated in terms of the distribution H_t, the quantity $\lambda_{ij}(x)$ and its limit can be evaluated for any one-factor Lévy model.

2.3 The Large Homogeneous Portfolio Approximation

We are interested in approximating the distribution function of the portfolio loss fraction $L_{T,n}^{HP}$. This proves to be possible if the homogeneous portfolio gets very large (i.e., $n \to \infty$).

When conditioned on the systematic factor X_ρ, the default variables $\mathbf{1}_{\{A_i(T) \leq K(T)\}}$, $i = 1, \ldots, n$ become independent. Hence, by the strong law of large numbers, we obtain

$$P\left(\lim_{n \to \infty} L_{T,n}^{HP} = (1 - R)\, p(X_\rho; T)\,\middle|\, X_\rho\right) = 1 \quad \text{a.s.,}$$

and taking expectations on both sides gives

$$L_{T,n}^{HP} \xrightarrow{a.s.} (1 - R)\, p(X_\rho; T) \quad \text{as } n \to \infty.$$

For large homogeneous portfolios, we then make the approximation

$$F_{T,n}^{HP}(x) := P(L_{T,n}^{HP} \leq x) = F_T^{HP}\left(\frac{x}{1 - R}\right), \quad x \in [0, 1 - R], \tag{3}$$

where F_T^{HP} denotes the distribution function of $p(X_\rho; T)$, that turns out to be the loss fraction with a zero recovery rate (or equivalently the fraction of defaults) on the limiting portfolio at time T. We easily compute

$$
\begin{aligned}
F_T^{HP}(x) &= P(H_{1-\rho}(K(T) - X_\rho) \leq x) \\
&= P(X_\rho \geq K(T) - H_{1-\rho}^{[-1]}(x)) \\
&= 1 - H_\rho\left(H_1^{[-1]}(p(T)) - H_{1-\rho}^{[-1]}(x)\right), \quad x \in [0, 1], \tag{4}
\end{aligned}
$$

so that we obtain an explicit handy expression for the distribution $F_{T,n}^{HP}$ of the portfolio fractional loss.

Note that if the portfolio contains a moderately large number of credits, the approximation (3) turns out to be remarkably good.

3 Examples of One-Factor Lévy Models for Correlated Defaults

3.1 Based on the Normal Distribution—Brownian Motion

The Vasicek [22] one-factor model assumes the following dynamics.

- $A_i(T) = \sqrt{\rho}\, Y + \sqrt{1 - \rho}\, X_i$, $i = 1, \ldots, n$,
- Y, X_i, $i = 1, \ldots, n$ are i.i.d. standard normal random variables with common distribution function Φ.

Table 2. Mean, variance, skewness, and kurtosis of the Gamma(a, b) distribution.

	Gamma(a,b)
Mean	a/b
Variance	a/b^2
Skewness	$2/\sqrt{a}$
Kurtosis	$3(1 + 2/a)$

This model can be cast in the above general Lévy framework. The mother infinitely divisible distribution is here the standard normal distribution and the associated Lévy process is the standard Brownian motion $W = \{W_t, t \in [0, 1]\}$. Indeed, note that W_ρ follows an $\mathcal{N}(0, \rho)$ distribution as does $\sqrt{\rho}\, Y$; similarly $W_{1-\rho}^{(i)}$ follows an $\mathcal{N}(0, 1 - \rho)$ distribution as does $\sqrt{1 - \rho}\, X_i$. Adding these independent random variables leads to a standard normally distributed random variable.

Using the classical properties of normal random variables, the distribution function (4) of the fraction of defaulted securities in the limiting portfolio at time T transforms into

$$
\begin{aligned}
F_T^{\mathrm{HP}}(x) &= 1 - \Phi\left(\frac{\Phi^{[-1]}(p(T)) - \sqrt{1 - \rho}\, \Phi^{[-1]}(x)}{\sqrt{\rho}} \right) \\
&= \Phi\left(\frac{\sqrt{1 - \rho}\, \Phi^{[-1]}(x) - \Phi^{[-1]}(p(T))}{\sqrt{\rho}} \right), \quad x \in [0, 1].
\end{aligned}
$$

3.2 Based on the Shifted Gamma Process

The density function of the gamma distribution Gamma(a, b) with parameters $a > 0$ and $b > 0$ is given by

$$
f_{Gamma}(x; a, b) = \frac{b^a}{\Gamma(a)}\, x^{a-1}\, e^{-xb}, \quad x > 0.
$$

The distribution function of the Gamma(a, b) distribution is denoted $\mathcal{H}_G(x; a, b)$ and has the following characteristic function,

$$
\phi_{Gamma}(u; a, b) = (1 - iu/b)^{-a}, \quad u \in \mathbb{R}.
$$

Clearly, this characteristic function is infinitely divisible. The gamma process $X^{(G)} = \{X_t^{(G)}, t \geq 0\}$ with parameters $a, b > 0$ is defined as the stochastic process that starts at zero and has stationary independent gamma-distributed increments such that $X_t^{(G)}$ follows a Gamma(at, b) distribution.

The properties of the Gamma(a, b) distribution given in Table 2 can be easily derived from its characteristic function. Now, if X is Gamma(a, b)-distributed and $c > 0$, then cX is Gamma$(a, b/c)$-distributed. Furthermore, if $X \sim$ Gamma(a_1, b) is independent of $Y \sim$ Gamma(a_2, b), then $X + Y \sim$ Gamma$(a_1 + a_2, b)$.

Let us start with a gamma process $G = \{G_t, t \in [0,1]\}$ with parameters $a > 0$ and $b = \sqrt{a}$, so that $E[G_1] = \sqrt{a}$ and $\text{Var}[G_1] = 1$. As driving Lévy process, we take the shifted gamma process $X = \{X_t, t \in [0,1]\}$ defined as

$$X_t = \sqrt{a}t - G_t, \quad t \in [0,1].$$

The interpretation in terms of firm value is that there is a deterministic up trend ($\sqrt{a}t$) with random downward shocks (G_t).

The one-factor shifted gamma–Lévy model is then in the form of (1),

$$A_i(T) = X_\rho + X_{1-\rho}^{(i)}, \quad i = 1, \ldots, n,$$

where X_ρ, $X_{1-\rho}^{(i)}$, $i = 1, \ldots, n$ are independent shifted gamma random variables defined as $X_\rho = \sqrt{a}\rho - G_\rho$ and $X_{1-\rho}^{(i)} = \sqrt{a}(1-\rho) - G_{1-\rho}$. By construction, each $A_i(T)$ follows the same distribution as X_1 and as such has zero mean and unit variance.

As derived in general, we have that the distribution of the limiting portfolio fractional loss with a zero recovery rate at time T is as in (4). The distribution function $H_t(x; a)$ of X_t, $t \in [0,1]$, can be easily obtained from the gamma distribution function. Indeed,

$$\begin{aligned} H_t(x; a) &= P(\sqrt{a}t - G_t \le x) \\ &= 1 - P(G_t < \sqrt{a}t - x) \\ &= 1 - \mathcal{H}_G(\sqrt{a}t - x; at, \sqrt{a}), \quad x \in (-\infty, \sqrt{a}t). \end{aligned}$$

For the inverse function, we have the following relation for each $t \in [0,1]$:

$$H_t^{[-1]}(y; a) = \sqrt{a}t - \mathcal{H}_G^{[-1]}(1 - y; at, \sqrt{a}), \quad y \in [0,1].$$

3.3 Based on the Shifted IG Process

The inverse Gaussian IG(a,b) law with parameters $a > 0$ and $b > 0$ has characteristic function

$$\phi_{IG}(u; a, b) = \exp\left(-a\left(\sqrt{-2iu + b^2} - b\right)\right), \quad u \in \mathbb{R}.$$

The IG distribution is infinitely divisible and we define the IG process $X^{(IG)} = \{X_t^{(IG)}, t \ge 0\}$ with parameters $a, b > 0$ as the process that starts at zero and has stationary independent IG-distributed increments such that

$$E[\exp(iuX_t^{(IG)})] = \phi_{IG}(u; at, b) = \exp\left(-at\left(\sqrt{-2iu + b^2} - b\right)\right), \quad u \in \mathbb{R}$$

meaning that $X_t^{(IG)}$ follows an IG(at, b) distribution.

Table 3. Mean, variance, skewness, and kurtosis of the IG(a, b) distribution.

	IG(a,b)
Mean	a/b
Variance	a/b^3
Skewness	$3/\sqrt{ab}$
Kurtosis	$3(1 + 5(ab)^{-1})$

The density function of the IG(a, b) law is given by

$$f_{IG}(x; a, b) = \frac{a\, e^{ab}}{\sqrt{2\pi}} x^{-3/2} \exp\left(-(a^2 x^{-1} + b^2 x)/2\right), \quad x > 0,$$

and we denote its distribution function $\mathcal{H}_{IG}(x; a, b)$. The characteristics of the IG distribution given in Table 3 can be easily obtained. Now, if X is IG(a, b)-distributed, then cX is IG$(a\sqrt{c}, b/\sqrt{c})$-distributed for any positive c. Furthermore, if $X \sim$ IG(a_1, b) is independent of $Y \sim$ IG(a_2, b), then $X + Y \sim$ IG$(a_1 + a_2, b)$.

Let us start with an IG process $I = \{I_t, t \in [0, 1]\}$ with parameters $a > 0$ and $b = a^{1/3}$, so that $E[I_1] = a^{2/3}$ and $\text{Var}[I_1] = 1$. As driving Lévy process, we take the shifted IG process $X = \{X_t, t \in [0, 1]\}$ defined as

$$X_t = a^{2/3}t - I_t, \quad t \in [0, 1].$$

The interpretation in terms of firm value is that there is a deterministic up trend $(a^{2/3}t)$ with random downward shocks (I_t).

The one-factor shifted IG-Lévy model is then

$$A_i(T) = X_\rho + X_{1-\rho}^{(i)}, \quad i = 1, \ldots, n,$$

where X_ρ, $X_{1-\rho}^{(i)}$, $i = 1, \ldots, n$ are independent shifted IG random variables defined as $X_\rho = a^{2/3}\rho - I_\rho$ and $X_{1-\rho}^{(i)} = a^{2/3}(1 - \rho) - I_{1-\rho}$. Each $A_i(T)$ follows the same distribution as X_1 and as such has zero mean and unit variance.

The distribution of the fraction of defaulted securities in the limiting portfolio at time T is therefore as in (4). The distribution function $H_t(x; a)$ of X_t, $t \in [0, 1]$, can be easily obtained from the IG distribution function as follows.

$$\begin{aligned}
H_t(x; a) &= P(a^{2/3}t - I_t \le x) \\
&= 1 - P(I_t < a^{2/3}t - x) \\
&= 1 - \mathcal{H}_{IG}(a^{2/3}t - x; at, a^{1/3}), \quad x \in (-\infty, a^{2/3}t).
\end{aligned}$$

For the inverse function, we have the following relation for each $t \in [0, 1]$,

$$H_t^{[-1]}(y; a) = a^{2/3}t - \mathcal{H}_{IG}^{[-1]}(1 - y; at, a^{1/3}), \quad y \in [0, 1].$$

Table 4. Mean, variance, skewness, and kurtosis of the $VG(\sigma, \nu, \theta, \mu)$ distribution.

	$VG(\sigma, \nu, \theta, \mu)$	$VG(\sigma, \nu, 0, \mu)$
Mean	$\theta + \mu$	μ
Variance	$\sigma^2 + \nu\theta^2$	σ^2
Skewness	$\theta\nu(3\sigma^2 + 2\nu\theta^2)/(\sigma^2 + \nu\theta^2)^{3/2}$	0
Kurtosis	$3(1 + 2\nu - \nu\sigma^4(\sigma^2 + \nu\theta^2)^{-2})$	$3(1 + \nu)$

3.4 Based on the VG Process

The Variance-Gamma (VG) distribution with parameters $\sigma > 0$, $\nu > 0$, $\theta \in \mathbb{R}$, and $\mu \in \mathbb{R}$, denoted $VG(\sigma, \nu, \theta, \mu)$, is infinitely divisible with characteristic function

$$\phi_{VG}(u; \sigma, \nu, \theta, \mu) = e^{iu\mu} \left(1 - iu\theta\nu + u^2\sigma^2\nu/2\right)^{-1/\nu}, \quad u \in \mathbb{R}.$$

We can then define the VG process $X^{(VG)} = \{X_t^{(VG)}, t \geq 0\}$ with parameters $\sigma, \nu > 0$ and $\theta, \mu \in \mathbb{R}$ as the process that starts at zero and has stationary independent VG-distributed increments such that $X_t^{(VG)}$ follows a $VG(\sigma\sqrt{t}, \nu/t, \theta t, \mu t)$ distribution. Note that a VG process may also be defined as a Brownian motion with drift time-changed by a gamma process (e.g., Schoutens [20]).

The density function of the $VG(\sigma, \nu, \theta, \mu)$ distribution is given by

$$f_{VG}(x; \sigma, \nu, \theta, \mu) = \frac{(GM)^C}{\sqrt{\pi}\Gamma(C)} \exp\left(\frac{(G-M)(x-\mu)}{2}\right)$$

$$\times \left(\frac{|x-\mu|}{G+M}\right)^{C-1/2} K_{C-1/2}\left((G+M)|x-\mu|/2\right), \quad x \in \mathbb{R},$$

where C, G, M are positive constants given by

$$C = 1/\nu,$$

$$G = \left(\sqrt{\frac{\theta^2\nu^2}{4} + \frac{\sigma^2\nu}{2}} - \frac{\theta\nu}{2}\right)^{-1},$$

$$M = \left(\sqrt{\frac{\theta^2\nu^2}{4} + \frac{\sigma^2\nu}{2}} + \frac{\theta\nu}{2}\right)^{-1}.$$

In Table 4, we give the values of the mean, variance, skewness, and kurtosis of the $VG(\sigma, \nu, \theta, \mu)$ distribution (the case $\theta = 0$ is also included). This distribution is symmetric around μ if $\theta = 0$, whereas negative values of θ result in negative skewness. Also, the parameter ν primarily controls the kurtosis.

The $VG(\sigma, \nu, \theta, \mu)$ distribution satisfies the following scaling property. If $X \sim VG(\sigma, \nu, \theta, \mu)$, then $cX \sim VG(c\sigma, \nu, c\theta, c\mu)$ for all $c > 0$. Also, we have the following convolution property. If $X \sim VG(\sigma\sqrt{\rho}, \nu/\rho, \theta\rho, \mu\rho)$ is independent of $Y \sim \mathrm{VG}(\sigma\sqrt{1-\rho}, \nu/(1-\rho), \theta(1-\rho), \mu(1-\rho))$, then $X+Y \sim \mathrm{VG}(\sigma, \nu, \theta, \mu)$ under the constraint $\rho \in (0,1)$.

The class of VG distributions was introduced by Madan and Seneta [9]. A number of papers have developed the VG model for asset returns and its implications for option pricing. In Madan and Seneta [9; 10] and Madan and Milne [8], the symmetric case $(\theta = 0)$ is considered. In Madan et al. [11], the general case with skewness is treated. In equity and interest rate modeling, the VG process has already proven its capabilities (see, e.g., Schoutens [20]).

Moosbrucker assumes in [12] a one-factor VG model where the asset value of obligor $i = 1, \ldots, n$ is of the form

$$A_i(T) = cY + \sqrt{1 - c^2}\, X_i,$$

where Y, X_i, $i = 1, \ldots, n$ are independently VG-distributed random variables with $X_i \sim \mathrm{VG}(\sqrt{1 - \nu\theta^2}, \nu/(1-c^2), \theta\sqrt{1-c^2}, -\theta\sqrt{1-c^2})$ for all i and $Y \sim \mathrm{VG}\left(\sqrt{1-\nu\theta^2}, \nu/c^2, \theta c, -\theta c\right)$. In this setting, the random variable $A_i(T)$ is $\mathrm{VG}(\sqrt{1-\nu\theta^2}, \nu, \theta, -\theta)$-distributed. All these random variables have indeed zero mean and unit variance but there is a constraint on the parameters, namely $\nu\theta^2 < 1$. Furthermore, we point out that the fractional loss distribution obtained by Moosbrucker [12, p. 19, Eq.(13)] with the LHP method under the above one-factor model approach is not correct. It should be in his notation:

$$F_{\text{portfolio loss}}(x) = 1 - F_M\left(\frac{C - \sqrt{1-c^2}F_{Z_i}^{-1}(x)}{c}\right),$$

because the VG distribution function is not an even function if $(\theta, \mu) \neq (0,0)$.

For a variant of the VG model, extended with an additional normal factor, we refer to Baxter [3].

Many variations are possible. For example, one could start with a zero mean $\mathrm{VG}(\kappa\sigma, \nu, \kappa\theta, -\kappa\theta)$ distribution for $A_i(T)$, with $\kappa = 1/\sqrt{\sigma^2 + \nu\theta^2}$ in order to force unit variance. The one-factor VG-Lévy model is then in the form of (1)

$$A_i(T) = X_\rho + X_{1-\rho}^{(i)}, \qquad i = 1, \ldots, n,$$

where X_ρ, $X_{1-\rho}^{(i)}$, $i = 1, \ldots, n$ are independent VG random variables with the following parameters. The common factor X_ρ follows a distribution $\mathrm{VG}(\kappa\sqrt{\rho}\sigma, \nu/\rho, \kappa\rho\theta, -\kappa\rho\theta)$ and the idiosyncratic risks $X_{1-\rho}^{(i)}$ all follow a distribution $\mathrm{VG}(\kappa\sqrt{1-\rho}\sigma, \nu/(1-\rho), \kappa(1-\rho)\theta, -\kappa(1-\rho)\theta)$.

3.5 Based on the NIG Process

The normal inverse Gaussian (NIG) distribution with parameters $\alpha > 0$, $\beta \in (-\alpha, \alpha)$, $\delta > 0$, and $\mu \in \mathbb{R}$, denoted $\mathrm{NIG}(\alpha, \beta, \delta, \mu)$, has a characteristic function given by

Table 5. Mean, variance, skewness, and kurtosis of the $\mathrm{NIG}(\alpha, \beta, \delta, \mu)$ distribution.

	$\mathrm{NIG}(\alpha, \beta, \delta, \mu)$	$\mathrm{NIG}(\alpha, 0, \delta, \mu)$
Mean	$\mu + \delta\beta/\sqrt{\alpha^2 - \beta^2}$	μ
Variance	$\alpha^2\delta(\alpha^2 - \beta^2)^{-3/2}$	δ/α
Skewness	$3\beta\alpha^{-1}\delta^{-1/2}(\alpha^2 - \beta^2)^{-1/4}$	0
Kurtosis	$3\left(1 + \frac{\alpha^2 + 4\beta^2}{\delta\alpha^2\sqrt{\alpha^2 - \beta^2}}\right)$	$3(1 + \delta^{-1}\alpha^{-1})$

$$\phi_{NIG}(u; \alpha, \beta, \delta, \mu) = \exp\left(iu\mu - \delta\left(\sqrt{\alpha^2 - (\beta + iu)^2} - \sqrt{\alpha^2 - \beta^2}\right)\right), \quad u \in \mathbb{R}.$$

We clearly see that this characteristic function is infinitely divisible. Hence, we can define the NIG process $X^{(NIG)} = \{X_t^{(NIG)}, t \geq 0\}$ with parameters $\alpha > 0$, $\beta \in (-\alpha, \alpha)$, $\delta > 0$, and $\mu \in \mathbb{R}$ as the process that starts at zero and has stationary independent NIG-distributed increments such that $X_t^{(NIG)}$ is $\mathrm{NIG}(\alpha, \beta, \delta t, \mu t)$-distributed.

The NIG distribution was introduced by Barndorff-Nielsen [1]; see also Barndorff-Nielsen [2] and Rydberg [13–15]. Note that the density function of the $\mathrm{NIG}(\alpha, \beta, \delta, \mu)$ distribution is given for any $x \in \mathbb{R}$ by

$$f_{NIG}(x; \alpha, \beta, \delta, \mu) = \frac{\alpha\delta}{\pi} \exp\left(\delta\sqrt{\alpha^2 - \beta^2} + \beta(x - \mu)\right) \frac{K_1(\alpha\sqrt{\delta^2 + (x - \mu)^2})}{\sqrt{\delta^2 + (x - \mu)^2}}.$$

If a random variable X is $\mathrm{NIG}(\alpha, \beta, \delta, \mu)$-distributed and $c > 0$, then cX is $\mathrm{NIG}(\alpha/c, \beta/c, c\delta, c\mu)$-distributed. If $\beta = 0$, the distribution is symmetric around μ. This can be seen from the characteristics of the NIG distribution given in Table 5. Furthermore, if $X \sim \mathrm{NIG}(\alpha, \beta, \delta_1, \mu_1)$ is independent of $Y \sim \mathrm{NIG}(\alpha, \beta, \delta_2, \mu_2)$, then $X + Y \sim \mathrm{NIG}(\alpha, \beta, \delta_1 + \delta_2, \mu_1 + \mu_2)$.

Guégan and Houdain propose in [6] a factor model based on a NIG-distributed common factor but with standard normal idiosyncratic risks.

Kalemanova et al. [7] define a one-factor NIG model in the following way,

$$A_i(T) = aY + \sqrt{1 - a^2}X_i, \qquad i = 1, \ldots, n,$$

where Y, X_i, $i = 1, \ldots, n$ are independent NIG random variables with the following parameters. The common factor Y follows a $\mathrm{NIG}(\alpha, \beta, \alpha, -(\alpha\beta/\sqrt{\alpha^2 - \beta^2}))$ distribution and the idiosyncratic risks X_i all follow a $\mathrm{NIG}(\alpha\sqrt{1 - a^2}/a, \beta\sqrt{1 - a^2}/a, a\sqrt{1 - a^2}/a, -(\alpha\beta/\sqrt{\alpha^2 - \beta^2})\sqrt{1 - a^2}/a)$ distribution. This leads to a $\mathrm{NIG}(\alpha/a, \beta/a, \alpha/a, -(\alpha\beta/(a\sqrt{\alpha^2 - \beta^2})))$ distribution for $A_i(T)$. Note that this distribution does not have unit variance whereas the parameters δ and μ are fixed to obtain a zero mean distribution. Incidentally, we point out that the distribution of the fractional loss obtained by Kalemanova et al. [7, p.10] with the LHP approximation under the above one-factor model approach is not correct. It should be in their notation:

$$F_\infty(x) = 1 - F_{\mathcal{NIG}(1)}\left(\frac{C - \sqrt{1-a^2}F^{-1}_{\mathcal{NIG}\sqrt{1-a^2}/a}(x)}{a}\right),$$

because the NIG distribution function is not an even function if $(\beta, \mu) \neq (0,0)$.

Using our methodology, one can set up similar NIG models. For example, let $X = \{X_t, t \in [0,1]\}$ be a NIG process, where X_1 follows a distribution $\text{NIG}(\alpha, \beta, (\alpha^2 - \beta^2)^{3/2}/\alpha^2, -(\alpha^2 - \beta^2)\beta/\alpha^2)$. Note that the parameters δ and μ are chosen such that X_1 has zero mean and unit variance. The one-factor NIG–Lévy model is then

$$A_i(T) = X_\rho + X^{(i)}_{1-\rho}, \qquad i = 1, \dots, n,$$

where $X_\rho, X^{(i)}_{1-\rho}, i = 1, \dots, n$ are independently NIG-distributed random variables with $X^{(i)}_{1-\rho} \sim \text{NIG}(\alpha, \beta, (1-\rho)(\alpha^2 - \beta^2)^{3/2}/\alpha^2, -(1-\rho)(\alpha^2 - \beta^2)\beta/\alpha^2)$ for all i and $X_\rho \sim \text{NIG}(\alpha, \beta, \rho(\alpha^2 - \beta^2)^{3/2}/\alpha^2, -\rho(\alpha^2 - \beta^2)\beta/\alpha^2)$. By construction, each $A_i(T)$ follows the same distribution as X_1.

3.6 Based on the Meixner Process

The density function of the Meixner distribution $(\text{Meixner}(\alpha, \beta, \delta, \mu))$ is given for any $x \in \mathbb{R}$ by

$$f_{Meixner}(x; \alpha, \beta, \delta, \mu) = \frac{(2\cos(\beta/2))^{2\delta}}{2\alpha\pi\,\Gamma(2\delta)}\exp\left(\frac{\beta(x-\mu)}{\alpha}\right)\left|\Gamma\left(\delta + \frac{i(x-\mu)}{\alpha}\right)\right|^2,$$

where $\alpha > 0$, $\beta \in (-\pi, \pi)$, $\delta > 0$, and $\mu \in \mathbb{R}$.

The characteristic function of the $\text{Meixner}(\alpha, \beta, \delta, \mu)$ distribution is

$$\phi_{Meixner}(u; \alpha, \beta, \delta, \mu) = e^{iu\mu}\left(\frac{\cos(\beta/2)}{\cosh((\alpha u - i\beta)/2)}\right)^{2\delta}, \quad u \in \mathbb{R}.$$

The $\text{Meixner}(\alpha, \beta, \delta, \mu)$ distribution being infinitely divisible, we can then associate with it a Lévy process which we call the Meixner process. More precisely, a Meixner process $X^{(Meixner)} = \{X^{(Meixner)}_t, t \geq 0\}$ with parameters $\alpha > 0$, $\beta \in (-\pi, \pi)$, $\delta > 0$, and $\mu \in \mathbb{R}$ is a stochastic process that starts at zero and has stationary independent Meixner-distributed increments such that $X^{(Meixner)}_t$ is $\text{Meixner}(\alpha, \beta, \delta t, \mu t)$-distributed.

The Meixner process was introduced in Schoutens and Teugels [21]; see also Schoutens [17]. It was suggested to serve for fitting stock returns in Grigelionis [5]. This application in finance was worked out in Schoutens [18; 19].

In Table 6, we give some relevant quantities for the general case and the symmetric case around μ, that is, with $\beta = 0$. Note that the kurtosis of any Meixner distribution is greater than that of the normal distribution. Now, if X

Table 6. Mean, variance, skewness, and kurtosis of the Meixner($\alpha, \beta, \delta, \mu$) distribution.

	Meixner($\alpha, \beta, \delta, \mu$)	Meixner($\alpha, 0, \delta, \mu$)
Mean	$\mu + \alpha\delta\tan(\beta/2)$	μ
Variance	$(\cos^{-2}(\beta/2))\alpha^2\delta/2$	$\alpha^2\delta/2$
Skewness	$\sin(\beta/2)\sqrt{2/\delta}$	0
Kurtosis	$3 + (2 - \cos(\beta))/\delta$	$3 + 1/\delta$

is Meixner($\alpha, \beta, \delta, \mu$)-distributed and $c > 0$, then cX is Meixner($c\alpha, \beta, \delta, c\mu$)-distributed. Furthermore, if $X \sim$ Meixner($\alpha, \beta, \delta_1, \mu_1$) is independent of $Y \sim$ Meixner($\alpha, \beta, \delta_2, \mu_2$), then $X + Y \sim$ Meixner($\alpha, \beta, \delta_1 + \delta_2, \mu_1 + \mu_2$).

Using our methodology, one can easily set up a Meixner model. For example, let $X = \{X_t, t \in [0,1]\}$ be a Meixner process, where X_1 follows a distribution Meixner($\alpha, \beta, 2\cos^2(\beta/2)/\alpha^2, -\sin(\beta)/\alpha$). Note that again the parameters δ and μ are chosen such that X_1 has zero mean and unit variance. The one-factor Meixner–Lévy model is then

$$A_i(T) = X_\rho + X_{1-\rho}^{(i)}, \qquad i = 1, \ldots, n,$$

where X_ρ, $X_{1-\rho}^{(i)}$, $i = 1, \ldots, n$ are independent Meixner random variables with the following parameters. The common factor X_ρ follows a Meixner($\alpha, \beta,$ $2\rho\cos^2(\beta/2)/\alpha^2, -\rho\sin(\beta)/\alpha$) distribution and the idiosyncratic risks $X_{1-\rho}^{(i)}$ all follow a Meixner($\alpha, \beta, 2(1 - \rho)\cos^2(\beta/2)/\alpha^2, -(1 - \rho)\sin(\beta)/\alpha$) distribution. By construction, each $A_i(T)$ follows the same distribution as X_1.

3.7 Other Candidate Models

We hope the idea is clear and invite the reader to set up the CGMY, Generalized Hyperbolic (GH), Generalized z (GZ), and other Lévy-based models in a similar way. For definitions, we refer to Schoutens [20].

4 Fair Pricing of a Synthetic CDO Tranche

In this section, we explain the procedure for valuing the tranches of synthetic CDOs. Consider a synthetic CDO tranche on a given reference portfolio of n names defined by an interval $[K_1, K_2]$ of loss fractions on the total portfolio notional N for which the tranche investor is responsible. The endpoints K_1 and K_2 of the interval are called attachment and detachment points, respectively. The tranche investor receives periodic spread payments from the CDO issuer (the *premium leg*) and makes payments to the CDO issuer when defaults affect the tranche (the *protection leg*). Note that for a synthetic CDO, any default corresponds to a credit event under a CDS in the reference portfolio. It turns out that the fair price of the tranche $[K_1, K_2]$ can be calculated using the same idea as for the pricing of a CDS, that is, by setting the fair premium s such that the expected present values of the premium leg and the protection leg are equal.

The loss fraction on the portfolio notional at time t is given by

$$L_{t,n} := \frac{1}{N} \sum_{i=1}^{n} 1_{\{\tau_i \le t\}}(1 - R_i)N_i,$$

where τ_i, R_i, and N_i denote the default time, the recovery rate, and the notional amount of name i, respectively, $i = 1, \ldots, n$. Under the factor model (1), we have $\{\tau_i \le t\} = \{A_i(t) \le K_i(t)\}$ with $K_i(t) = H_1^{[-1]}(p_i(t))$, $i = 1, \ldots, n$.

The loss fraction on the CDO tranche $[K_1, K_2]$ at time t is simply expressed by means of $L_{t,n}$ as

$$L_{t,n}(K_1, K_2) := \frac{\max\{\min(L_{t,n}, K_2) - K_1, 0\}}{K_2 - K_1}.$$

We assume that the payments (premium and protection legs) occur on periodic payment dates $t_1, \ldots, t_m$. Furthermore, we assume that the CDO issuer receives compensation at the next scheduled payment date after a default has occurred. Note that payments are only made as long as the effective notional of the tranche at time t_i is positive. Denote $\tau_{(1)}, \ldots, \tau_{(n)}$ the order statistics, arranged in increasing order, of the random sample $\tau_1, \ldots, \tau_n$ of default times. Put $t_0 := 0$, $\tau_{(0)} := 0$, and $L_{0,n} := 0$. In what follows, expectations are taken under a risk-neutral measure, that is, risk-adjusted expectations.

The expected present value of the premium leg of the tranche is the present value of all spread payments the tranche investor expects to receive

$$\begin{aligned} \mathrm{PL}(s) = s\,E\Bigg[\sum_{j=1}^{m}\Bigg\{ &(t_j - t_{j-1})\,D(0,t_j)\left(1 - L_{t_j,n}(K_1,K_2)\right)\\ &+ \sum_{i=1}^{n} 1_{\{t_{j-1}<\tau_{(i)}<t_j\}}\left(\tau_{(i)} - t_{j-1}\right)D(0,\tau_{(i)})\\ &\times \left(L_{\tau_{(i)},n}(K_1,K_2) - L_{\tau_{(i-1)},n}(K_1,K_2)\right)\Bigg\}\Bigg], \end{aligned}$$

where $D(0,t_j)$ is the risk-free discount factor for payment date t_j and s is the spread per annum paid to the tranche investor. The term $1 - L_{t_j,n}(K_1,K_2)$ is the fraction of the tranche notional outstanding on payment date t_j and reflects the decline in notional as defaults affect the tranche. The second term in the sum over j corresponds to the discounted sum of accrual payments the tranche investor receives when defaults occur between payment dates. They are paid at the next payment date and are based on the previous effective tranche notional.

The expected present value of the protection leg of the tranche is the discounted sum of the expected payments the tranche investor must make when defaults affect the tranche:

$$\text{LL} = E\left[\sum_{j=1}^{m} D(0,t_j) \sum_{i=1}^{n} \mathbf{1}_{\{t_{j-1}<\tau_{(i)}<t_j\}} \left(L_{\tau_{(i)},n}(K_1,K_2) - L_{\tau_{(i-1)},n}(K_1,K_2)\right)\right].$$

The fair tranche premium s_{par} is then the spread s solving $\text{PL}(s) = \text{LL}$. As a consequence, the periodic payment received by the tranche investor from the CDO issuer in return for bearing the risk of losses is equal to s_{par} times the effective outstanding notional of the tranche.

Now, assume a homogeneous portfolio. Recall that all obligors have the same default barrier $K_i(t) = K(t)$, the same recovery rate $R_i = R$, and the same notional amount $N_i = N/n$. Each of the n obligors either takes no loss or a loss of $(1-R)(N/n)$ so that multiplying the number of defaults by $(1-R)(N/n)$ gives losses. Clearly, $L_{t,n}$ reduces to the loss fraction $L_{t,n}^{\text{HP}}$ defined in (2). It follows that the expected loss fraction on the portfolio at time t is

$$E[L_{t,n}^{\text{HP}}] = \frac{1-R}{n} \sum_{k=0}^{n} k\, P\left(L_{t,n}^{\text{HP}} = \frac{k(1-R)}{n}\right),$$

and, denoting $L_{t,n}^{\text{HP}}(K_1,K_2) := L_{t,n}(K_1,K_2)$, that the expected loss fraction on the CDO tranche $[K_1, K_2]$ at time t is

$$E[L_{t,n}^{\text{HP}}(K_1,K_2)]$$
$$= \frac{1}{K_2 - K_1} \sum_{k=0}^{n} P\left(L_{t,n}^{\text{HP}} = \frac{k(1-R)}{n}\right) \max\left\{\min\left(\frac{k(1-R)}{n}, K_2\right) - K_1, 0\right\},$$

where $P\left(L_{t,n}^{\text{HP}} = (k(1-R))/n\right)$ is the probability that exactly k defaults occur by time t, $k = 0, \ldots, n$.

As soon as the expected loss fraction on the tranche is calculated, the computation of the tranche premium becomes easy. Unfortunately, the derivation of the fractional loss distribution on the reference portfolio is not trivial. However, under the above homogeneity assumptions, we know from Subsection 2.3 that this distribution can be approximated using the LHP approximation method. As a consequence, this method provides an easy tool to compute both the fractional loss distribution and the expected loss fraction on the tranche over different time horizons.

With the LHP approximation given in (3), that is, $F_{t,n}^{\text{HP}}(x) = F_t^{\text{HP}}(x/(1-R))$ for $x \in [0, 1-R]$, we compute the expected loss fraction on the CDO tranche $[K_1, K_2]$ at time t as

$$E[L_{t,n}^{\text{HP}}(K_1,K_2)]$$
$$= E\left[\frac{\max\{\min(L_{t,n}^{\text{HP}}, K_2) - K_1, 0\}}{K_2 - K_1}\right]$$

Table 7. Pricing of iTraxx tranches of May 4, 2006 with LHP Lévy models.

Model/Quotes	0–3% (bp)	3–6% (bp)	6–9% (bp)	9–12% (bp)	12–22% (bp)	Absolute Error (bp)
Market	17	44.0	12.8	6.0	2.0	
Gaussian	17	105.7	22.4	5.7	0.7	73.7
Shifted Gamma	17	44.0	19.7	11.9	6.0	16.8
Shifted IG	17	44.0	19.8	12.2	6.5	17.7
VG	17	43.9	21.8	14.1	7.8	23.0
NIG	17	44.0	24.1	17.1	11.7	32.1

$$= E\left[\frac{\max(L_{t,n}^{\mathrm{HP}} - K_1, 0) - \max(L_{t,n}^{\mathrm{HP}} - K_2, 0)}{K_2 - K_1}\right]$$

$$= \frac{1}{K_2 - K_1}\left(\int_{K_1}^{1-R}(x - K_1)\,dF_{t,n}^{\mathrm{HP}}(x) - \int_{K_2}^{1-R}(x - K_2)\,dF_{t,n}^{\mathrm{HP}}(x)\right)$$

$$= \frac{1-R}{K_2 - K_1}\left(\int_{K_1/(1-R)}^{1}\left(x - \frac{K_1}{1-R}\right)dF_t^{\mathrm{HP}}(x)\right.$$

$$\left. - \int_{K_2/(1-R)}^{1}\left(x - \frac{K_2}{1-R}\right)dF_t^{\mathrm{HP}}(x)\right).$$

Similarly, the expected loss fraction on the last tranche $[K, 1]$ at time t is given by

$$E[L_{t,n}^{\mathrm{HP}}(K, 1)] = E\left[\frac{\max(L_{t,n}^{\mathrm{HP}} - K, 0)}{1 - K}\right]$$

$$= \frac{1}{1-K}\int_{K}^{1-R}(x - K)\,dF_{t,n}^{\mathrm{HP}}(x)$$

$$= \frac{1-R}{1-K}\int_{K/(1-R)}^{1}\left(x - \frac{K}{1-R}\right)dF_t^{\mathrm{HP}}(x).$$

Finally, we report on a small calibration exercise of the Gaussian, shifted gamma, shifted IG, VG, and NIG cases. We calibrate the model to the iTraxx of the 4th of May 2006. In Table 7, one finds the market quotes together with the calibrated model quotes for the different tranches. Note that for the 0–3% tranche, the upfront is quoted with a 500 bp running.

Acknowledgments

The authors thank Nomura and ING for providing the data. Special thanks go to Martin Baxter and Jürgen Tistaert. Hansjörg Albrecher is supported by the Austrian Science Fund Project P18392. Sophie A. Ladoucette is supported by the grant BDB-B/04/03 from the Catholic University of Leuven.

References

1. O.E. Barndorff-Nielsen. Normal inverse Gaussian distributions and the modelling of stock returns. Research Report No. 300, Department of Theoretical Statistics, Aarhus University, 1995.
2. O.E. Barndorff-Nielsen. Normal inverse Gaussian distributions and stochastic volatility modelling. *Scandinavian Journal of Statistics*, 24(1):1–13, 1997.
3. M. Baxter. Dynamic modelling of single-name credits and CDO tranches. Working Paper - Nomura Fixed Income Quant Group, 2006.
4. J. Bertoin. *Lévy Processes*. Cambridge Tracts in Mathematics 121. Cambridge University Press, 1996.
5. B. Grigelionis. Processes of Meixner type. *Lithuanian Mathematical Journal*, 39(1):33–41, 1999.
6. D. Guégan and J. Houdain. Collateralized debt obligations pricing and factor models: A new methodology using normal inverse Gaussian distributions. Research Report IDHE-MORA No. 07-2005, ENS Cachan, 2005.
7. A. Kalemanova, B. Schmid, and R. Werner. The normal inverse Gaussian distribution for synthetic CDO pricing. Technical Report, 2005.
8. D.B. Madan and F. Milne. Option pricing with VG martingale components. *Mathematical Finance*, 1(4):39–55, 1991.
9. D.B. Madan and E. Seneta. Chebyshev polynomial approximations and characteristic function estimation. *Journal of the Royal Statistical Society, Series B*, 49(2):163–169, 1987.
10. D.B. Madan and E. Seneta. The Variance-Gamma (V.G.) model for share market returns. *Journal of Business*, 63(4):511–524, 1990.
11. D.B. Madan, P.P. Carr, and E.C. Chang. The variance gamma process and option pricing. *European Finance Review*, 2:79–105, 1998.
12. T. Moosbrucker. Pricing CDOs with correlated variance gamma distributions. Research Report, Department of Banking, University of Cologne, 2006.
13. T. Rydberg. Generalized hyperbolic diffusions with applications towards finance. Research Report No. 342, Department of Theoretical Statistics, Aarhus University, 1996.
14. T. Rydberg. The normal inverse Gaussian Lévy process: Simulations and approximation. Research Report No. 344, Department of Theoretical Statistics, Aarhus University, 1996.
15. T. Rydberg. A note on the existence of unique equivalent martingale measures in a Markovian setting. *Finance and Stochastics*, 1:251–257, 1997.
16. K. Sato. *Lévy Processes and Infinitely Divisible Distributions*. Cambridge Studies in Advanced Mathematics 68. Cambridge University Press, 2000.
17. W. Schoutens. *Stochastic Processes and Orthogonal Polynomials*. Lecture Notes in Statistics 146. Springer-Verlag, 2000.
18. W. Schoutens. The Meixner process in finance. EURANDOM Report 2001-002, EURANDOM, Eindhoven, 2001.
19. W. Schoutens. Meixner processes: Theory and applications in finance. EURANDOM Report 2002-004, EURANDOM, Eindhoven, 2002.
20. W. Schoutens. *Lévy Processes in Finance - Pricing Financial Derivatives*. John Wiley & Sons, 2003.
21. W. Schoutens and J.L. Teugels. Lévy processes, polynomials and martingales. *Communications in Statistics - Stochastic Models*, 14(1–2):335–349, 1998.
22. O. Vasicek. Probability of loss on loan portfolio. Technical Report, KMV Corporation, 1987.

Utility Valuation of Credit Derivatives: Single and Two-Name Cases

Ronnie Sircar[1] and Thaleia Zariphopoulou[2]

[1] ORFE Department
Princeton University
E-Quad
Princeton, NJ 08544, USA
sircar@princeton.edu

[2] Department of Mathematics and Information
Department of Risk and Operations Management
The University of Texas at Austin
Austin, TX 78712, USA
zariphop@math.utexas.edu

Summary. We study the effect of risk aversion on the valuation of credit derivatives. Using the technology of utility-indifference valuation in intensity-based models of default risk, we analyze resulting yield spreads for single-name defaultable bonds and a simple representative two-name credit derivative. The impact of risk averse valuation on prices and yield spreads is expressed in terms of "effective correlation."

Key words: Credit derivatives; indifference pricing; reaction–diffusion equations.

1 Introduction

In this article, we analyze the impact of risk aversion on the valuation of defaultable bonds and a simple multiname credit derivative. Our approach is to work within intensity-based models, as initiated by, among others, Artzner and Delbaen [1], Madan and Unal [26], Lando [23], and Jarrow and Turnbull [19]. In these models, a firm's default time is modeled directly by a totally inaccessible stopping time, typically the first jump of a Cox process. However, rather than pricing using no-arbitrage arguments, we apply the utility-based valuation methodology, which entails analysis of portfolio optimization problems under default risk.

A major limitation of many traditional approaches is the inability to capture and explain high premiums observed in credit derivatives markets for unlikely events, for example, the spreads quoted for senior tranches of collaterized debt obligations (CDOs) written on investment grade firms. The approach explored here, and in our related work [29], aims to explain such phenomena as a consequence of tranche holders' risk aversion, and to quantify this through the mechanism of utility-indifference valuation. For a general introduction to credit risk, including other approaches to default, such as structural models, we refer, for example, to the books [5; 13; 24; 27].

1.1 Valuation Mechanisms

In complete financial market environments, such as in the classical Black–Scholes model, the payoffs of derivative securities can be replicated by trading strategies in the underlying securities, and their prices are naturally deduced from the value of these associated portfolios. However, once nontraded risks, such as unpredictable defaults, are considered, the possibility of replication and, therefore, risk elimination breaks down and alternative ways are needed for the quantification of risk and assignation of price. One approach is to use market derivatives data, when available, to identify which of the many feasible arbitrage-free pricing measures is consistent with market prices. In a different direction, valuation of claims involving nontradable risks can be based on optimality of decisions once this claim is incorporated in the investor's portfolio. Naturally, the risk attitude of the individual needs to be taken into account, and this is typically modeled by a concave and increasing utility function U. In a static framework, prices are determined through the *certainty equivalent*, otherwise known as the principle of equivalent utility [6; 17]. The utility-based value of the claim, written on the risk Y and yielding payoff $C(Y)$, is $\nu(C) = U^{-1}\{E_P[U(C(Y))]\}$. Note that the arbitrage-free price and the certainty-equivalent price are very different. The first is linear and uses the risk-neutral measure. The certainty-equivalent price is nonlinear and uses the historical assessment of risks.

Prompted by the ever-increasing number of applications (event risk-sensitive claims, insurance plans, mortgages, weather derivatives, etc.), considerable effort has been put into analyzing the utility-based valuation mechanism. Due to the prevalence of instruments dependent on nonmarket risks (such as default), there is a great need for building new dynamic pricing rules. These rules should identify and price unhedgeable risks and, at the same time, build optimal risk-monitoring policies. In this direction, a dynamic utility-based pricing theory has been developed producing the so-called *indifference prices*. The approach is based on finding the amount at which the buyer of the claim is indifferent, in terms of maximum expected utility, between holding or not holding the derivative. Specification of the indifference price requires understanding how investors act optimally with or without the derivative at hand. These issues are naturally addressed through stochastic optimization problems of expected utility maximization. We refer to [21; 22] and [8] as classical references in this area. The indifference approach was initiated for European claims by Hodges and Neuberger [18] and further extended by Davis et al. [10].

1.2 Credit Derivatives

As well as single-name securities, such as credit default swaps (CDSs), in which there is a relatively liquid market, basket, or multiname, products have generated considerable over-the-counter activity. Popular cases are collateralized debt obligations, whose payoffs depend on the default events of a basket portfolio of up to 300 firms, over a five-year period. As long as there are no defaults, investors in CDO tranches enjoy high yields, but, as defaults start occurring, they affect first the high-yield equity tranche, then the mezzanine tranches, and, perhaps, the senior and super-senior tranches. See Davis and Lo [9] or Elizalde [15] for a concise introduction to these products.

The focus of modeling in the credit derivatives industry has been on correlation between default times. Partly, this is due to the adoption of the one-factor Gaussian copula model as industry standard and the practice (up until recently) of analyzing tranche prices through implied correlation. This revealed that traded prices of senior tranches could only be realized through these models with an implausibly high correlation parameter, the so-called correlation smile.

Rather than focusing on models with "enough correlation" to reproduce market observations via traditional no-arbitrage pricing, our goal is to understand the effects of risk aversion on valuation of single- and two-name credit derivatives. Questions of interest are (i) how does risk aversion affect the value of portfolios that are sensitive to the potential default of a number of firms, and so to correlation between these events, and (ii) does the nonlinearity of the indifference pricing mechanism enhance the impact of correlation. It seems natural that some of the prices, or spreads, seen in credit markets are due mainly to "crash-o-phobia" in a relatively illiquid market, with the effect enhanced nonlinearly in baskets. When super-senior tranches offer nontrivial spreads (albeit a few basis points) for protection against the default risk of 15–30% of investment grade U.S. firms over the next five years, they are ascribing a seemingly large probability to "the end of the world as we know it." We seek to capture this directly as an effect of risk aversion leading to effective or perceived correlation, in contrast to a mechanism of high direct correlation.

Taking the opposite angle, the method of indifference valuation should be attractive to participants in this still quite illiquid over-the-counter (OTC) market. It is a direct way for them to quantify the default risks they face in a portfolio of complex instruments, when calibration data are scarce. Unlike well-developed equity and fixed-income derivatives markets, where the case for traditional arbitrage-free valuation is more compelling, the potential for utility valuation to account for high dimensionality in a way that is consistent with investors' fears of a cascade of defaults is a case for its application here.

For applications of indifference valuation to credit risk, see also Collin-Dufresne and Hugonnier [7], Bielecki et al. [4; 3], and Shouda [28].

1.3 Outline

In Section 2, we introduce the model used to value a single-name defaultable bond. We study there the two utility optimization problems that lead to the indifference price. The first is the usual Merton optimal investment problem, and the second is the portfolio problem where the investor also holds the corporate bond. We present the analysis under stochastic default intensity, and give bounds on the indifference value. In the case of constant intensity, we give the explicit formula and study the implied term structure of yield spreads.

In Section 3, we take a first step towards extending the analysis to multi-name credit derivatives, by looking at a simple two-name example. This highlights the key role of a diversity coefficient in indifference valuation. In the case of constant intensities, we study the indifference value as a function of the maturity, and describe the correlating effect of the utility valuation mechanism. We conclude in Section 4.

2 Indifference Valuation: Single Name

We start with single-name defaultable bonds to illustrate the approach. We work within models incorporating information from the firm's stock price S, but unlike in a traditional structural approach, default occurs at a nonpredictable stopping time τ with stochastic intensity process λ, which is correlated with the firm's stock price. These are sometimes called *hybrid* models (see, e.g., [25]). The process S could alternatively be taken as the price of another firm or index used to hedge the default risk. Of course, the choice of the investment opportunity set affects the ensuing indifference price.

The stock price S is taken to be a geometric Brownian motion. The intensity process is $\lambda(Y_t)$, where $\lambda(\cdot)$ is a nonnegative, locally Lipschitz, smooth, and bounded function, and Y is a correlated diffusion. The dynamics of S and Y are

$$dS_t = \mu S_t \, dt + \sigma S_t \, dW_t^{(1)}, \qquad S_0 = S > 0,$$
$$dY_t = b(Y_t) \, dt + a(Y_t) \left(\rho \, dW_t^{(1)} + \sqrt{1 - \rho^2} \, dW_t^{(2)} \right), \qquad Y_0 = y \in \mathbb{R}.$$

The coefficients a and b are taken to be Lipschitz functions with sublinear growth. The processes $W^1 = (W_t^{(1)})$ and $W^2 = (W_t^{(2)})$ are independent standard Brownian motions defined on a probability space $(\Omega, \mathcal{F}, P)$, and we denote $\mathcal{F}_t$ the augmented σ-algebra generated by $((W_u^{(1)}, W_u^{(2)}); 0 \le u \le t)$. The parameter $\rho \in (-1, 1)$ measures the instantaneous correlation between shocks to the stock price S and shocks to the intensity-driving process Y. In applications, it is natural to expect that $\lambda(\cdot)$ and ρ are specified in a way such that the intensity tends to rise when the stock price falls. Random fluctuations in the default intensity may be due to economywide factors, as well as firm-

or industry-specific issues that cause yield spreads to change. They may also incorporate the effects of ratings changes. We refer to [13, Chapter 3] for a detailed discussion.

The default time τ of the firm is defined by

$$\tau = \inf \left\{ t \geq 0 : \int_0^t \lambda(Y_s)\, ds = \xi \right\},$$

the first time the cumulated intensity reaches the standard exponentially distributed random variable ξ, which is independent of the Brownian motions.

2.1 Maximal Expected Utility Problem

Let $T < \infty$ denote our finite fixed horizon, chosen later to coincide with the expiration date of the derivative contract of interest. We consider the portfolio problem of an investor who can trade the stock S and has access to a riskless money market account that pays interest at rate r. The investor's control process is π_t, representing the dollar amount held in the stock at time t, until $\tau \wedge T$. For $t < \tau \wedge T$, her wealth process $X = (X_t)$ follows

$$dX_t = \pi_t \frac{dS_t}{S_t} + r(X_t - \pi_t)\, dt$$

$$= (rX_t + \pi_t(\mu - r))\, dt + \sigma \pi_t\, dW_t^{(1)}.$$

The control $\pi = (\pi_t)$ is called admissible if it is $\mathcal{F}_t$-measurable and satisfies the integrability constraint $E[\int_0^T \pi_s^2\, ds] < \infty$. The set of admissible policies is denoted $\mathcal{A}$.

If the default event occurs before T, the investor can no longer trade the firm's stock. She has to liquidate holdings in the stock and deposit in the bank account, so the effect is to reduce her investment opportunities. For simplicity, we assume she receives full predefault market value on her stock holdings on liquidation, although one might extend to consider some loss, or jump downwards in the stock price at the default time (see, e.g., [25] for such a model). Therefore, given that $\tau < T$, for $\tau \leq t \leq T$, we have

$$X_t = X_\tau e^{r(t-\tau)},$$

as the bank account is the only remaining investment.

We work with exponential utility of discounted (to time zero) wealth. We are first interested in the optimal investment problem up to time T of the investor who does not hold any derivative security. At time zero, the maximum expected utility payoff then takes the form

$$\sup_{\pi \in \mathcal{A}} E\left[-e^{-\gamma(e^{-rT}X_T)} \mathbf{1}_{\{\tau > T\}} + (-e^{-\gamma(e^{-r\tau}X_\tau)}) \mathbf{1}_{\{\tau \leq T\}} \right].$$

We switch to the discounted variable $X_t \mapsto e^{-rt}X_t$ and excess growth rate $\mu \mapsto \mu - r$; with a slight abuse, we use the same notation.

Next, we consider the stochastic control problem initiated at time $t \leq T$, and define the default time τ_t by

$$\tau_t = \inf\left\{s \geq t : \int_t^s \lambda(Y_u)\,du = \xi\right\},$$

where ξ is an independent standard exponential random variable.

In the absence of the defaultable claim, the investor's value function is given by

$$M(t,x,y) = \sup_{\pi \in \mathcal{A}} E_{t,x,y}\left[-e^{-\gamma X_T}\mathbf{1}_{\{\tau_t > T\}} + (-e^{-\gamma X_{\tau_t}})\mathbf{1}_{\{\tau_t \leq T\}}\right], \qquad (1)$$

where $E_{t,x,y}[\cdot] = E[\cdot \mid X_t = x, Y_t = y]$.

Proposition 1. *The value function* $M : [0,T] \times \mathbb{R} \times \mathbb{R} \to \mathbb{R}^-$ *is the unique viscosity solution in the class of functions that are concave and increasing in x, and uniformly bounded in y of the Hamilton–Jacobi–Bellman (HJB) equation*

$$M_t + \mathcal{L}_y M + \max_\pi \left\{\frac{1}{2}\sigma^2\pi^2 M_{xx} + \pi(\rho\sigma a(y)M_{xy} + \mu M_x)\right\}$$

$$+ \lambda(y)(-e^{-\gamma x} - M) = 0, \qquad (2)$$

with $M(T,x,y) = -e^{-\gamma x}$ *and*

$$\mathcal{L}_y = \frac{1}{2}a(y)^2\frac{\partial^2}{\partial y^2} + b(y)\frac{\partial}{\partial y}.$$

Proof. The proof follows by extension of the arguments used in Theorem 4.1 of Duffie and Zariphopoulou [14] and is omitted. $\qquad\square$

2.2 Bondholder's Problem and Indifference Price

We now consider the same problem from the point of view of an investor who owns a defaultable bond of the firm. The bond pays \$1 on date T if the firm has survived until then. Defining $c = e^{-rT}$, we have the bondholder's value function

$$H(t,x,y) = \sup_{\pi \in \mathcal{A}} E_{t,x,y}\left[-e^{-\gamma(X_T+c)}\mathbf{1}_{\{\tau_t > T\}} + (-e^{-\gamma X_{\tau_t}})\mathbf{1}_{\{\tau_t \leq T\}}\right]. \qquad (3)$$

As in Proposition 1 for the plain investor's value function M, we have the following HJB characterization.

Proposition 2. *The value function* $H : [0, T] \times \mathbb{R} \times \mathbb{R} \to \mathbb{R}^-$ *is the unique viscosity solution in the class of functions that are concave and increasing in* x, *and uniformly bounded in* y *of the HJB equation*

$$H_t + \mathcal{L}_y H + \max_\pi \left\{ \frac{1}{2}\sigma^2\pi^2 H_{xx} + \pi(\rho\sigma a(y)H_{xy} + \mu H_x) \right\} + \lambda(y)(-e^{-\gamma x} - H) = 0,$$

$$(4)$$

with $H(T, x, y) = -e^{-\gamma(x+c)}$.

The indifference value of the defaultable bond, from the point of view of the bondholder, is the reduction in his initial wealth level such that his maximum expected utility H is the same as the plain investor's value function M.

Definition 1. *The buyer's indifference price* $p_0(T)$ *(at time zero) of a defaultable bond with expiration date* T *is defined by*

$$M(0, x, y) = H(0, x - p_0, y).$$

$$(5)$$

The indifference price at times $0 < t < T$ can be defined similarly, with minor modifications to the previous calculations (in particular, with quantities discounted to time t dollars.)

2.3 Variational Results

In this section, we present some simple bounds for the value functions and the indifference price introduced above.

Proposition 3. *The value functions* M *and* H *satisfy, respectively,*

$$-e^{-\gamma x} \leq M(t, x, y) \leq -e^{-\gamma x - (\mu^2/2\sigma^2)(T-t)},$$

$$(6)$$

and

$$-e^{-\gamma x} + (e^{-\gamma x} - e^{-\gamma(x+c)})P(\tau_t > T \mid Y_t = y) \leq H(t, x, y)$$

$$\leq -e^{-\gamma(x+c)-(\mu^2/2\sigma^2)(T-t)}. (7)$$

Proof. We start with establishing (6). We first observe that the function $\tilde{M}(t, x, y) = -e^{-\gamma x}$ is a subsolution of the HJB equation (2). Moreover, $\tilde{M}(T, x, y) = M(T, x, y)$. The lower bound then follows from the comparison principle. □

Similarly, testing the function

$$\tilde{M}(t, x, y) = -e^{-\gamma x - (\mu^2/2\sigma^2)(T-t)}$$

yields

$$\tilde{M}_t + \mathcal{L}_y \tilde{M} + \max_\pi \left\{ \frac{1}{2}\sigma^2 \pi^2 \tilde{M}_{xx} + \pi(\rho\sigma a(y)\tilde{M}_{xy} + \mu\tilde{M}_x) \right\}$$
$$+ \lambda(y)\left(-e^{-\gamma x} + e^{-\gamma x - (\mu^2/2\sigma^2)(T-t)}\right)$$
$$= \lambda(y)e^{-\gamma x}\left(e^{-(\mu^2/2\sigma^2)(T-t)} - 1\right) \le 0.$$

Therefore, $\tilde{M}$ is a supersolution, with $\tilde{M}(T,x,y) = M(T,x,y)$, and the upper bound follows.

Next, we establish (7). To obtain the lower bound, we follow the sub-optimal policy of investing exclusively in the default-free bank account (i.e., taking $\pi \equiv 0$). Then

$$H(t,x,y) \ge E_{t,x,y}\left[-e^{-\gamma(x+c)}\mathbf{1}_{\{\tau_t > T\}} + (-e^{-\gamma x})\mathbf{1}_{\{\tau_t \le T\}} \right]$$
$$= -e^{-\gamma(x+c)}P(\tau_t > T \mid Y_t = y) + (-e^{-\gamma x})P(\tau_t \le T \mid Y_t = y)$$
$$= -e^{-\gamma x} + (e^{-\gamma x} - e^{-\gamma(x+c)})P(\tau_t > T \mid Y_t = y),$$

and the lower bound follows. The upper bound is established by testing the function

$$\tilde{H}(t,x,y) = -e^{-\gamma(x+c)-(\mu^2/2\sigma^2)(T-t)}$$

in the HJB equation (4) for H, and showing that $\tilde{H}$ is a supersolution.   □

Remark 1. The bounds given above reflect that, in the presence of default, the value functions are bounded between the solutions of two extreme cases. For example, the lower bounds correspond to a degenerate market (only the bank account available for trading in $[0,T]$), and the upper bounds correspond to the standard Merton case with no default risk.

2.4 Reduction to Reaction–Diffusion Equations

The HJB equation (2) can be simplified by the familiar distortion scaling

$$M(t,x,y) = -e^{-\gamma x}u(t,y)^{1/(1-\rho^2)}, \tag{8}$$

with $u : [0,T] \times \mathbb{R} \to \mathbb{R}^+$ solving the reaction–diffusion equation

$$u_t + \tilde{\mathcal{L}}_y u - (1-\rho^2)\left(\frac{\mu^2}{2\sigma^2} + \lambda(y)\right)u + (1-\rho^2)\lambda(y)u^{-\theta} = 0, \tag{9}$$
$$u(T,y) = 1,$$

where

$$\theta = \frac{\rho^2}{1-\rho^2},$$

and

$$\widetilde{\mathcal{L}}_y = \mathcal{L}_y - \frac{\rho\mu}{\sigma}a(y)\frac{\partial}{\partial y}. \tag{10}$$

Similar equations arise in other utility problems in incomplete markets, for example, in portfolio choice with recursive utility [30], valuation of mortgage-backed securities [31], and life-insurance problems [2]. One might work first with (9) and then provide the verification results for the HJB equation (2), because the solutions of (2) and (9) are related through (8). It is worth noting, however, that the reaction–diffusion equation (9) does not belong to the class of such equations with Lipschitz reaction term. Therefore, more detailed analysis is needed for directly establishing existence, uniqueness, and regularity results. In the context of a portfolio choice problem with stochastic differential utilities, the analysis can be found in [30]. The equation at hand is slightly more complicated than the one analyzed there, in that the reaction term has the multiplicative intensity factor. Because $\lambda(\cdot)$ is taken to be bounded and Lipschitz, an adaptation of the arguments in [30] can be used to show that the reaction–diffusion problem (9) has a unique bounded and smooth solution. Furthermore, using (8) and the bounds obtained for M in Proposition 3, we have

$$e^{-(1-\rho^2)(\mu^2/2\sigma^2)(T-t)} \le u(t,y) \le 1.$$

For the bondholder's value function, the transformation

$$H(t,x,y) = -e^{-\gamma(x+c)}w(t,y)^{1/(1-\rho^2)}$$

reduces to

$$w_t + \widetilde{\mathcal{L}}_y w - (1-\rho^2)\left(\frac{\mu^2}{2\sigma^2} + \lambda(y)\right)w + (1-\rho^2)e^{\gamma c}\lambda(y)w^{-\theta} = 0, \tag{11}$$

$$w(T,y) = 1,$$

which is a similar reaction–diffusion equation as (9). The only difference is the coefficient $e^{\gamma c} > 1$ in front of the reaction term. Existence of a unique smooth and bounded solution follows similarly.

The following lemma gives a relationship between u and w.

Lemma 1. *Let u and w be solutions of the reaction–diffusion problems (9) and (11). Then*

$$u(t,y) \le w(t,y) \qquad for\ (t,y) \in [0,T] \times \mathbb{R}.$$

Proof. We have $u(T,y) = w(T,y) = 1$. Moreover, because $e^{\gamma c} > 1$ and $\lambda > 0$,

$$(1-\rho^2)e^{\gamma c}\lambda(y)w^{-\theta} > (1-\rho^2)\lambda(y)w^{-\theta},$$

which yields

$$w_t + \widetilde{\mathcal{L}}_y w - (1-\rho^2)\left(\frac{\mu^2}{2\sigma^2} + \lambda(y)\right)w + (1-\rho^2)\lambda(y)w^{-\theta} < 0.$$

Therefore, w is a supersolution of (9), and the result follows. $\qquad\square$

From this, we easily obtain the following sensible bounds on the indifference value of the defaultable bond, and the yield spread.

Proposition 4. *The indifference bond price* p_0 *in (5) is given by*

$$p_0(T) = e^{-rT} - \frac{1}{\gamma(1-\rho^2)} \log\left(\frac{w(0,y)}{u(0,y)}\right), \qquad (12)$$

and satisfies $p_0(T) \le e^{-rT}$. *The yield spread defined by*

$$\mathcal{Y}_0(T) = -\frac{1}{T} \log(p_0(T)) - r$$

is nonnegative for all $T > 0$.

Remark 2. We denote the seller's indifference price by $\tilde{p}_0(T)$. In order to construct it, we replace c by $-c$ in the definition (3) of the value function H and in the ensuing transformations. If $\tilde{w}$ is the solution of

$$\tilde{w}_t + \tilde{\mathcal{L}}_y \tilde{w} - (1-\rho^2)\left(\frac{\mu^2}{2\sigma^2} + \lambda(y)\right)\tilde{w} + (1-\rho^2)e^{-\gamma c}\lambda(y)\tilde{w}^{-\theta} = 0, \qquad (13)$$

with $\tilde{w}(T,y) = 1$, then

$$\tilde{p}_0(T) = e^{-rT} - \frac{1}{\gamma(1-\rho^2)} \log\left(\frac{u}{\tilde{w}}\right).$$

Using comparison results, we obtain $u > \tilde{w}$ as $e^{-\gamma c} < 1$. Therefore $\tilde{p}_0(T) \le e^{-rT}$, and the seller's yield spread is nonnegative for all $T > 0$.

2.5 Connection with Relative Entropy Minimization

For completeness, we connect the HJB equations characterizing the primal optimal investment problem that we study with the dual problem of relative entropy minimization. The reader can skip this section without affecting the understanding of the rest of the paper.

Let G be the bounded $\mathcal{F}_T$-measurable payoff of a credit derivative, and let $\mathcal{P}$ denote the primal problem's value (for simplicity, at time zero):

$$\mathcal{P} = \sup_{\pi \in \mathcal{A}} E\left[-e^{-\gamma(X_{\tau \wedge T} + G\mathbf{1}_{\{\tau > T\}})}\right].$$

In our problem (3), we have $G = c$.

As is well known, under quite general conditions, we have the duality relation

$$\mathcal{P} = -e^{-\gamma x - \gamma \mathcal{D}},$$

where x is the initial wealth and $\mathcal{D}$ is the value of the dual optimization problem:

$$D = \inf_{Q \in P_f} \left(E^Q[G] + \frac{1}{\gamma} \mathcal{H}(Q|P) \right). \tag{14}$$

Here $\mathcal{H}(Q|P)$ is the relative entropy between Q and P, namely,

$$\mathcal{H}(Q|P) = \begin{cases} E\left[\frac{dQ}{dP} \log\left(\frac{dQ}{dP} \right) \right], & Q \ll P, \\ \infty, & \text{otherwise.} \end{cases} \tag{15}$$

In (14), P_f denotes the set of absolutely continuous local martingale measures with finite relative entropy with respect to P. We refer the reader to [11; 20] for full details.

We now derive the related HJB equation for the dual problem. This approach is taken in [3]. Under $Q \in P_f$, the stock price S is a local martingale, but the intensity-driving process Y need not be. Under mild regularity conditions, the measure change from P to Q is parameterized by a pair of adapted processes, ψ_t and $\phi_t \geq 0$, with

$$E^Q\left[\int_0^T \psi_t^2 \, dt \right] < \infty, \quad \text{and} \quad \int_0^T \phi_t \lambda(Y_t) \, dt < \infty \text{ a.s.},$$

such that

$$dS_t = \sigma S_t \, dW_t^{Q(1)},$$
$$dY_t = \left(b(Y_t) - \frac{\rho \mu}{\sigma} a(Y_t) - \psi_t \rho' a(Y_t) \right) dt + a(Y_t) \left(\rho \, dW_t^{Q(1)} + \rho' \, dW_t^{Q(2)} \right),$$

and the intensity is

$$\lambda_t^Q = \phi_t \lambda(Y_t).$$

Here, $W^{Q(1)}$ and $W^{Q(2)}$ are independent Q-Brownian motions, and $\rho' = \sqrt{1 - \rho^2}$. The control ψ can be interpreted as a risk premium for the non-traded component of Y, whereas ϕ directly affects the stochastic intensity. The Radon–Nikodym derivative is given by (cf., e.g., [12, Appendix E])

$$\log \frac{dQ}{dP} = \mathbf{1}_{\{\tau < T\}} \log \phi_\tau - \frac{1}{2} \int_0^{\tau \wedge T} \left(\frac{\mu^2}{\sigma^2} + \psi_t^2 \right) dt - \int_0^{\tau \wedge T} \frac{\mu}{\sigma} \, dW_t^{(1)}$$
$$- \int_0^{\tau \wedge T} \psi_t \, dW_t^{(2)} + \int_0^{\tau \wedge T} (1 - \phi_t) \lambda(Y_t) \, dt.$$

Therefore, we have the expression

$$\mathcal{H}(Q|P) = E^Q\left[\mathbf{1}_{\{\tau < T\}} \log \phi_\tau \right] + \int_0^{\tau \wedge T} \left(\frac{\mu^2}{2\sigma^2} + \frac{1}{2} \psi_t^2 + (1 - \phi_t) \lambda(Y_t) \right) dt.$$

In passing to the associated HJB equation, we use the fact that if ϕ is bounded and adapted to the filtration generated by the two Brownian motions, then τ retains the so-called "doubly stochastic" property under Q. This means that, conditioned on the path of Y, the distribution of τ under Q is given by

$$P^Q\left(\tau > t \mid (Y_s)_{0 \le s \le t}\right) = e^{-\int_0^t \phi_s \lambda(Y_s)\, ds}.$$

We henceforth assume G is bounded and of European-type, in the sense that $G = G(Y_T)$ (note that the payoff does not depend on the stock). In the defaultable bond case of interest, G is a constant.

We are thus led to define the stochastic optimization problem

$$J(t,y) = \inf_{\psi; \phi \ge 0} E_{t,y}^Q \left[\gamma \mathcal{E}_{t,T} G \right.$$

$$\left. + \int_t^T \left(\frac{\mu^2}{2\sigma^2} + \frac{1}{2}\psi_s^2 + (1 - \phi_s)\lambda(Y_s) + \phi_s \lambda(Y_s) \log \phi_s \right) \mathcal{E}_{t,s}\, ds \right],$$

where

$$\mathcal{E}_{t,s} = e^{-\int_t^s \phi_u \lambda(Y_u)\, du}.$$

The associated HJB equation is

$$J_t + \widetilde{\mathcal{L}}_y J + \frac{\mu^2}{2\sigma^2} + \lambda(y) + \inf_\psi \left(\frac{1}{2}\psi^2 - \psi \rho' a(y) J_y \right)$$

$$+ \inf_{\phi \ge 0} \left(\phi \lambda(y) \log \phi - (J + 1)\phi \lambda(y) \right) = 0,$$

with $J(T,y) = \gamma G(y)$, and $\widetilde{\mathcal{L}}$ as in (10). In turn, the optimizing ϕ is given by $\phi^* = e^J$, so we have

$$J_t + \widetilde{\mathcal{L}}_y J + \frac{\mu^2}{2\sigma^2} - \frac{1}{2}(1 - \rho^2)a(y)^2 J_y^2 + \lambda(y)(1 - e^J) = 0.$$

Finally, setting $G = c = e^{-rT}$ and making the transformation

$$J = \gamma c - \frac{1}{(1 - \rho^2)} \log w$$

recovers the reaction–diffusion equation (11).

2.6 Intensity Bounds

We next investigate the behavior of the prices with respect to the intensity process. Specifically, we assume that, for $y \in \mathbb{R}$,

$$0 < \underline{\lambda} \le \lambda(y) \le \bar{\lambda} < \infty. \tag{16}$$

Proposition 5. *Let*

$$\bar{\alpha} = \frac{\mu^2}{2\sigma^2} + \bar{\lambda}, \quad and \quad \underline{\alpha} = \frac{\mu^2}{2\sigma^2} + \underline{\lambda}.$$

Then, under assumption (16), the value functions M and H satisfy for $x, y \in \mathbb{R}$,

$$-\left[\left(1-\frac{\bar\lambda}{\bar\alpha}\right)e^{-\bar\alpha(T-t)}+\frac{\bar\lambda}{\bar\alpha}\right]\le e^{\gamma x}M(t,x,y)\le-\left[\left(1-\frac{\underline\lambda}{\underline\alpha}\right)e^{-\underline\alpha(T-t)}+\frac{\underline\lambda}{\underline\alpha}\right],\tag{17}$$

and

$$-\left[\left(1-\frac{\bar\lambda e^{\gamma c}}{\bar\alpha}\right)e^{-\bar\alpha(T-t)}+\frac{\bar\lambda e^{\gamma c}}{\bar\alpha}\right]\le e^{\gamma(x+c)}H(t,x,y)$$

$$\le-\left[\left(1-\frac{\underline\lambda e^{\gamma c}}{\underline\alpha}\right)e^{-\underline\alpha(T-t)}+\frac{\underline\lambda e^{\gamma c}}{\underline\alpha}\right].\tag{18}$$

Proof. To show (17), we introduce the function

$$\bar M(t,x,y)=-e^{-\gamma x}\left[\left(1-\frac{\bar\lambda}{\bar\alpha}\right)e^{-\bar\alpha(T-t)}+\frac{\bar\lambda}{\bar\alpha}\right].$$

Direct calculations show that, for $x\in\mathbb{R},t\in[0,T]$,

$$\bar M(t,x,y)\ge-e^{-\gamma x},$$

and that

$$\bar M_t+\mathcal{L}_y\bar M+\max_\pi\left\{\frac{1}{2}\sigma^2\pi^2\bar M_{xx}+\pi(\rho\sigma a(y)\bar M_{xy}+\mu\bar M_x)\right\}+\lambda(y)(-e^{-\gamma x}-\bar M)\ge0.$$

$$\bar M_t+\mathcal{L}_y\bar M+\max_\pi\left\{\frac{1}{2}\sigma^2\pi^2\bar M_{xx}+\pi(\rho\sigma a(y)\bar M_{xy}+\mu\bar M_x)\right\}$$

$$+\lambda(y)(-e^{-\gamma x}-\bar M)\ge0.$$

Moreover, $\bar M(T,x,y)=-e^{-\gamma x}$. We easily conclude using the comparison principle. The other bounds are obtained similarly. □

Proposition 6. *The indifference price satisfies*

$$-\frac{1}{\gamma}\log\left(\frac{\left(1-\frac{\bar\lambda e^{\gamma c}}{\bar\alpha}\right)e^{-\bar\alpha(T-t)}+\frac{\bar\lambda e^{\gamma c}}{\bar\alpha}}{\left(1-\frac{\underline\lambda}{\underline\alpha}\right)e^{-\underline\alpha(T-t)}+\frac{\underline\lambda}{\underline\alpha}}\right)\le(p_0-e^{-rT})$$

$$\le-\frac{1}{\gamma}\log\left(\frac{\left(1-\frac{\underline\lambda e^{\gamma c}}{\underline\alpha}\right)e^{-\underline\alpha(T-t)}+\frac{\underline\lambda e^{\gamma c}}{\underline\alpha}}{\left(1-\frac{\bar\lambda}{\bar\alpha}\right)e^{-\bar\alpha(T-t)}+\frac{\bar\lambda}{\bar\alpha}}\right).$$

Proof. The assertion follows from the definition of the indifference price and the inequalities (17) and (18). □

2.7 Constant Intensity Case

We study explicitly the case of constant intensity, when the default time τ is independent of the level of the firm's stock price S, and is simply an exponential random variable with parameter λ. This simplified structure is employed in the multiname models that we analyze for CDO valuation in [29].

Proposition 7. *When λ is constant, the indifference price $p_0(T)$ (at time zero) of the defaultable bond expiring on date T is given by*

$$p_0(T) = e^{-rT} - \frac{1}{\gamma} \log \left(\frac{e^{-\alpha T} + \frac{\lambda}{\alpha} e^{\gamma c} \left(1 - e^{-\alpha T} \right)}{e^{-\alpha T} + \frac{\lambda}{\alpha} \left(1 - e^{-\alpha T} \right)} \right), \tag{19}$$

where

$$\alpha = \frac{\mu^2}{2\sigma^2} + \lambda.$$

Proof. We construct the explicit solutions of the HJB equations solved by the two value functions M and H. When λ is constant, the value functions M and H do not depend on y, and the HJB equation (2) reduces to

$$M_t - \frac{\mu^2}{2\sigma^2} \frac{M_x^2}{M_{xx}} + \lambda(-e^{-\gamma x} - M) = 0, \tag{20}$$

with $M(T, x) = -e^{-\gamma x}$. Substituting $M(t, x) = -e^{-\gamma x} m(t)$, we obtain

$$m' - \alpha m + \lambda = 0,$$

with $m(T) = 1$, and α as above. The unique solution is

$$m(t) = e^{-\alpha(T-t)} + \frac{\lambda}{\alpha} \left(1 - e^{-\alpha(T-t)} \right).$$

Similarly, the defaultable bondholder's value function $H(t, x)$ satisfies the same equation as M, but with terminal condition $H(T, x) = -e^{-\gamma(x+c)}$. Substituting $H(t, x) = -e^{-\gamma(x+c)} h(t)$, we obtain $h' - \alpha h + \lambda e^{\gamma c} = 0$, with $h(T) = 1$. The unique solution is

$$h(t) = e^{-\alpha(T-t)} + \frac{\lambda e^{\gamma c}}{\alpha} \left(1 - e^{-\alpha(T-t)} \right).$$

We easily deduce that the indifference price of the defaultable bond at time zero is given by

$$p_0(T) = e^{-rT} - \frac{1}{\gamma} \log \left(\frac{h(0)}{m(0)} \right),$$

leading to formula (19).   □

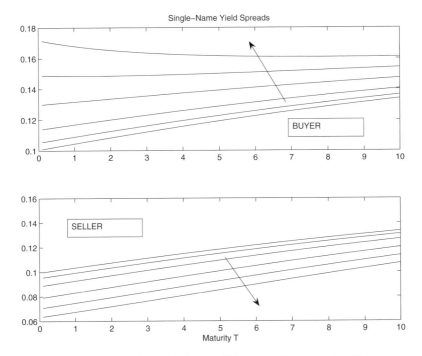

Fig. 1. Single-name buyer's and seller's indifference yield spreads. The parameters are $\lambda = 0.1$, along with $\mu = 0.09, r = 0.03$, and $\sigma = 0.15$. The curves correspond to different risk aversion parameters γ and the arrows show the direction of increasing γ over the values $(0.01, 0.1, 0.25, 0.5, 0.75, 1)$.

Remark 3. The seller's indifference price is given by

$$\tilde{p}_0(T) = e^{-rT} + \frac{1}{\gamma} \log \left(\frac{e^{-\alpha T} + \frac{\lambda}{\alpha} e^{-\gamma c} \left(1 - e^{-\alpha T} \right)}{e^{-\alpha T} + \frac{\lambda}{\alpha} \left(1 - e^{-\alpha T} \right)} \right).$$

A plot of the yield spreads $\mathcal{Y}_0(T) = -(1/T) \log(p_0(T)/e^{-rT})$ for the buyer, and similarly $\tilde{\mathcal{Y}}_0(T)$ for the seller, for various risk aversion coefficients, is shown in Figure 1. Observe that both spread curves are, in general, sloping, so the spreads are not flat even though we started with a constant intensity model. Although the seller's curve is upward sloping, the buyer's may become downward sloping when the risk aversion is large enough. Upward sloping yield spreads are commonly observed in market data; see, for example, [13, Figure 6.8] and [16, Figures 7, 8]. The short-term limit of the yield spread is nonzero, as we would expect in the presence of nonpredictable defaults. For the buyer's yield spread, we have

$$\lim_{T \downarrow 0} \mathcal{Y}_0(T) = \frac{(e^\gamma - 1)}{\gamma} \lambda,$$

which is larger than λ because $\gamma > 0$. This is amplified as γ becomes larger. In other words, the buyer values the claim as though the intensity were larger than the historically estimated value λ. The seller, on the other hand, values short-term claims as though the intensity were lower, because

$$\lim_{T\downarrow 0} \tilde{\mathcal{Y}}_0(T) = \frac{(1-e^{-\gamma})}{\gamma}\lambda \leq \lambda.$$

The long-term limit for both buyer's and seller's spread is simply α,

$$\lim_{T\to\infty} \mathcal{Y}_0(T) = \lim_{T\to\infty} \tilde{\mathcal{Y}}_0(T) = \frac{\mu^2}{2\sigma^2} + \lambda,$$

which is always larger than λ. Both long-term yield spreads converge to the intensity plus a term proportional to the square of the Sharpe ratio of the firm's stock.

3 A Two-Name Credit Derivative

To illustrate how the nonlinearity of the utility indifference valuation mechanism affects basket or multiname claims, we look at the case of two ($N = 2$) firms. The more realistic application to CDOs, where N might be on the order of a hundred, is studied in [29]. As with other approaches to these problems, particularly copula models, it becomes necessary to make substantial simplifications, typically involving some sort of symmetry assumption, in order to be able to handle the high-dimensional computational challenge. For example, in the one-factor copula model, the default times have the same pairwise correlation. Often, the single-name default probabilities and the losses-given-default are assumed to be identical, so the basket is homogeneous, or exchangeable: it does not specify which names default, just how many of them.

We assume throughout this section that intensities are constant or, equivalently, that the default times of the firms are independent.

The firms' stock prices processes $(S^{(i)})$ follow geometric Brownian motions:

$$\frac{dS_t^{(i)}}{S_t^{(i)}} = (r + \mu_i)\,dt + \sigma_i\,dW_t^{(i)}, \quad i = 1, 2,$$

where $(W^{(i)})$ are Brownian motions with instantaneous correlation coefficient $\rho \in (-1, 1)$ and volatilities $\sigma_i > 0$. The two firms have independent exponentially distributed default times τ_1 and τ_2, with intensities λ_1 and λ_2, respectively. We let $\tau = \min(\tau_1, \tau_2)$, and value a claim that pays \$1 if both firms survive until time T, and zero otherwise.

While both firms are alive, investors can trade the two stocks and the risk-free money market. When one defaults, its stock can no longer be traded, and if the second also subsequently defaults, the portfolio is invested entirely in the bank account. As in the single-name case, we work with discounted wealth X, and μ_i denotes the excess growth rate of the ith stock. The control processes $\pi^{(i)}$ are the dollar amount held in each stock, and the discounted wealth process X evolves according to

$$dX_t = \sum_{i=1}^{2} \pi_t^{(i)} \mathbf{1}_{\{\tau_i > t\}} \mu_i \, dt + \sum_{i=1}^{2} \pi_t^{(i)} \mathbf{1}_{\{\tau_i > t\}} \sigma_i \, dW_t^{(i)}.$$

When both firms are alive, the investor's objective is to maximize her expected utility from terminal wealth,

$$\sup_{\pi^{(1)}, \, \pi^{(2)}} E\left[-e^{-\gamma X_T}\right].$$

To solve the problem, we need to recursively deal with the cases when there are no firms left, when there is one firm left, and, finally, when both firms are present. Let $M^{(j)}(t,x)$ denote the value function of the investor who starts at time $t \le T$, with wealth x, when there are $j \in \{0,1,2\}$ firms available to invest in. In the case $j = 1$, we denote by $M_i^{(j)}(t,x)$ the subcases when it is the firm $i \in \{1,2\}$ that is alive.

When there are no firms left, we have the value function

$$M^{(0)}(t,x) = -e^{-\gamma x}. \tag{21}$$

When only firm i is alive, we have the single-name value functions computed in the proof of Proposition 7, namely,

$$M_i^{(1)}(t,x) = -e^{-\gamma x} v_i^{(1)}(t),$$

where

$$v_i^{(1)}(t) = e^{-\alpha_i(T-t)} + \frac{\lambda_i}{\alpha_i}\left(1 - e^{-\alpha_i(T-t)}\right), \tag{22}$$

and

$$\alpha_i = \frac{\mu_i^2}{2\sigma_i^2} + \lambda_i.$$

When both firms are alive, the value function $M^{(2)}(t,x)$ solves

$$M_t^{(2)} - \frac{1}{2}D_2\frac{(M_x^{(2)})^2}{M_{xx}^{(2)}} + \sum_{i=1}^{2}\lambda_i(M_i^{(1)} - M^{(2)}) = 0, \tag{23}$$

with $M^{(2)}(T,x) = -e^{-\gamma x}$. Here, the diversity coefficient D_2 is given by

$$D_2 = \mu^T A^{-1}\mu,$$

where

$$\mu = \begin{pmatrix} \mu_1 \\ \mu_2 \end{pmatrix} \quad \text{and} \quad A = \begin{pmatrix} \sigma_1^2 & \rho\sigma_1\sigma_2 \\ \rho\sigma_1\sigma_2 & \sigma_2^2 \end{pmatrix}.$$

Evaluating D_2 yields

$$D_2 = \frac{\mu_1^2}{\sigma_1^2} + \frac{1}{(1-\rho^2)} \left(\rho \frac{\mu_1}{\sigma_1} + \frac{\mu_2}{\sigma_2} \right)^2. \tag{24}$$

Remark 4. In the single-name case (with constant intensity), the analogue of (23) is (20). It is clear that the analogue of D_2 would be $D_1 = \mu_1^2/\sigma_1^2$ (if it is firm 1, e.g., in question), the square of the Sharpe ratio of the stock $S^{(1)}$. The formula (24) implies $D_2 \geq \mu_1^2/\sigma_1^2$, and, by interchanging subscripts, $D_2 \geq \mu_2^2/\sigma_2^2$. Therefore, it is natural to think of D_2 as a measure of the improved investment opportunity set offered by the diversity of having two stocks in which to invest. This idea may be naturally extended to $N > 2$ dimensions (see [29]).

Proposition 8. *The value function $M^{(2)}(t, x)$, solving (23), is given by*

$$M^{(2)}(t, x) = -e^{-\gamma x} v^{(2)}(t), \tag{25}$$

where

$$v^{(2)}(t) = e^{-\alpha_{1,2}(T-t)} + \sum_{i=1}^{2} \lambda_i \left[\left(1 - \frac{\lambda_1}{\alpha_1} \right) \frac{1}{(\alpha_{1,2} - \alpha_i)} \right.$$
$$\left. \left(e^{-\alpha_i(T-t)} - e^{-\alpha_{1,2}(T-t)} \right) + \frac{\lambda_i}{\alpha_i \alpha_{1,2}} \left(1 - e^{-\alpha_{1,2}(T-t)} \right) \right], \tag{26}$$

and $\alpha_{1,2} = D_2 + \lambda_1 + \lambda_2$.

Proof. Inserting (25) into (23) gives the following ODE for $v^{(2)}(t)$,

$$\frac{d}{dt} v^{(2)} - \alpha_{1,2} v + \sum_{i=1}^{2} \lambda_i v_i^{(1)}(t) = 0,$$

with $v^{(2)}(T) = 1$. Using the formula (22) for $v_i^{(1)}$ and solving the ODE leads to (26). ☐

We next consider the investment problem for the holder of the basket claim that pays \$1 if both firms survive up to time T. With $c = e^{-rT}$ as before, the value function $H^{(2)}(t, x)$ for the claimholder, starting with wealth x at time $t \leq T$ when both firms are still alive, solves

$$H_t^{(2)} - \frac{1}{2} D_2 \frac{(H_x^{(2)})^2}{H_{xx}^{(2)}} + \sum_{i=1}^{2} \lambda_i (M_i^{(1)} - H^{(2)}) = 0, \tag{27}$$

with $H^{(2)}(T, x) = -e^{-\gamma(x+c)}$. Notice that in the case of this simple claim, we do not have to consider separately the case of one or no firm left because the claim pays nothing in these cases. Once one firm defaults, the bond holder's problem reduces to the previous case of no claim. In general, however, for a more complicated claim, there will be a chain of value functions $H^{(j)}$.

Working as above, we can show the following.

Proposition 9. *The value function $H^{(2)}(t,x)$, solution of (27), is given by*

$$H^{(2)}(t,x) = -e^{-\gamma(x+c)}w^{(2)}(t), \tag{28}$$

where

$$w^{(2)}(t) = e^{-\alpha_{1,2}(T-t)} + \sum_{i=1}^{2} \lambda_i e^{\gamma c} \left[\left(1 - \frac{\lambda_1}{\alpha_1}\right) \frac{1}{(\alpha_{1,2} - \alpha_i)} \right. \tag{29}$$

$$\left. \left(e^{-\alpha_i(T-t)} - e^{-\alpha_{1,2}(T-t)}\right) + \frac{\lambda_i}{\alpha_i \alpha_{1,2}} \left(1 - e^{-\alpha_{1,2}(T-t)}\right) \right].$$

Finally, the buyer's indifference price at time zero of the claim with maturity T is given by

$$p_0(T) = c + \frac{1}{\gamma} \log \left(\frac{v^{(2)}(0)}{w^{(2)}(0)} \right).$$

Next, we collect the analogous formulas for the seller of the claim, which are found by straightforward calculations, in the following proposition.

Proposition 10. *The value function of the seller is given by*

$$\tilde{H}^{(2)}(t,x) = -e^{-\gamma(x-c)}\tilde{w}^{(2)}(t),$$

where

$$\tilde{w}^{(2)}(t) = e^{-\alpha_{1,2}(T-t)} + \sum_{i=1}^{2} \lambda_i e^{-\gamma c} \left[\left(1 - \frac{\lambda_1}{\alpha_1}\right) \frac{1}{(\alpha_{1,2} - \alpha_i)} \right.$$

$$\left. \left(e^{-\alpha_i(T-t)} - e^{-\alpha_{1,2}(T-t)}\right) + \frac{\lambda_i}{\alpha_i \alpha_{1,2}} \left(1 - e^{-\alpha_{1,2}(T-t)}\right) \right].$$

The seller's indifference price of the claim with maturity T at time zero is given by

$$p_0(T) = c - \frac{1}{\gamma} \log \left(\frac{v^{(2)}(0)}{\tilde{w}^{(2)}(0)} \right).$$

A plot of the yield spreads,

$$\mathcal{Y}_0(T) = -\frac{1}{T} \log(p_0(T)/e^{-rT})$$

for the buyer, and similarly for the seller, for various risk aversion coefficients, is shown in Figure 2. As in the single-name case, both spread curves are, in general, sloping, so the spreads are not flat even though we started with a constant intensity model. Although the seller's curve is upward sloping, the buyer's may become downward sloping when the risk aversion is large enough. The long-term limit of both buyer's and seller's yield spread is simply $\alpha_{1,2}$:

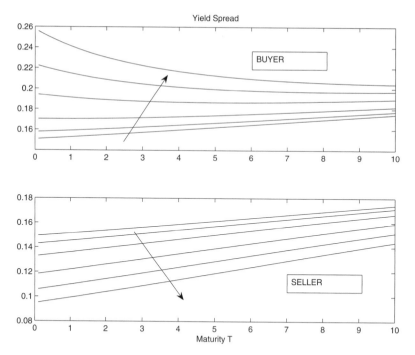

Fig. 2. Buyer's and seller's indifference yield spreads for two-name survival claim. The curves correspond to different risk aversion parameters γ and the arrows show the direction of increasing γ over the values $(0.01, 0.1, 0.25, 0.5, 0.75, 1)$. The other parameters are: excess growth rates $(0.04, 0.06)$, volatilities $(0.2, 0.15)$, correlation $\rho = 0.2$, and intensities $(0.05, 0.1)$.

$$\lim_{T \to \infty} \mathcal{Y}_0(T) = D_2 + \lambda_1 + \lambda_2,$$

which dominates the actual joint survival probability's hazard rate $\lambda_1 + \lambda_2$.

Another way to express the correlating effect of utility indifference valuation is through the linear correlation coefficient. Let $p_1(T)$ and $p_2(T)$ denote the indifference prices of the (single-name) defaultable bonds for firms 1 and 2, respectively, computed as in Section 2.7. Let $p_{12}(T)$ denote the value of the two-name survival claim, as in Proposition 9. Then we define the linear correlation coefficient (see [27, Section 10.1]):

$$\varrho(T) = \frac{p_{12}(T) - p_1(T)p_2(T)}{\sqrt{p_1(T)(1 - p_1(T))p_2(T)(1 - p_2(T))}}.$$

This is plotted for different maturities, risk aversions, and for buyer and seller in Figure 3. We observe that the correlating effect is enhanced by the more risk averse buyer, and reduced by the risk averse seller. For both, the effect increases over short to medium maturities, before plateauing, or dropping off slightly.

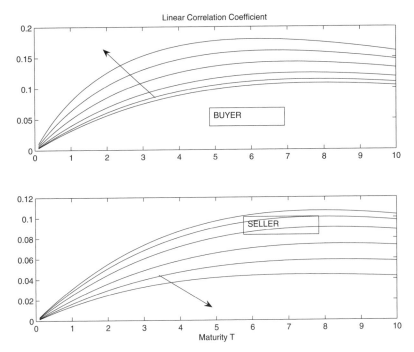

Fig. 3. Linear correlation coefficient $\varrho(T)$ from buyer's and seller's indifference values for single-name and two-name survival claims. The curves correspond to different risk aversion parameters γ and the arrows show the direction of increasing γ over the values $(0.01, 0.1, 0.25, 0.5, 0.75, 1)$. The other parameters are as in Figure 2.

4 Conclusions

The preceding analysis demonstrates that utility valuation produces nontrivial yield spreads and "effective correlations" within even the simplest of intensity-based models of default. They are able to incorporate equity market information (growth rates, volatilities of the nondefaulted firms) as well as investor risk aversion to provide a relative value mechanism for credit derivatives.

Here we have studied single-name defaultable bonds, whose valuation under a stochastic (diffusion) intensity process leads naturally to the study of reaction–diffusion equations. However, even with constant intensity, the yield spreads due to risk aversion are striking. The subsequent analysis of the simple two-name claim demonstrates how nonlinear pricing can be interpreted as high "effective correlation." The impact on more realistic multiname basket derivatives, such as CDOs, is investigated in detail in [29].

Acknowledgments

Ronnie Sircar's work is partially supported by NSF grant DMS-0456195. Thaleia Zariphopoulou's work is partially supported by NSF grants DMS-0456118 and DMS-0091946. This work was presented at conferences at the CRM Montreal, Newton Institute Cambridge, SAMSI North Carolina, the SIAM Annual Meeting in New Orleans, the INFORMS Annual Meeting in San Francisco, and at a seminar at UC Santa Barbara. We are grateful to the participants for fruitful comments.

References

1. P. Artzner and F. Delbaen. Default risk insurance and incomplete markets. *Mathematical Finance*, 5:187–195, 1995.
2. D. Becherer. Rational hedging and valuation of integrated risks under constant absolute risk aversion. *Insurance: Mathematics and Economics*, 33(1):1–28, 2003.
3. T. Bielecki and M. Jeanblanc. Indifference pricing of defaultable claims. In R. Carmona, editor, *Indifference Pricing*. Princeton University Press, 2006.
4. T. Bielecki, M. Jeanblanc, and M. Rutkowski. Hedging of defaultable claims. *Paris-Princeton Lectures on Mathematical Finance*, ed. R. Carmona, Springer, 2004.
5. T. Bielecki and M. Rutkowski. *Credit Risk*. Springer-Verlag, 2001.
6. H. Buhlmann. *Mathematical Methods in Risk Theory*. Springer-Verlag, 1970.
7. P. Collin-Dufresne and J. Hugonnier. Event risk, contingent claims and the temporal resolution of uncertainty. Carnegie Mellon University Working Paper, 2001.
8. J. Cvitanić, W. Schachermayer, and H. Wang. Utility maximization in incomplete markets with random endowment. *Finance and Stochastics*, 5(2):259–272, 2001.
9. M. Davis and V. Lo. Infectious defaults. *Quantitative Finance*, 1(4):382-387, 2001.
10. M.H.A. Davis, V. Panas, and T. Zariphopoulou. European option pricing with transaction costs. *SIAM J. Control and Optimization*, 31:470–93, 1993.
11. F. Delbaen, P. Grandits, T. Rheinländer, D. Samperi, M. Schweizer, and C. Stricker. Exponential hedging and entropic penalties. *Mathematical Finance*, 12(2):99–123, 2002.
12. D. Duffie. Credit risk modeling with affine processes. *J. Banking and Finance*, 29:2751-2802, 2005.
13. D. Duffie and K. Singleton. *Credit Risk*. Princeton University Press, 2003.
14. D. Duffie and T. Zariphopoulou. Optimal investment with undiversifiable income risk. *Mathematical Finance*, 3(2):135–148, 1993.
15. A. Elizalde. Credit risk models IV: Understanding and pricing CDOs. *www.abelelizalde.com*, 2005.
16. J.P. Fouque, R. Sircar, and K. Sølna. Stochastic volatility effects on defaultable bonds. *Applied Finance*, 13(3), 215–244, 2006.

17. H. Gerber. *An Introduction to Mathematical Risk Theory*. Huebner Foundation for Insurance Education, Wharton School, University of Pennsylvania, 1979.

18. S.D. Hodges and A. Neuberger. Optimal replication of contingent claims under transaction costs. *Review of Futures Markets*, 8:222–239, 1989.

19. R. Jarrow and S. Turnbull. Pricing options on financial securities subject to credit risk. *J. Finance*, 50:53–85, 1995.

20. Y. M. Kabanov and C. Stricker. On the optimal portfolio for the exponential utility maximization: Remarks to the six-author paper. *Mathematical Finance*, 12(2):125–34, 2002.

21. I. Karatzas and S. Shreve. *Methods of Mathematical Finance*. Springer-Verlag, 1998.

22. D. Kramkov and W. Schachermayer. The asymptotic elasticity of utility functions and optimal investment in incomplete markets. *Annals of Applied Probability*, 9(3):904–950, 1999.

23. D. Lando. On Cox processes and credit risky securities. *Review of Derivatives Research*, 2:99–120, 1998.

24. D. Lando. *Credit Risk Modeling: Theory and Applications*. Princeton Series in Finance. Princeton University Press, 2004.

25. V. Linetsky. Pricing equity derivatives subject to bankruptcy. *Mathematical Finance*, 16(2):255–282, 2006.

26. D. Madan and H. Unal. Pricing the risks of default. *Review of Derivatives Research*, 2:121–160, 1998.

27. P. Schönbucher. *Credit Derivatives Pricing Models*. Wiley, 2003.

28. T. Shouda. The indifference price of defaultable bonds with unpredictable recovery and their risk premiums. Preprint, Hitosubashi University, Tokyo, 2005.

29. R. Sircar and T. Zariphopoulou. Utility valuation of credit derivatives and application to CDOs. Submitted, 2006.

30. C. Tiu. *On the Merton Problem in Incomplete Markets*. Ph.D. thesis, The University of Texas at Austin, 2002.

31. T. Zhou. Indifference valuation of mortgage-backed securities in the presence of payment risk. Preprint, The University of Texas at Austin, 2006.

Investment and Valuation Under Backward and Forward Dynamic Exponential Utilities in a Stochastic Factor Model

Marek Musiela[1] and Thaleia Zariphopoulou[2]

[1] BNP Paribas
London, United Kingdom
marek.musiela@bnpparibas.com

[2] Department of Mathematics and Information
Department of Risk and Operations Management
The University of Texas at Austin
Austin, TX 78712, USA
zariphop@math.utexas.edu

Summary. We introduce a new class of dynamic utilities that are generated forward in time. We discuss the associated value functions, optimal investments, and indifference prices and we compare them with their traditional counterparts, implied by backward dynamic utilities.

Key words: Forward dynamic utilities; time invariance; indifference valuation; minimal martingale measure; minimal entropy measure; incomplete markets.

1 Introduction

This paper is a contribution to integrated portfolio management in incomplete markets. Incompleteness stems from a correlated stochastic factor affecting the dynamics of the traded risky security (stock). The investor trades between a riskless bond and the stock, and may incorporate in his or her portfolio derivatives and liabilities. The optimal investment problem is embedded into a partial equilibrium one that can be solved by the so-called utility-based pricing approach. The optimal portfolios can be, in turn, constructed as the sum of the policy of the plain investment problem and the indifference hedging strategy of the associated claim.

In a variety of applications, the investment horizon and the maturities of the claims do not coincide. This misalignment might cause price discrepancies if the current optimal expected utility is not correctly specified. The focus herein is in exploring which classes of utilities preclude such pathological situations.

In the traditional framework of expected utility from terminal wealth, the correct dynamic utility is easily identified, namely, it is given by the implied value function. Such a utility is, then, called self-generating in that it is indistinguishable from the value function it produces. This is an intuitively clear consequence of the dynamic programming principle. There are, however, two important underlying ingredients. Firstly, the risk preferences are a priori specified at a future time, say T, and, secondly, the utility, denoted $U_t^B(x;T)$, is generated at previous times $(0 \leq t \leq T)$. Herein, T denotes the end of the investment horizon and x represents the wealth argument. Due to the backward in time generation, T is called the backward normalization point and $U_t^B(x;T)$ the backward dynamic utility.

Despite their popularity, the traditional backward dynamic utilities considerably constrain the set of claims that can be priced to the ones that expire before the normalization point T. Moreover, even in the absence of payoffs and liabilities, utilities of terminal wealth do not seem to capture very accurately changes in the risk attitude as the market environment evolves. In many aspects, in the familiar utility framework utilities "move" backwards in time and market shocks are revealed forward in time.

Motivated by such considerations, the authors recently introduced the notion of forward dynamic utilities (see Musiela and Zariphopoulou [23; 24]) in a simple multiperiod incomplete binomial model. These utilities, as with their backward counterparts, are created via an expected criterion, but in contrast, they evolve forward in time. Specifically, they are determined today, say at s, and they are generated for future times, via a self-generating criterion. In other words, the forward dynamic utility $U_t^F(x;s)$ is normalized at the present time and not at the end of the generic investment horizon.

In this paper, we extend the notion of forward dynamic utilities in a diffusion model with a correlated stochastic factor. For simplicity, we assume that the utility data, at both backward and forward normalization points, are taken to be of exponential type with constant risk aversion. This assumption can be relaxed without losing the fundamental properties of the dynamic utilities. We also concentrate our analysis on European-type liabilities so that closed-form variational results can be obtained.

The two classes of dynamic utilities, as well as the emerging prices and investment strategies, have similarities but also striking differences. As mentioned above, both utilities are self-generating and, therefore, price discrepancies are precluded in the associated backward and forward indifference pricing systems. A consequence of self-generation is that an investor endowed with backward and forward utilities receives the same dynamic utility across different investment horizons. It is worth noting that although the backward dynamic utility is unique, the forward one might not be.

The associated indifference prices have very distinct characteristics. Backward indifference prices depend on the backward normalization point (and, thus, implicitly on the trading horizon) even if the claim matures before T. However, forward indifference prices are not affected by the choice of the forward normalization point. For the class of European claims examined herein, backward prices are represented as nonlinear expectations associated with the minimal relative entropy measure. On the other hand, forward prices are also represented as nonlinear expectations but with respect to the minimal martingale measure. The two prices coincide when the market is complete. This is a direct consequence of the fact that the internal market incompleteness—coming from the stochastic factor—is processed by the backward and forward dynamic utilities in a very distinct manner.

The portfolio strategies related to the backward and forward utilities also have very different characteristics. The optimal backward investments consist of the myopic portfolio, the backward indifference deltas, and the excess risky demand. The latter policy reflects, in contrast to the myopic portfolio, the incremental changes in the optimal behavior due to the movement of the stochastic factor. The forward optimal investments have the same structure as their backward counterparts but do not include the excess risky demand.

The paper is organized as follows. In Section 2, we introduce the investment model and its dynamic utilities. In Section 3, we provide some auxiliary technical results related to the minimal martingale and minimal entropy measures. In Sections 4 and 5, respectively, we construct the backward and forward dynamic utilities, the associated prices, and the optimal investments. We conclude in Section 6, where we provide a comparative study for integrated portfolio problems under the two classes of dynamic risk preferences.

2 The Model and Its Dynamic Utilities

Two securities are available for trading, a riskless bond and a risky stock whose price solves

$$dS_s = \mu(Y_s) S_s ds + \sigma(Y_s) S_s dW_s^1, \tag{1}$$

for $s \geq 0$ and $S_0 = S > 0$. The bond offers zero interest rate. The case of (deterministic) nonzero interest rate may be handled by straightforward scaling arguments and is not discussed.

The process Y, referred to as the stochastic factor, is assumed to satisfy

$$dY_s = b(Y_s) ds + a(Y_s) dW_s, \tag{2}$$

for $s \geq 0$ and $Y_0 = y \in \mathcal{R}$.

The processes W^1 and W are standard Brownian motions defined on a probability space $(\Omega, \mathcal{F}, (\mathcal{F}_s), P)$ with $\mathcal{F}_s$ being the augmented σ-algebra. We assume that the correlation coefficient $\rho \in (-1, 1)$ and, thus, we may write

$$dW_s = \rho dW_s^1 + \sqrt{1-\rho^2} dW_s^{1,\perp}, \tag{3}$$

with $W^{1,\perp}$ being a standard Brownian motion on $(\Omega, \mathcal{F}, (\mathcal{F}_s), P)$ orthogonal to W^1. For simplicity, we assume that the dynamics in (1) and (2) are autonomous. We denote the stock's Sharpe ratio process by

$$\lambda_s = \lambda(Y_s) = \frac{\mu(Y_s)}{\sigma(Y_s)}. \tag{4}$$

The following assumption is assumed to hold throughout.

Assumption 4 *The market coefficients μ, σ, a, and b are assumed to be $C^2(\mathcal{R})$ functions that satisfy $|f(y)| \leq C(1 + |y|)$ for $y \in \mathcal{R}$ and $f = \mu, \sigma, a$ and b, and are such that (1) and (2) have a unique strong solution satisfying $S_s > 0$ a.e. for $s \geq 0$. There also exists $\varepsilon > 0$ such that $g(y) > \varepsilon$, for $y \in \mathcal{R}$ and $g = \sigma$ and a.*

Next, we consider an arbitrary trading horizon $[0, T]$, and an investor who starts, at time $t_0 \in [0, T]$, with initial wealth $x \in \mathcal{R}$ and trades between the two securities. His or her current wealth X_s, $t_0 \leq s \leq T$, satisfies the budget constraint $X_s = \pi_s^0 + \pi_s$ where π_s^0 and π_s are self-financing strategies representing the amounts invested in the bond and the stock accounts. Direct calculations, in the absence of intermediate consumption, yield the evolution of the wealth process

$$dX_s = \mu(Y_s)\pi_s ds + \sigma(Y_s)\pi_s dW_s, \tag{5}$$

with $X_{t_0} = x \in \mathcal{R}$. The set $\mathcal{A}$ of admissible strategies is defined as $\mathcal{A} = \left\{ \pi : \pi \text{ is } \mathcal{F}_s\text{-measurable, self-financing, and } E_P\left[\int_0^T \sigma^2(Y_s)\pi_s^2 ds\right] < \infty \right\}$.

Further constraints might be binding due to the specific application and/or the form of the involved utility payoffs. In order, however, to keep the exposition simple and to concentrate on the new notions and insights, we choose to abstract from such constraints. We denote $\mathcal{D}$ the generic spatial solvency domain for (x, y).

We start with an informal motivational discussion for the upcoming notions of backward and forward dynamic utilities. In the traditional economic model of expected utility from terminal wealth, a utility datum is assigned at a given time, representing the end of the investment horizon. We denote the utility datum $u(x)$ and the time at which it is assigned T. At $t_0 \geq 0$, the investor starts trading between the available securities until T. At intermediate times $t \in [t_0, T]$, the associated value function $v : \mathcal{D} \times [0, T] \to \mathcal{R}$ is defined as the maximal expected (conditional on $\mathcal{F}_t$) utility that the agent achieves from investment. For the model at hand, v takes the form

$$v(x, y, t) = \sup_{\mathcal{A}} E_P\left[u(X_T) \mid X_t = x, Y_t = y\right], \quad t \in [t_0, T], \tag{6}$$

with the wealth and stochastic factor processes X, Y solving (5) and (2).

The scope is to specify v and to construct the optimal control policies. The duality approach can be applied to general market models and provides characterization results for the value function, but limited results for the optimal portfolios. The latter can be constructed via variational methods for certain classes of diffusion models. To date, although there is a rich body of work for the value function, very little is understood about how investors adjust their portfolios in terms of their risk preferences, trading horizon, and the market environment.

2.1 Integrated Models of Portfolio Choice

In a more realistic setting, the investor might be interested in incorporating in his or her portfolio derivative securities, liabilities, proceeds from additional assets, labor income, and so on. Given that such situations arise frequently in practice, it is important to develop an approach that accommodates integrated investment problems and yields quantitative and qualitative results for the optimal portfolios. This is the aim of the study below.

To simplify the presentation, we assume, for the moment, that the investor faces a liability at T, represented by a random variable $C_T \in \mathcal{F}_T$. We recall that $\mathcal{F}_T$ is generated by both the traded stock and the stochastic factor, and that the investor uses only self-financing strategies.

In a complete market setup (e.g., when the processes S and Y are perfectly correlated) the optimal strategy for this generalized portfolio choice model is as follows. At initiation t_0, the investor splits the initial wealth, say x, into the amounts $E_Q\left[C_T | \mathcal{F}_{t_0}\right]$ and $\tilde{x} = x - E_Q\left[C_T | \mathcal{F}_{t_0}\right]$, with $E_Q\left[C_T | \mathcal{F}_{t_0}\right]$ being the arbitrage-free price of C_T. The residual amount $\tilde{x}$ is used for investment as if there were no liability. The dynamic optimal strategy is, then, the sum of the optimal portfolio, corresponding to initial endowment $\tilde{x}$ and the hedging strategy, denoted $\delta_s\left(C_T\right)$, of a European-type contingent claim written on the traded stock, maturing at T and yielding C_T. Using $*$ to denote optimal policies, we may write

$$\pi_s^{x,*} = \pi_s^{\tilde{x},*} + \delta_s\left(C_T\right) \qquad \text{with } \tilde{x} = x - E_Q\left[C_T | \mathcal{F}_{t_0}\right]. \qquad (7)$$

This can be established either by variational methods or duality. This remarkable additive structure, arising in the highly nonlinear utility setting, is a direct consequence of the ability to replicate the liability. Note that for a fixed choice of wealth units, the second portfolio component is not affected by the risk preferences.

When the market is incomplete, similar argumentation can be developed by formulating the problem as a partial equilibrium one and, in turn, using results from the utility-based valuation approach. The liability may be, then, viewed as a derivative security and the optimal portfolio choice problem is embedded into an indifference valuation one. Using payoff decomposition results (see, e.g., Musiela and Zariphopoulou [20] and [21], Stoikov and Zariphopoulou [32], and Monoyios [19]), we associate to the liability an indifference hedging strategy, say $\Delta_s\left(C_T\right)$, that is the incomplete market counterpart of its arbitrage-free replicating portfolio. Denoting the relevant indifference price by $\nu_t\left(C_T\right)$, we obtain an analogous to (7) decomposition of the optimal investment strategy in the stock account, namely,

$$\pi_s^{x,*} = \pi_s^{\tilde{x},*} + \Delta_s\left(C_T\right) \quad \text{with } \tilde{x} = x - \nu_{t_0}\left(C_T\right).\tag{8}$$

Because the indifference valuation approach incorporates the investor's risk preferences, the choice of utility will influence—in contrast to the complete market case—both components of the emerging optimal investment strategy. Note, however, that due to the dynamic nature of the problem, utility effects evolve both with time and market information. In order to correctly quantify these effects, it is imperative to be able to specify the dynamic value of our investment strategies across horizons, maturities, and units. As the analysis below indicates, the cornerstone of this endeavor is the specification of a dynamic utility structure that yields consistent valuation results and investment behavior across optimally chosen self-financing strategies.

Before we introduce the dynamic utilities, we first recall the auxiliary concept of indifference value. To preserve simplicity, we consider the aforementioned single liability C_T. To calculate its indifference price $\nu_t\left(C_T\right)$ for $t \in [0, T]$, we look at the investor's modified utility,

$$v^{C_T}\left(x, y, t\right) = \sup_{\mathcal{A}} E_P\left[u\left(X_T - C_T\right)| X_t = x, Y_t = y\right],\tag{9}$$

and, subsequently, impose the equilibrium condition

$$v\left(x - \nu_t\left(C_T\right), y, t\right) = v^{C_T}\left(x, y, t\right).\tag{10}$$

The optimal policy is given by (8) and can be retrieved in closed form for special cases. For example, when the utility is exponential, $u\left(x\right) = -e^{-\gamma x}$ with $\gamma > 0$, the stock's Sharpe ratio is constant and $C_T = G\left(Y_T\right)$, for some bounded function G, variational arguments yield the optimal investment representation

$$\pi_s^{x,*} = \pi_s^{\tilde{x},*} + \rho\frac{a\left(Y_s\right)}{\sigma\left(Y_s\right)}\left(\left.\frac{\partial g\left(y, t\right)}{\partial y}\right|_{y=Y_s, t=s}\right),$$

with $\tilde{x} = x - g\left(y, t_0\right)$, and $g : \mathcal{R} \times [0, T] \to \mathcal{R}$ solving a quasilinear PDE, of quadratic gradient nonlinearities, with terminal condition $g\left(y, T\right) = G\left(y\right)$.

2.2 Liabilities and Payoffs of Shorter Maturities

Consider a liability to be paid before the fixed horizon T, say at $T_0 < T$. There are two ways to proceed. The first alternative is to work with portfolio choice in the initial investment horizon $[t_0, T]$. In this case, the utility (9) becomes

$$v^{C_{T_0}}\left(x, y, t\right) = \sup_{\mathcal{A}} E_P\left[u\left(X_T - C_{T_0}\right)| X_t = x, Y_t = y\right],\tag{11}$$

where we took into consideration that the riskless interest rate is zero. The indifference value is, then, calculated by the pricing condition (10), for $t \in [t_0, T_0]$. However, such arguments might not be easily implemented, if at all, as it is the case of liabilities and payoffs of random maturity and/or various exotic characteristics.

The second alternative is to derive the indifference value by considering the investment opportunities, with and without the liability, up to the claim's maturity T_0. For this, we first need to correctly specify the value functions, denoted, respectively, $\bar{v}$ and $\bar{v}^{C_{T_0}}$, that correspond to optimality of investments in the shorter investment horizon $[t_0, T_0]$. Working along the lines that their long-horizon counterparts v and v^{C_T} were defined, let us, hypothetically, assume that we are given a utility datum for the point T_0. We denote this datum $\bar{u}(x, y, T_0)$. We henceforth use the $-$ notation for all quantities, that is, utilities, investments, and indifference values, associated with the shorter horizon. For $t \in [t_0, T_0]$,

$$\bar{v}(x, y, t) = \sup_{\mathcal{A}} E_P\left[\bar{u}(X_{T_0}, Y_{T_0}, T_0) \mid X_t = x, Y_t = y\right],$$

and

$$\bar{v}^{C_{T_0}}(x, y, t) = \sup_{\mathcal{A}} E_P\left[\bar{u}(X_{T_0} - C_{T_0}, Y_{T_0}, T_0) \mid X_t = x, Y_t = y\right].$$

The associated indifference value $\bar{\nu}_t(C_{T_0})$ is then, given by

$$\bar{v}(x - \bar{\nu}_t(C_{T_0}), y, t) = \bar{v}^{C_{T_0}}(x, y, t).$$

Clearly, in order to have a well-specified valuation system, we must have, for all $C_{T_0} \in \mathcal{F}_{T_0}$ and $t \in [t_0, T_0]$,

$$\nu_t(C_{T_0}) = \bar{\nu}_t(C_{T_0}),$$

which strongly suggests that the utility datum $\bar{u}(x, y, T_0)$ cannot be exogenously assigned in an arbitrary manner.

Such issues, related to the correct specification and alignment of intermediate utilities, and their value functions, with the claims' possibly different and/or random maturities, were first discussed in Davis and Zariphopoulou [7] in the context of utility-based valuation of American claims in markets with transaction costs. Recall that when early exercise is allowed, the first alternative computational step (cf. (11)) cannot be implemented because T_0 is not a priori known. For the same class of early exercise claims, but when incompleteness comes exclusively from a nontraded asset (which does not affect the dynamics of the stock), and the claim is written on both the traded and nontraded assets, further analysis on the specification of preferences across exercise times, was provided in Kallsen and Kuehn [15], Oberman and Zariphopoulou [25], and Musiela and Zariphopoulou [22]. In the latter papers, the related intermediate utilities and valuation condition took, respectively, the forms

$$\bar{u}(x, S, t) = v(x, S, t) \quad \text{and} \quad \bar{v}(x, S, t) = \bar{u}(x, S, t),$$

$$\bar{v}^{C_\tau}(x, S, z, t) = \sup_{\mathcal{A} \times \mathcal{T}} E_P\left[v\left(X_\tau - C\left(S_\tau, Z_\tau\right), S_\tau, \tau\right)\middle| X_t = x, S_t = S, Z_t = z\right],$$

where $\mathcal{T}$ is the set of stopping times in $[t_0, T]$. The processes S and Z represent the traded and nontraded assets and X the wealth process. The early exercise indifference price of C_τ is, then, given by

$$\bar{v}(x, S, t) = \bar{v}^{C_\tau}\left(x + \bar{\nu}_t\left(C\left(S_\tau, Z_\tau\right)\right), S, z, t\right).$$

Although the above calculations might look pedantic when a single exogenous cash flow (liability or payoff) is incorporated, the arguments get much more involved when a family of claims is considered and arbitrary, or stochastic, maturities are allowed. Naturally, the related difficulties disappear when the market is complete. However, when perfect replication is not viable and a utility-based approach is implemented for valuation, discrepancies leading to arbitrage might arise if we fail to properly incorporate in our model dynamic risk preferences that process and price the market incompleteness in a consistent manner. This issue was exposed by the authors in Musiela and Zariphopoulou [23] and [24], who initiated the construction of indifference pricing systems based on the so-called backward and forward dynamic exponential utilities. In these papers, indifference valuation of arbitrary claims and specification of integrated optimal policies were studied in an incomplete binomial case. Even though this model setup was rather simple, it offered a starting point in exploring the effects of the evolution of risk preferences to prices and investments. What follows is, to a great extent, a generalization of the theory developed therein.

2.3 Utility Measurement Across Investment Times

Let us now see how a dynamic utility can be introduced and incorporated in the stochastic factor model in which we are interested. We recall that the standing assumptions are: (i) the trading horizon $[0, T]$ is preassigned, (ii) a utility datum is given for T, and (iii) T dominates the maturities of all claims and liabilities in consideration.

We next assume that instead of having the single static measurement of utility u at expiration, the investor is endowed with a dynamic utility $u_t(x, y; T)$, $t \in [0, T]$. Being vague, for the moment, we view this utility as a functional, at each intermediate time t, of his current wealth and the level of the stochastic factor. Obviously, we must have

$$u_T(x, y; T) = u(x),$$

in which case, we say that $u_t(x, y; T)$ is *normalized* at T. As a consequence, we refer to T as the *normalization* point. For reasons that are apparent in the sequel, we choose to carry T in our notation.

If a maximal expected criterion is involved, the associated value function, denoted with a slight abuse of notation by v_t, will naturally take the form

$$v_t\left(x, y, \bar{T}\right) = \sup_{\mathcal{A}} E_P\left[u_T\left(X_{\bar{T}}, Y_{\bar{T}}; T\right) \middle| X_t = x, Y_t = y\right],$$

in an arbitrary subhorizon $[t, \bar{T}] \in [t_0, T]$ and with X, Y solving (5) and (2).

Let us now see how the generic liability $C_{T_0} \in \mathcal{F}_{T_0}$ would be valued under such a utility structure. For $t \in [t_0, T_0]$, the relevant maximal expected dynamic utility will be

$$v_t^{C_{T_0}}\left(x, y, T_0\right) = \sup_{\mathcal{A}} E_P\left[u_{T_0}\left(X_{T_0} - C_{T_0}, Y_{T_0}; T\right) \middle| X_t = x, Y_t = y\right].$$

Respectively, the indifference value $\nu_t\left(C_{T_0}; T\right)$ must satisfy, for $t \in [t_0, T_0]$,

$$v_t\left(x - \nu_t\left(C_{T_0}; T\right), y, T_0\right) = v_t^{C_{T_0}}\left(x, y, T_0\right).$$

Observe that because u_t is normalized at T, the associated value functions will depend on the normalization point. The latter will also affect the indifference price $\nu_t\left(C_{T_0}; T\right)$, even though the claim matures at an earlier time.

So far, the above formulation seems convenient and flexible enough for the valuation of claims with arbitrary maturities, as long as these maturities are shorter than the time at which risk preferences are normalized. However, as the next two examples show, it is wrong to assume that a dynamic utility can be introduced in an ad hoc way.

In both examples, it is assumed that the terminal utility datum is of exponential type, and independent of the level of the stochastic factor,

$$u_T\left(x, y; T\right) = -e^{-\gamma x} \tag{12}$$

with $(x, y) \in \mathcal{D}$ and γ being a given positive constant. It is also assumed that there is a single claim to be priced. Its payoff is taken to be of the form $C_{T_0} = G\left(Y_{T_0}\right)$, for some bounded function $G : \mathcal{R} \to \mathcal{R}^+$. Albeit the fact that in the model considered herein such a payoff is, to a certain extent, artificial, we, nevertheless, choose to work with it because explicit formulae can be obtained and the exposition is, thus, considerably facilitated.

Example 1. Consider a dynamic utility of the form

$$u_t\left(x, y; T\right) = \begin{cases} -e^{-\gamma x}, & \bar{T} < t \leq T, \\ -e^{-\bar{\gamma} x}, & 0 < t \leq \bar{T}, \end{cases}$$

with $\bar{T} > T_0$, γ as in (12) and $\bar{\gamma} \neq \gamma$.

Let us now see how C_{T_0} will be valued under the above choice of dynamic utility. If the investor chooses to trade in the original horizon $[t_0, T]$, the associated intermediate utilities are

$$v_t^{0, C_{T_0}}\left(x, y, T\right) = \sup_{\mathcal{A}} E_P\left[u_T\left(X_T - C_{T_0}, Y_T; T\right) \middle| X_t = x, Y_t = y\right]$$

$$= \sup_{\mathcal{A}} E_P \left[-e^{-\gamma(X_T - C_{T_0})} \middle| X_t = x, Y_t = y \right].$$

Obviously, the discontinuity, with regard to the risk aversion coefficient of u_t will not alter the above value functions. Following the results of Sircar and Zariphopoulou [31] yields

$$\nu_t\left(C_{T_0}\right) = \frac{1}{\gamma\left(1 - \rho^2\right)} \ln E_{Q^{me}} \left[e^{\gamma\left(1 - \rho^2\right)G\left(Y_{T_0}\right)} \middle| Y_t = y \right], \tag{13}$$

with Q^{me} being the minimal relative entropy martingale measure (see the next section for the relevant technical arguments).

If, however, the investor chooses to trade solely in the shorter horizon $\left[t_0, \bar{T}\right]$, analogous argumentation yields

$$\bar{v}_t^{0, C_{T_0}}\left(x, y, \bar{T}\right) = \sup_{\mathcal{A}} E_P \left[-e^{-\bar{\gamma}\left(X_{\bar{T}} - G\left(Y_{T_0}\right)\right)} \middle| X_t = x, Y_t = y \right],$$

where we used the $-$ notation to denote the shorter horizon choice. The associated indifference price is

$$\bar{\nu}_t\left(C_{T_0}\right) = \frac{1}{\bar{\gamma}\left(1 - \rho^2\right)} \ln E_{Q^{me}} \left[e^{\bar{\gamma}\left(1 - \rho^2\right)G\left(Y_{T_0}\right)} \middle| Y_t = y \right],$$

and we easily deduce that, in general,

$$\nu_t\left(C_{T_0}\right) \neq \bar{\nu}_t\left(C_{T_0}\right),$$

an obviously wrong result.

Note that even if we, naively, allow $\gamma = \bar{\gamma}$ price discrepancies will still emerge.

Example 2. Consider the dynamic utility

$$u_t\left(x, y; T\right) = -e^{-\gamma x - F(y, t; T)},$$

with γ as in (12) and

$$F\left(y, t; T\right) = E_P \left[\int_t^T \frac{1}{2} \lambda^2\left(Y_s\right) ds \middle| Y_t = y \right],$$

where P is the historical measure. If the agent chooses to invest in the longer horizon $[t_0, T]$, the indifference value remains the same as in (13). However, if she chooses to invest exclusively till the liability is met, we have, for $t \in [t_0, T_0]$,

$$\bar{v}_t^{0, C_{T_0}}\left(x, y, T_0\right) = \sup_{\mathcal{A}} E_P \left[u_{T_0}\left(X_{T_0} - G\left(Y_{T_0}\right), Y_{T_0}; T\right) \middle| X_t = x, Y_t = y \right]$$

$$= \sup_{\mathcal{A}} E_P \left[-e^{-\gamma\left(X_{T_0} - G\left(Y_{T_0}\right)\right) - F\left(Y_{T_0}, T_0; T\right)} \middle| X_t = x, Y_t = y \right].$$

Setting

$$Z_{T_0} = \frac{1}{\gamma} F(Y_{T_0}, T_0; T),$$

we deduce that, in the absence of the liability, the current utility is

$$\bar{v}_t^0(x, y, T_0) = \sup_{\mathcal{A}} E_P\left[-e^{-\gamma(X_{T_0} + Z_{T_0})} \,\middle|\, X_t = x, Y_t = y\right].$$

Note that, by definition, $Z_{T_0} \in \mathcal{F}_{T_0}$. Therefore, we may interpret $\bar{v}_t^0$ as a buyer's value function for the claim Z_{T_0}, in a traditional (nondynamic) exponential utility setting of constant risk aversion γ and investment horizon $[0, T_0]$. Then,

$$\bar{v}_t^0(x, y, T_0) = -e^{-\gamma(x + \bar{\mu}_t(Z_{T_0})) - \tilde{H}(y,t;T_0)},$$

with $\tilde{H}$ being the aggregate entropy function (see Equation (24) in the next section) and

$$\bar{\mu}_t(Z_{T_0}) = -\frac{1}{\gamma(1 - \rho^2)} \ln E_{Q^{me}}\left[e^{-\gamma(1-\rho^2)Z_{T_0}} \,\middle|\, Y_t = y\right].$$

Proceeding similarly, we deduce

$$\bar{v}_t^{C_{T_0}}(x, y, T_0) = \sup_{\mathcal{A}} E_P\left[-e^{-\gamma(X_{T_0} + (Z_{T_0} - G(Y_{T_0})))} \,\middle|\, X_t = x, Y_t = y\right]$$

$$= -e^{-\gamma(x + \bar{\mu}_t(Z_{T_0} - G(Y_{T_0}))) - \tilde{H}(y,t;T_0)},$$

with

$$\bar{\mu}_t(Z_{T_0} - G(Y_{T_0})) = -\frac{1}{\gamma(1 - \rho^2)} \ln E_{Q^{me}}\left[e^{-\gamma(1-\rho^2)(Z_{T_0} - G(Y_{T_0}))} \,\middle|\, Y_t = y\right].$$

Applying the definition of the indifference value, we deduce that, with regard to the shorter horizon,

$$\bar{\nu}_t(G(Y_{T_0})) = \bar{\mu}_t(Z_{T_0}) - \bar{\mu}_t(Z_{T_0} - G(Y_{T_0}))$$

$$= \frac{1}{\gamma(1 - \rho^2)} \ln \frac{E_{Q^{me}}\left[e^{-\gamma(1-\rho^2)(Z_{T_0} - G(Y_{T_0}))} \,\middle|\, Y_t = y\right]}{E_{Q^{me}}\left[e^{-\gamma(1-\rho^2)(Z_{T_0})} \,\middle|\, Y_t = y\right]},$$

which, in general, does not coincide with $\nu_t(G(Y_{T_0}))$, given in (13).

2.4 Backward and Forward Dynamic Exponential Utilities

The above examples expose that an ad hoc choice of dynamic utility might lead to price discrepancies. This, in view of the structural form of the optimal policy for the integrated model (cf. (8)) would, in turn, yield wrongly specified investment policies. It is, thus, important to investigate which classes of dynamic utilities preclude such pathological situations.

For the simple examples above, the correct choice of the dynamic utility is essentially obvious, namely,

$$
u_t\left(x, y; T\right) = \begin{cases} u\left(x\right), & t = T, \\ v\left(x, y, t\right), & t \in [t_0, T], \end{cases}
$$

with v as in (6).

This simple observation indicates the following. First, observe that if u_t is the candidate dynamic utility, then, in all trading subhorizons, say $[t, \tilde{t}]$, the associated dynamic value function v_t will be

$$
v_t\left(x, y; T\right) = \sup_{\mathcal{A}} E_P\left[u_{\tilde{t}}\left(X_{\tilde{t}}, Y_{\tilde{t}}; T\right)\middle| X_t = x, Y_t = y\right].
$$

Discrepancies in prices will be, then, precluded if at all intermediate times the dynamic utility coincides with the dynamic value function it generates,

$$
u_t\left(x, y; T\right) = v_t\left(x, y; T\right).
$$

We then say that the dynamic utility is self-generating.

Building on this concept, we are led to two classes of dynamic utilities, the *backward* and *forward* ones. Their definitions are given below. Because the applications herein are concentrated on exponential preferences, we work with such utility data. Throughout, we take the risk aversion coefficient to be a positive constant γ.

Although the backward dynamic utility is essentially the traditional value function, the concept of forward utility is, to the best of our knowledge, new. As mentioned earlier, it was recently introduced by the authors in an incomplete binomial setting (see Musiela and Zariphopoulou [23] and [24]) and it is extended herein to the diffusion case.

We continue with the definition of the backward dynamic utility. This utility takes the name *backward* because it is first specified at the normalization point T and is then generated at previous times.

Definition 1. *Let $T > 0$. An $\mathcal{F}_t$-measurable stochastic process $U_t^B\left(x; T\right)$ is called a* backward dynamic utility (BDU), *normalized at T, if for all $t, \tilde{T}$ it satisfies the stochastic optimality criterion*

$$
U_t^B\left(x; T\right) = \begin{cases} -e^{-\gamma x}, & t = T, \\ \sup_{\mathcal{A}} E_P\left[U_{\tilde{T}}^B\left(X_{\tilde{T}}; T\right)\middle| \mathcal{F}_t\right], & 0 \le t \le \tilde{T} \le T, \end{cases} \tag{14}
$$

with X given by (5) and $X_t = x \in \mathcal{R}$.

The above equation provides the dynamic law for the backward dynamic utility. Note that even though this dynamic utility coincides with the familiar value function, its notion was created from a very different point of view and scope. Under mild regularity assumptions on the coefficients of the state processes, it is easy to deduce that the above problem has a solution that is unique. There is ample literature on the value function and, thus, on the backward dynamic utility (see, e.g., Kramkov and Schachermayer [17], Rouge and El Karoui [28], Delbaen et al. [8], and Kabanov and Stricker [14]).

The fact that U_t^B is self-generating is immediate. Indeed, in an arbitrary subhorizon $[t, \bar{T}]$, the associated value function V_t^B, given by

$$V_t^B\left(x, \bar{T}; T\right) = \sup_{\mathcal{A}} E_P\left[U_{\bar{T}}^B\left(X_{\bar{T}}; T\right)\middle| \mathcal{F}_t\right],$$

coincides with its associated dynamic utility

$$U_t^B\left(x; T\right) = V_t^B\left(x, \bar{T}; T\right)$$

by Definition 1.

A consequence of self-generation is that the investor receives the same dynamic utility across different investment horizons. This is seen by the fact that, for $\bar{T} \leq \bar{T}'$, self-generation yields

$$V_t^B\left(x, \bar{T}; T\right) = U_t^B\left(x; T\right),$$

and

$$V_t^B\left(x, \bar{T}'; T\right) = U_t^B\left(x; T\right),$$

and the horizon invariance

$$V_t^B\left(x, \bar{T}; T\right) = V_t^B\left(x, \bar{T}'; T\right)$$

follows.

In most of the existing utility models, the dynamic utility (or, equivalently, its associated value function) is generated backwards in time. The form of the utility might be more complex, as it is in the case of recursive utilities where dynamic risk preferences are generated by an aggregator. Nevertheless, the features of utility prespecification at a future fixed point in time and generation at previous times are still prevailing.

One might argue that an ad hoc specification of utility at a future time is, to a certain extent, nonintuitive, given that our risk attitude might change with the way the market environment enfolds from one time period to the next. Note that changes in the investment opportunities and losses/gains are revealed forward in time and the traditional value function appears to process this information backwards in time. Such issues have been considered in prospect theory where, however, utility normalization at a given future point is still present.

From the valuation perspective, working with utilities normalized in the future severely constrains the class of claims that can be priced. Indeed, their maturities must be always dominated by the time at which the backward utility is normalized. This precludes opportunities related to claims arriving at a later time and/or maturing beyond the normalization point.

In order to be able to accommodate claims of arbitrary maturities, one might propose to work in an infinite horizon framework and to employ either discounted at optimal growth utility functionals or utilities allowing for intermediate consumption. The perpetual nature of these problems, however, might not be appropriate for a variety of applications in which the agent faces defaults, constraints due to reporting periods, and other "real-time" issues.

Motivated by these considerations, the authors recently introduced the concept of *forward dynamic utilities*. Their main characteristic is that they are determined at present time and, as their name indicates, are generated, via their constitutive equation, forward in time.

Definition 2. *Let $s \geq 0$. An $\mathcal{F}_t$-measurable stochastic process $U_t^F(x; s)$ is called a* forward dynamic exponential utility *(FDU), normalized at s, if, for all t, T, with $s \leq t \leq T$, it satisfies the stochastic optimization criterion*

$$U_t^F(x; s) = \begin{cases} -e^{-\gamma x}, & t = s, \\ \sup_{\mathcal{A}} E_P\left[U_T^F(X_T; s) \middle| \mathcal{F}_t\right], & t \geq s. \end{cases} \tag{15}$$

Observe that by construction, there is no constraint on the length of the trading horizon.

As with its backward dynamic counterpart, the forward dynamic utility is self-generating and makes the investor indifferent across distinct investment horizons. Indeed, self-generation, that is,

$$V_t^F(x, T; s) = U_t^F(x; s),$$

with

$$V_t^F(x, T; s) = \sup_{\mathcal{A}} E_P\left[U_T^F(X_T; s) \middle| \mathcal{F}_t\right],$$

is an immediate consequence of the above definition. For the horizon invariance, it is enough to observe that in distinct investment subhorizons, say $[t, T]$ and $[t, \bar{T}]$,

$$V_t^F(x, T; s) = \sup_{\mathcal{A}} E_P\left[U_T^F(X_T; s) \middle| \mathcal{F}_t\right],$$

and

$$V_t^F(x, \bar{T}; s) = \sup_{\mathcal{A}} E_P\left[U_{\bar{T}}^F(X_{\bar{T}}; s) \middle| \mathcal{F}_t\right],$$

and, therefore,

$$V_t^F\left(x, T; s\right) = V_t^F\left(x, \bar{T}; s\right).$$

We stress that, in contrast to their backward dynamic counterparts, forward dynamic utilities might not be unique. In general, the problem of existence and uniqueness is an open one. This issue is discussed in Section 5. Determining a natural class of forward utilities in which uniqueness is established is a challenging and, in our view, interesting question.

3 Auxiliary Technical Results

In the upcoming sections, two equivalent martingale measures are used, namely, the minimal martingale and the minimal entropy ones. They are denoted, respectively, Q^{mm} and Q^{me} and are defined as the minimizers of the entropic functionals

$$\mathcal{H}^0\left(Q^{mm}\mid P\right) = \min_{Q \in \mathcal{Q}_e} E_P\left[-\ln \frac{dQ}{dP}\right],$$

and

$$\mathcal{H}\left(Q^{me}\mid P\right) = \min_{Q \in \mathcal{Q}_e} E_P\left[\frac{dQ}{dP}\ln \frac{dQ}{dP}\right],$$

where $\mathcal{Q}_e$ stands for the set of equivalent martingale measures. There is ample literature on these measures and on their role in valuation and optimal portfolio choice in the traditional framework of exponential utility; see, respectively, Foellmer and Schweizer [10], Schweizer [29] and [30], Bellini and Frittelli [2] and Frittelli [11], Rouge and El Karoui [28], Arai [1], Delbaen et al. [8], and Kabanov and Stricker [14].

For arbitrary $T > 0$, the restrictions of Q^{mm} and Q^{me} on the σ-algebra $\mathcal{F}_T = \sigma\left\{\left(W_u^1, W_u\right) : 0 \leq u \leq T\right\}$, can be explicitly constructed as discussed next. We remark that, with a slight abuse of notation, the restrictions of the two measures are denoted as their original counterparts.

The density of the minimal martingale measure is given by

$$\frac{dQ^{mm}}{dP} = \exp\left(-\int_0^T \lambda_s dW_s^1 - \int_0^T \frac{1}{2}\lambda_s^2 ds\right), \tag{16}$$

with λ being the Sharpe ratio process (4).

Calculating the density of the minimal relative entropy measure is more involved and we refer the reader to Rheinlander [27] (see, also, Grandits and Rheinlander [12]) for a concise treatment. For the diffusion case considered herein, the density can be found through variational arguments and is represented by

$$\frac{dQ^{me}}{dP} = \exp\left(-\int_0^T \lambda_s dW_s^1 - \int_0^T \hat{\lambda}_s dW_s^{1,\perp} - \int_0^T \frac{1}{2}\left(\lambda_s^2 + \hat{\lambda}_s^2\right) ds\right), \tag{17}$$

with $W^{1,\perp}$ as in (3). The process $\hat{\lambda}$ is given by

$$\hat{\lambda}_s = \hat{\lambda}(Y_s, s; T), \tag{18}$$

with Y solving (2) and $\hat{\lambda} : \mathcal{R} \times [0, T] \to \mathcal{R}^+$ defined as

$$\hat{\lambda}(y, t; T) = -\frac{1}{\sqrt{1 - \rho^2}} a(y) \frac{f_y(y, t; T)}{f(y, t; T)}, \tag{19}$$

where $f : \mathcal{R} \times [0, T] \to \mathcal{R}^+$ is the unique $C^{1,2}(\mathcal{R} \times [0, T])$ solution of the terminal value problem

$$\begin{cases} f_t + \frac{1}{2}a^2(y) f_{yy} + (b(y) - \rho\lambda(y) a(y)) f_y = \frac{1}{2}(1 - \rho^2) \lambda^2(y) f, \\ \\ f(y, T) = 1. \end{cases} \tag{20}$$

The proof can be found in Benth and Karslen [3] (see, also, Stoikov and Zariphopoulou [32] and Monoyios [19]).

The dependence of f and $\hat{\lambda}$ on the end of the horizon T is highlighted due to the role that it will play in the upcoming dynamic utilities.

It easily follows that the aggregate, relative to the historical measure, entropies of Q^{mm} and Q^{me} are, respectively,

$$\mathcal{H}(Q^{mm}|P) = E_P\left[\frac{dQ^{mm}}{dP} \ln \frac{dQ^{mm}}{dP}\right] = E_{Q^{mm}}\left[\int_0^T \frac{1}{2}\lambda_s^2 ds\right],$$

and

$$\mathcal{H}(Q^{me}|P) = E_P\left[\frac{dQ^{me}}{dP} \ln \frac{dQ^{me}}{dP}\right] = E_{Q^{me}}\left[\int_0^T \frac{1}{2}\left(\lambda_s^2 + \hat{\lambda}_s^2\right) ds\right].$$

When the market becomes complete, the two measures Q^{mm} and Q^{me} coincide with the unique risk neutral measure. In general, they differ and their respective relative entropies are related in a nonlinear manner. This was explored in Stoikov and Zariphopoulou [32, Corollary 3.1], where it was shown that

$$\mathcal{H}(Q^{me}|P) = \mathcal{E}_{Q^{mm}}\left[\int_0^T \frac{1}{2}\lambda_s^2 ds \,|\mathcal{F}_0\right]. \tag{21}$$

The conditional nonlinear expectation $\mathcal{E}_Q$ of a generic random variable $Z \in \mathcal{F}_T$ and measure Q on $(\Omega, \mathcal{F}_T)$ is defined, for $t \in [0, T]$ and $\gamma \in \mathcal{R}^+$, by

$$\mathcal{E}_Q[Z|\mathcal{F}_t; \gamma] = -\frac{1}{\gamma(1 - \rho^2)} \ln E_Q\left[e^{-\gamma(1-\rho^2)Z} |\mathcal{F}_t\right]. \tag{22}$$

The aggregate entropy $\mathcal{H}(Q^{me}|P)$ is then the nonlinear expectation of the random variable $Z_T = \int_0^T \frac{1}{2}\lambda_s^2 ds$, for $Q = Q^{me}$ and $\gamma = 1$.

Next, we introduce two quantities that facilitate our analysis. Namely, for $0 \leq t \leq \tilde{T} \leq T$, we define the aggregate relative entropy process

$$H\left(t, \tilde{T}\right) = E_{Q^{me}}\left[\int_t^{\tilde{T}} \frac{1}{2}\left(\lambda^2\left(Y_s\right) + \hat{\lambda}^2\left(Y_s, s; T\right)\right) ds \middle| \mathcal{F}_t\right], \qquad (23)$$

and the function $\tilde{H} : \mathcal{R} \times \left[0, \tilde{T}\right] \to \mathcal{R}^+$,

$$\tilde{H}\left(y, t; \tilde{T}\right) = E_{Q^{me}}\left[\int_t^{\tilde{T}} \frac{1}{2}\left(\lambda^2\left(Y_s\right) + \hat{\lambda}^2\left(Y_s, s; T\right)\right) ds \middle| Y_t = y\right], \qquad (24)$$

for λ, $\hat{\lambda}$ defined in (4) and (18). We also introduce the linear operators

$$\mathcal{L}^Y = \frac{1}{2}a^2\left(y\right)\frac{\partial^2}{\partial y^2} + b\left(y\right)\frac{\partial}{\partial y}, \qquad (25)$$

$$\mathcal{L}^{Y,mm} = \frac{1}{2}a^2\left(y\right)\frac{\partial^2}{\partial y^2} + \left(b\left(y\right) - \rho\lambda\left(y\right)a\left(y\right)\right)\frac{\partial}{\partial y}, \qquad (26)$$

and

$$\mathcal{L}^{Y,me} = \frac{1}{2}a^2\left(y\right)\frac{\partial^2}{\partial y^2} + \left(b\left(y\right) - \rho\lambda\left(y\right)a\left(y\right)\right)\frac{\partial}{\partial y} \qquad (27)$$

$$+ a^2\left(y\right)\frac{f_y\left(y, t; T\right)}{f\left(y, t; T\right)}\frac{\partial}{\partial y},$$

with f solving (20).

The following results follow directly from the definition of $\tilde{H}$, and (19) and (20).

Lemma 1. *For $\tilde{T} \leq T$, the function $\tilde{H} : \mathcal{R} \times [0, \tilde{T}] \to \mathcal{R}^+$ solves the quasilinear equation*

$$\tilde{H}_t + \mathcal{L}^{Y,mm}\tilde{H} - \frac{1}{2}(1 - \rho^2)a\left(y\right)^2\tilde{H}_y^2 + \frac{\lambda^2(y)}{2} = 0,$$

or, equivalently, the linear equation

$$\tilde{H}_t + \mathcal{L}^{Y,me}\tilde{H} + \frac{\lambda^2(y) + \hat{\lambda}^2\left(y, t; T\right)}{2} = 0,$$

with $\tilde{H}\left(y, \tilde{T}; T\right) = 0$, and $\mathcal{L}^{Y,mm}$, $\mathcal{L}^{Y,me}$ as in (26) and (27).

4 Investment and Valuation Under Backward Dynamic Exponential Utilities

In this section, we provide an analytic representation of the backward dynamic exponential utility (cf. Definition 1) and construct the agent's optimal investment in an integrated portfolio choice problem. We recall that the investment horizon is fixed, the utility is normalized at this horizon's end, and that no liabilities, or cash flows, are allowed beyond the normalization point. For convenience, we occasionally rewrite some of the quantities introduced in earlier sections.

Proposition 1. *Let Q^{me} be the minimal relative entropy martingale measure and $H(t, T)$ the aggregate relative entropy process (cf. (23)),*

$$H\left(t, \tilde{T}\right) = E_{Q^{me}}\left[\int_t^{\tilde{T}} \frac{1}{2}\left(\lambda^2\left(Y_s\right) + \hat{\lambda}^2\left(Y_s, s; T\right)\right) ds \,\bigg|\, \mathcal{F}_t\right],$$

with λ and $\hat{\lambda}$ as in (4) and (18). Then, for $x \in \mathcal{R}$, $t \in [0, T]$, the process $U_t^B \in \mathcal{F}_t$, given by

$$U_t^B\left(x; T\right) = -e^{-\gamma x - H(t, T)}, \tag{28}$$

is the backward dynamic exponential utility.

The proof is, essentially, a direct consequence of the dynamic programming principle and the results of Rouge and El Karoui [28]. For the specific technical arguments, related to the stochastic factor model we examine herein, we refer the reader to Stoikov and Zariphopoulou [32]. We easily deduce the following result.

Corollary 1. *The backward dynamic utility is given by*

$$U_t^B\left(x; T\right) = u\left(x, Y_t, t; T\right),$$

with $u : \mathcal{R} \times \mathcal{R}^+ \times [0, T] \to \mathcal{R}^-$ defined as

$$u\left(x, y, t; T\right) = -e^{-\gamma x - \tilde{H}(y, t; T)},$$

with

$$\tilde{H}\left(y, t; \tilde{T}\right) = E_{Q^{me}}\left[\int_t^{\tilde{T}} \frac{1}{2}\left(\lambda^2\left(Y_s\right) + \hat{\lambda}^2\left(Y_s, s; T\right)\right) ds \,\bigg|\, Y_t = y\right].$$

4.1 Backward Indifference Values

Next, we revisit the classical definition of indifference values but in the framework of backward dynamic utility. This framework allows for a concise valuation of claims and liabilities of arbitrary maturities, provided that these maturities occur before the normalization point. Due to self-generation, the notion of dynamic value function becomes redundant. Herein we concentrate on the indifference treatment of a liability, or, equivalently, on the optimal portfolio choice of the writer of a claim, yielding payoff equal to the liability at hand.

Definition 3. *Let T be the backward normalization point and consider a claim $C_{\bar{T}} \in \mathcal{F}_{\bar{T}}$, written at $t_0 \geq 0$ and maturing at $\bar{T} \leq T$. For $t \in [t_0, \bar{T}]$, the backward indifference value process (BIV) $\nu_t^B\left(C_{\bar{T}}; T\right)$ is defined as the amount that satisfies the pricing condition*

$$U_t^B\left(x - \nu_t^B\left(C_{\bar{T}}; T\right); T\right) = \sup_{\mathcal{A}} E_P\left[U_{\bar{T}}^B\left(X_{\bar{T}} - C_{\bar{T}}; T\right)\big| \mathcal{F}_t\right], \qquad (29)$$

for all $x \in \mathcal{R}$ and $X_t = x$.

We note that the backward indifference value coincides with the classical one, but it is constructed from a quite different point of view. The focus herein is not on rederiving previously known quantities, but, rather, in exploring how the backward indifference values are affected by the normalization point and the changes in the market environment, as well as how they differ from their forward dynamic counterparts.

We address these questions for the class of bounded European claims and liabilities, for which we can deduce closed-form variational expressions.

Proposition 2. *Let T be the backward normalization point and consider a European claim written at $t_0 \geq 0$ and maturing at $\bar{T} \leq T$, yielding payoff $C_{\bar{T}} = C\left(S_{\bar{T}}, Y_{\bar{T}}\right)$. For $t \in [t_0, \bar{T}]$, its backward indifference value process $\nu_t^B\left(C_{\bar{T}}; T\right)$ is given by*

$$\nu_t^B\left(C_{\bar{T}}; T\right) = p^B\left(S_t, Y_t, t\right),$$

where S and Y solve (1) and (2), and $p^F : \mathcal{R}^+ \times \mathcal{R} \times [0, \bar{T}] \to \mathcal{R}$ satisfies

$$\begin{cases} p_t^B + \mathcal{L}^{(S,Y),me} p^B + \frac{1}{2}\gamma\left(1 - \rho^2\right) a^2\left(y\right)\left(p_y^B\right)^2 = 0, \\ \\ p^B\left(S, y, \bar{T}\right) = C\left(S, y\right). \end{cases} \qquad (30)$$

Herein,

$$\mathcal{L}^{(S,Y),me} = \frac{1}{2}\sigma^2\left(y\right) S^2 \frac{\partial^2}{\partial S^2} + \rho\sigma\left(y\right) Sa\left(y\right) \frac{\partial^2}{\partial S \partial y} + \frac{1}{2}a^2\left(y\right) \frac{\partial^2}{\partial y^2} \qquad (31)$$

$$+ \left(b\left(y\right) - \rho\lambda\left(y\right) a\left(y\right) + a^2\left(y\right) \frac{f_y\left(y, t; T\right)}{f\left(y, t; T\right)}\right) \frac{\partial}{\partial y},$$

and f solves (cf. (20))

$$\begin{cases} f_t + \frac{1}{2}a^2\left(y\right) f_{yy} + \left(b\left(y\right) - \rho\lambda\left(y\right) a\left(y\right)\right) f_y = \frac{1}{2}\left(1 - \rho^2\right)\lambda^2\left(y\right) f, \\ \\ f\left(y, T\right) = 1. \end{cases}$$

Proof. For convenience, we recall the entropic quantities

$$H\left(t,t'\right) = E_{Q^{me}}\left[\int_t^{t'} \frac{1}{2}\left(\lambda^2\left(Y_s\right) + \hat{\lambda}^2\left(Y_s,s;T\right)\right)ds\,\Bigg|\,\mathcal{F}_t\right],$$

and

$$\tilde{H}\left(y,t;t'\right) = E_{Q^{me}}\left[\int_t^{t'} \frac{1}{2}\left(\lambda^2\left(Y_u\right) + \hat{\lambda}^2\left(Y_u,u;T\right)\right)du\,\Bigg|\,Y_t = y\right],$$

for $0 \le t \le t' \le \bar{T} \le T$. We first calculate the right-hand side of (29), which, in view of Proposition 1, becomes

$$\sup_{\mathcal{A}} E_P\left[-e^{-\gamma(X_{\bar{T}}-C_{\bar{T}})-H(\bar{T};T)}\,\Big|\,\mathcal{F}_t\right] = \sup_{\mathcal{A}} E_P\left[-e^{-\gamma(X_{\bar{T}}-G_{\bar{T}})}\,\Big|\,\mathcal{F}_t\right],$$

with

$$G_{\bar{T}} = C\left(S_{\bar{T}},Y_{\bar{T}}\right) - \frac{1}{\gamma}H\left(\bar{T};T\right).$$

One may, then, view this problem as a traditional indifference valuation one in which the trading horizon is $[t,\bar{T}]$ and the utility is the exponential function at $\bar{T}$. For the stochastic factor model we consider herein, we obtain (see Sircar and Zariphopoulou [31] and Grasselli and Hurd [13])

$$\sup_{\mathcal{A}} E_P\left[-e^{-\gamma(X_{\bar{T}}-G_{\bar{T}})}\,\Big|\,\mathcal{F}_t\right] = -e^{-\gamma(x-h(S_t,Y_t,t))-H(t;\bar{T})},$$

with $h : \mathcal{R}^+ \times \mathcal{R} \times [0,\bar{T}] \to \mathcal{R}$ solving

$$\begin{cases} h_t + \mathcal{L}^{(S,Y),me}h + \frac{1}{2}\gamma\left(1-\rho^2\right)a^2\left(y\right)h_y^2 = 0, \\[2mm] h\left(S,y,\bar{T}\right) = C\left(S,y\right) - \frac{1}{\gamma}\tilde{H}\left(y,\bar{T};T\right). \end{cases}$$

Next, we introduce the function $p^B : \mathcal{R}^+ \times \mathcal{R} \times [0,\bar{T}] \to \mathcal{R}$,

$$p^B\left(S,y,t\right) = h\left(S,y,t\right) + \frac{1}{\gamma}\left(\tilde{H}\left(y,t;T\right) - \tilde{H}\left(y,t;\bar{T}\right)\right).$$

Using the equation satisfied by h, we deduce that p^B solves (30). On the other hand, Corollary 1 and the above equalities yield

$$\sup_{\mathcal{A}} E_P\left[-e^{-\gamma(X_{\bar{T}}-G_{\bar{T}})}\,\Big|\,X_t = x, S_t = S, Y_t = y\right]$$

$$= -e^{-\gamma\left(x-p^B(S,y,t)\right)-\left(\tilde{H}\left(y,t;\bar{T}\right)+\tilde{H}\left(y,\bar{T};T\right)\right)} = -e^{-\gamma\left(x-p^B(S,y,t)\right)-\tilde{H}(y,t;T)},$$

and the assertion follows from Definition 3 and Proposition 1.   □

4.2 Optimal Portfolios Under Backward Dynamic Utility

Next, we construct the optimal portfolio strategies in the integrated portfolio problem. We start with the agent's optimal behavior in the absence of the liability/payoff. We concentrate our attention on optimal behavior in a shorter horizon. For simplicity, its end is taken to coincide with $\bar{T}$, the point at which the liability is met.

Proposition 3. *Let T be the backward normalization point and $[t, \bar{T}] \in [t, T]$ be the trading horizon of an investor endowed with the backward exponential dynamic utility U^B. The processes $\pi_s^{B,*}$ and $\pi_s^{B,0,*}$, representing the optimal investments in the risky and riskless asset, are given, for $s \in [t, \bar{T}]$, by*

$$\pi_s^{B,*} = \pi^{B,*}\left(X_s^{B,*}, Y_s, s\right) = \frac{\mu\left(Y_s\right)}{\gamma \sigma^2\left(Y_s\right)} - \rho \frac{a\left(Y_s\right)}{\sigma\left(Y_s\right)} \tilde{H}_y\left(Y_s, s; T\right) \qquad (32)$$

and

$$\pi_s^{B,0,*} = \pi^{B,0,*}\left(X_s^*, Y_s, s\right) = X_s^{B,*} - \pi_s^{B,*}.$$

Herein, $X_s^{B,*}$ solves (5) with $\pi_s^{B,*}$ being used, and $\tilde{H} : \mathcal{R} \times [0, T] \to \mathcal{R}^+$ satisfies

$$\tilde{H}_t + \mathcal{L}^{Y,me} \tilde{H} + \frac{\lambda^2(y) + \hat{\lambda}^2\left(y, t; T\right)}{2} = 0,$$

with terminal condition

$$\tilde{H}\left(y, \bar{T}; T\right) = E_{Q^{me}}\left[\left. \int_{\bar{T}}^T \frac{1}{2}\left(\lambda\left(Y_s\right)^2 + \hat{\lambda}\left(Y_s, s; T\right)^2\right) ds \right| Y_{\bar{T}} = y\right]. \qquad (33)$$

Given the diffusion nature of the model, the form of the utility data and the regularity assumptions on the market coefficients, optimality follows from classical verification results (see, among others, Duffie and Zariphopoulou [9], Zariphopoulou [35], Pham [26], and Touzi [33]).

Due to the stochasticity of the investment opportunity set, the optimal investment strategy in the stock account consists of two components, namely, the *myopic* portfolio and the so-called *excess risky demand*, given, respectively, by $\mu\left(Y_s\right)/\gamma \sigma^2\left(Y_s\right)$ and $-\rho((a\left(Y_s\right))/(\sigma\left(Y_s\right)))\tilde{H}_y\left(Y_s, s; T\right)$. The myopic component is what the investor would follow if the coefficients of the risky security remained constant across trading periods. The excess risky demand is the required investment that emerges from the local in time changes in the Sharpe ratio (see, among others, Kim and Omberg [16], Liu [18], Campbell and Viceira [4], Chacko and Viceira [6], Wachter [34], and Campell et al. [5]).

Note that even though the trading horizon $[t, \bar{T}]$ is shorter than the original one $[t, T]$, the optimal policies depend on the longer horizon because the dynamic risk preferences are normalized at T and not at $\bar{T}$.

Remark 1. The reader familiar with the representation of indifference prices might try to interpret the excess risky demand as the indifference hedging strategy of an appropriately chosen claim. Such questions were studied in Stoikov and Zariphopoulou [32] where the relevant claim was identified and priced.

We continue with the optimal strategies in the presence of a European-type liability $C_{\bar{T}}$, which, we recall, is taken to be bounded.

Proposition 4. *Let T be the backward normalization point and consider an investor endowed with the backward dynamic exponential utility U^B and facing a liability $C_{\bar{T}} = C\left(S_{\bar{T}}, Y_{\bar{T}}\right)$. The processes $\pi_s^{B,*}$ and $\pi_s^{B,0,*}$, representing the optimal investments in the risky and riskless asset, are given, for $s \in [t, \bar{T}]$, by*

$$\pi_s^{B,*} = \pi^{B,*}\left(X_s^{B,*}, S_s, Y_s, s\right) = \frac{\mu\left(Y_s\right)}{\gamma\sigma^2\left(Y_s\right)} - \rho\frac{a\left(Y_s\right)}{\sigma\left(Y_s\right)}\tilde{H}_y\left(Y_s, s; T\right) \quad (34)$$

$$+ S_s p_S^B\left(S_s, Y_s, s\right) + \rho\frac{a\left(Y_s\right)}{\sigma\left(Y_s\right)}p_y^B\left(S_s, Y_s, s\right)$$

and

$$\pi_s^{B,0,*} = \pi^{B,0,*}\left(X_s^*, S_s, Y_s, s\right) = X_s^{B,*} - \pi_s^{B,*}.$$

Herein, $X_s^{B,*}$ solves (5) with $\pi_s^{B,*}$ being used, $\tilde{H}$ as in Proposition 1, and p^B solves (30).

Proof. In the presence of the liability, we observe

$$\sup_{\mathcal{A}} E_P\left[U_{\bar{T}}^B\left(X_{\bar{T}} - C_{\bar{T}}; T\right)\middle|\mathcal{F}_t\right] = u^C\left(x, S_t, Y_t, t\right),$$

where $u^C : \mathcal{R} \times \mathcal{R}^+ \times \mathcal{R} \times [0, \bar{T}] \to \mathcal{R}^-$ solves the Hamilton–Jacobi–Bellman equation

$$u_t^C + \max_{\pi}\left\{\frac{1}{2}\sigma^2\left(y\right)\pi^2 u_{xx}^C + \pi\left(\sigma^2\left(y\right)Su_{xS}^C + \rho a(y)\sigma\left(y\right)u_{xy}^C + \mu\left(y\right)u_x^C\right)\right\}$$

$$+ \mathcal{L}^{(S,Y)}u^C = 0,$$

with

$$u^C\left(x, S, y, \bar{T}\right) = -e^{-\gamma(x - C(S,y)) - \tilde{H}\left(y, \bar{T}; T\right)}$$

and

$$\mathcal{L}^{(S,Y)} = \frac{1}{2}\sigma^2\left(y\right)S^2\frac{\partial^2}{\partial S^2} + \rho\sigma\left(y\right)Sa\left(y\right)\frac{\partial^2}{\partial S\partial y} + \frac{1}{2}a^2\left(y\right)\frac{\partial^2}{\partial y^2} \quad (35)$$

$$+ \mu\left(y\right)\frac{\partial}{\partial S} + b\left(y\right)\frac{\partial}{\partial y}.$$

Verification results yield that the optimal policy $\pi_s^{B,*}$ is given in the feedback form

$$\pi_s^{B,*} = \pi^{B,*}\left(X_s^{B,*}, S_s, Y_s, s\right),$$

with

$$\pi^{B,*}(x, S, y, t) = -\frac{\sigma^2(y) S u_{xS}^C + \rho a(y)\sigma(y) u_{xy}^C + \mu(y) u_x^C}{\sigma^2(y) u_{xx}^C}.$$

On the other hand, from Proposition 1,

$$u^C(x, S, y, t) = -e^{-\gamma\left(x - p^B(S,y,t)\right) - \tilde{H}(y,t;T)}.$$

Combining the above and the feedback form of $\pi^{B,*}(x, S, y, t)$ we conclude.

$\square$

5 Investment and Valuation Under Forward Dynamic Exponential Utilities

We now revert our attention to portfolio choice and pricing under the newly introduced class of forward dynamic utilities. We start with the analytic construction of such a utility. As mentioned in Section 2, general existence and uniqueness results for forward dynamic utilities are lacking. As a matter of fact, an alternative solution to (15) is presented in Example 3.

Proposition 5. *Let $s \geq 0$ be the forward normalization point. Define, for $t \geq s$, the process*

$$h(s, t) = \int_s^t \frac{1}{2}\lambda_u^2 du, \tag{36}$$

with λ being the Sharpe ratio (4). Then, the process $U_t^F(x; s)$ given, for $x \in \mathcal{R}$ and $t \geq s$, by

$$U_t^F(x; s) = -e^{-\gamma x + h(s,t)}, \tag{37}$$

is a forward dynamic exponential utility, normalized at s.

Proof. The fact that $U_t^F(x; s)$ is $\mathcal{F}_t$-measurable and normalized at s is immediate. It remains to show (15), namely, that for arbitrary $T \geq t$,

$$-e^{-\gamma x + h(s,t)} = \sup_{\mathcal{A}} E_P\left[-e^{-\gamma X_T + h(s,T)}\middle|\mathcal{F}_t\right].$$

Using (36), the above reduces to

$$-e^{-\gamma x} = \sup_{\mathcal{A}} E_P\left[-e^{-\gamma X_T + h(t,T)}\middle|\mathcal{F}_t\right]. \tag{38}$$

Next, we introduce the function $u : \mathcal{R} \times \mathcal{R} \times [0, T] \to \mathcal{R}^-$,

$$u(x, y, t) = \sup_{\mathcal{A}} E_P \left[-e^{-\gamma X_T + \int_t^T (1/2)\lambda^2 (Y_s) ds} \middle| X_t = x, Y_t = y \right].$$

Classical arguments imply that u solves the Hamilton–Jacobi–Bellman equation

$$u_t + \mathcal{L}^Y u + \frac{\lambda^2 (y)}{2} u$$

$$+ \max_\pi \left\{ \frac{1}{2}\sigma^2 (y) \pi^2 u_{xx} + \pi (\rho a(y)\sigma (y) u_{xy} + \mu (y) u_x) \right\} = 0,$$

with

$$u(x, y, T) = -e^{-\gamma x},$$

and $\mathcal{L}^Y$ as in (15). We deduce (see, e.g., Duffie and Zariphopoulou [9] and Pham [26]) that the above equation has a unique solution in the class of functions that are concave and increasing in x, and are uniformly bounded in y. We, then, see that the function $\breve{u}(x, y, t) = -e^{-\gamma x}$ is such a solution and, by uniqueness, it coincides with u. The rest of the proof follows easily.   □

We next present an alternative forward dynamic utility.

Example 3. Consider, for $x \in \mathcal{R}$ and $t \geq s$, the process

$$U_t^F (x; s) = -e^{-\gamma x - Z(s,t)},$$

with

$$Z(s, t) = \int_s^t \frac{1}{2}\lambda_s^2 ds + \int_s^t \lambda_s dW_s^1. \tag{39}$$

Observe that, for $X_t = x$, the forward stochastic criterion (cf. (15)),

$$-e^{-\gamma x - Z(s,t)} = \sup_{\mathcal{A}} E_P \left[-e^{-\gamma X_T - Z(s,T)} \middle| \mathcal{F}_t \right],$$

will hold if we establish

$$-e^{-\gamma x} = \sup_{\mathcal{A}} E_P \left[-e^{-\gamma X_T - Z(t,T)} \middle| \mathcal{F}_t \right],$$

or, equivalently,

$$-e^{-\gamma x} = \sup_{\mathcal{A}} E_P \left[-e^{-\gamma X_T - Z(t,T)} \middle| X_t = x, Y_t = y, \hat{Z}_t = 0 \right],$$

with $\hat{Z}_{t'} = z + Z(t, t')$. Defining $v : \mathcal{R} \times \mathcal{R} \times \mathcal{R} \times [0, T] \to \mathcal{R}^-$ by

$$v(x, y, z, t) = \sup_{\mathcal{A}} E_P \left[-e^{-\gamma X_T - \hat{Z}_T} \middle| X_t = x, Y_t = y, \hat{Z}_t = z \right],$$

we see that it solves the Hamilton–Jacobi–Bellman equation

$$v_t + \max_{\pi} \left\{ \frac{1}{2}\sigma^2(y)\pi^2 u_{xx} + \pi\left(\lambda(y)\sigma(y)u_{xz} + \rho a(y)\sigma(y)u_{xy} + \mu(y)u_x\right) \right\}$$

$$+ \frac{1}{2}\lambda^2(y)v_{zz} + \rho\lambda(y)a(y)v_{zy} + \frac{1}{2}a^2(y)v_{yy} + b(y)v_y + \frac{1}{2}\lambda^2(y)v_z,$$

with
$$v(x,y,z,T) = -e^{-\gamma x - z}.$$

Substituting above the function $\hat{v}(x,y,z,t) = -e^{-\gamma x - z}$, and after some calculations, yields

$$-\frac{(\lambda(y)\sigma(y)\hat{v}_{xz} + \mu(y)\hat{v}_x)^2}{2\sigma^2(y)\hat{v}_{xx}} + \frac{1}{2}\lambda^2(y)(\hat{v}_{zz} + \hat{v}_z) = 0.$$

We easily conclude that $\hat{v} \equiv v$, and the assertion follows.

5.1 Forward Indifference Values

Next, we introduce the concept of forward indifference value. As is its backward counterpart, it is defined as the amount that generates the same level of (forward) dynamic utility with and without incorporating the liability. Note, also, that in the definition below, it is only the forward dynamic utility that enters, eliminating the need to incorporate in the definition the forward dynamic value function. This allows for a concise treatment of payoffs and liabilities of arbitrary maturities. Finally, we remark, that the nomenclature "forward" does not refer to the terminology used in derivative valuation pertinent to wealth expressed in forward units. Rather, it refers to the forward in time manner that the dynamic utility evolves.

Although the concept of forward indifference value appears to be a straightforward extension of the backward one, it is important to observe that the maturities of the claims in consideration need not be bounded by any prespecified horizon. This is one of the striking differences between the classes of claims that can be priced by the two distinct dynamic utilities we consider herein.

Definition 4. *Let $s \geq 0$ be the forward normalization point and consider a claim $C_T \in \mathcal{F}_T$, written at $t_0 \geq s$ and maturing at T. For $t \in [t_0, T]$, the forward indifference value process (FIP) $v_t^F(C_T; s)$ is defined as the amount that satisfies the pricing condition*

$$U_t^F\left(x - v_t^F(C_T; s); s\right) = \sup_{\mathcal{A}} E_P\left[U_T^F(X_T - C_T; s)\big| \mathcal{F}_t\right], \qquad (40)$$

for all $x \in \mathcal{R}$ and $X_t = x$.

We continue with the valuation of a bounded European-type liability and we examine how its forward indifference value is affected by the choice of the normalization point. We show that even though both forward dynamic utilities, entering in (40) above, depend on the normalization point, the emerging forward price does not. This is another important difference between the backward and the forward indifference values.

Proposition 6. *Let* $s \geq 0$ *be the forward normalization point and consider a European claim written at* $t_0 \geq s$ *and maturing at* T *yielding payoff* $C_T = C\left(S_T, Y_T\right)$. *For* $t \in [t_0, T]$, *its forward indifference value* $\nu_t^F\left(C_T; s\right)$ *is given by*

$$\nu_t^F\left(C_T; s\right) = p^F\left(S_t, Y_t, t\right),$$

where S *and* Y *solve (1) and (2), and* $p^F : \mathcal{R}^+ \times \mathcal{R} \times [0, T] \to \mathcal{R}$ *satisfies*

$$\begin{cases} p_t^F + \mathcal{L}^{(S,Y),mm} p^F + \frac{1}{2}\gamma\left(1 - \rho^2\right) a^2\left(y\right)\left(p_y^F\right)^2 = 0, \\ \\ p^F\left(S, y, T\right) = C\left(S, y\right), \end{cases} \tag{41}$$

with

$$\mathcal{L}^{(S,Y),mm} = \frac{1}{2}\sigma^2\left(y\right) S^2 \frac{\partial^2}{\partial S^2} + \rho\sigma\left(y\right) Sa\left(y\right) \frac{\partial^2}{\partial S \partial y} + \frac{1}{2}a^2\left(y\right) \frac{\partial^2}{\partial y^2}$$

$$+ \left(b\left(y\right) - \rho\lambda\left(y\right) a\left(y\right)\right) \frac{\partial}{\partial y}.$$

Proof. We first note that

$$\sup_{\mathcal{A}} E_P\left[U_T^F\left(X_T - C_T; s\right) \middle| \mathcal{F}_t \right] = \sup_{\mathcal{A}} E_P\left[-e^{-\gamma(X_T - C_T) + \int_s^T (1/2)\lambda^2(Y_u) du} \middle| \mathcal{F}_t \right]$$

$$= e^{\int_s^t (1/2)\lambda^2(Y_u) du} \sup_{\mathcal{A}} E_P\left[-e^{-\gamma(X_T - C_T) + \int_t^T (1/2)\lambda^2(Y_u) du} \middle| \mathcal{F}_t \right],$$

where we used Proposition 5 and the measurability of the process h (cf. (36)).

Define $u^C : \mathcal{R} \times \mathcal{R}^+ \times \mathcal{R} \times [0, T] \to \mathcal{R}^-$,

$$u^C\left(x, S, y, t\right) = \sup_{\mathcal{A}} E_P\left[-e^{-\gamma(X_T - C_T) + \int_t^T (1/2)\lambda^2(Y_u) du} \middle| X_t = x, S_t = S, Y_t = y \right],$$

and observe that it solves the Hamilton–Jacobi–Bellman equation

$$u_t^C + \max_\pi \left\{ \frac{1}{2}\sigma^2\left(y\right) \pi^2 u_{yy}^C + \pi\left(\sigma^2\left(y\right) S u_{xS}^C + \rho a(y)\sigma\left(y\right) u_{xy}^C + \mu\left(y\right) u_x^C\right) \right\}$$

$$+ \mathcal{L}^{(S,Y)} u^C + \frac{\lambda^2\left(y\right)}{2} u^C = 0,$$

with

$$u^C\left(x, S, y, T\right) = -e^{-\gamma(x - C(S, y))},$$

and $\mathcal{L}^{(S,Y)}$ as in (35). Using the transformation
$$u^C(x, S, y, t) = -e^{-\gamma\left(x - p^F(S,y,t)\right)},$$

we deduce, after tedious but straightforward calculations, that the term $p^F(S, y, t)$ solves (41). We, then, easily, see that

$$\sup_{\mathcal{A}} E_P\left[U_T^F(X_T - C_T; s)\middle| \mathcal{F}_t\right]$$

$$= -e^{\int_s^t (1/2)\lambda^2(Y_u)du} u^C(x, S_t, Y_t, t),$$

and applying Proposition 5 and Definition 4 completes the proof.   □

5.2 Optimal Portfolios Under Forward Dynamic Utilities

We continue with the optimal investment policies under the forward dynamic risk preferences.

Proposition 7. *Let $s \geq 0$ be the forward normalization point and $[t, T]$ the trading horizon, with $s \leq t$. The processes $\pi^{F,*}$ and $\pi^{F,0,*}$, representing the optimal investments in the risky and riskless asset, are given, respectively, for $u \in [t, T]$, by*

$$\pi_u^{F,*} = \pi^{F,*}\left(X_u^{F,*}, Y_u, u\right) = \frac{\mu(Y_u)}{\gamma\sigma^2(Y_u)} \tag{42}$$

and

$$\pi_u^{F,0,*} = \pi^{F,0,*}\left(X_u^*, Y_u, u\right) = X_u^{F,*} - \pi_u^{F,*},$$

with $X_u^{F,*}$ solving (5) with $\pi_u^{F,*}$ being used.

Two important facts emerge. Firstly, both optimal investment policies $\pi^{F,*}$ and $\pi^{F,0,*}$ are independent of the spot normalization point. Secondly, the investment in the risky asset consists entirely of the myopic component. Indeed, the excess hedging demand, which emerges due to the presence of the stochastic factor, has vanished. The investor has processed the stochasticity of the (incomplete) market environment into her preferences, that are dynamically updated, following, forward in time, the market movements.

Proposition 8. *Let $s \geq 0$ be the forward normalization point and consider an investor endowed with the forward exponential dynamic utility U^F and facing a liability $C_T = C(S_T, Y_T)$. The processes $\pi_s^{F,*}$ and $\pi_s^{F,0,*}$, representing the optimal investments in the risky and riskless asset in the integrated portfolio choice problem, are given, for $u \in [t, T]$, by*

$$\pi_u^{F,*} = \pi^{F,*}\left(X_u^{F,*}, S_u, Y_u, u\right) = \frac{\mu(Y_u)}{\gamma\sigma^2(Y_u)}$$

$$+ S_u p_S^F(S_u, Y_u, u) + \rho\frac{a(Y_u)}{\sigma(Y_u)} p_y^F(S_u, Y_u, u) \tag{43}$$

and

$$\pi_u^{F,0,*} = \pi^{F,0,*}\left(X_u^*, S_u, Y_u, u\right) = X_u^{F,*} - \pi_u^{F,*}.$$

Herein, $X_s^{F,*}$ solves (5) with $\pi_s^{F,*}$ being used, and p^F satisfies (41).

6 Concluding Remarks: Forward Versus Backward Utilities and Their Associated Indifference Prices

In the previous two sections, we analyzed the investment and pricing problems of investors endowed with backward (BDU) and forward (FDU) dynamic exponential utilities. These utilities have similarities but also striking differences. These features are, in turn, inherited by the associated optimal policies, indifference prices, and risk monitoring strategies. Below, we provide a discussion on these issues.

We first observe that the backward and forward utilities are produced via a conditional expected criterion. They are both self-generating, in that they coincide with their implied value functions. Moreover, in the absence of exogenous cash flows, investors endowed with such utilities are indifferent to the investment horizons.

Backward and forward dynamic utilities are constructed in entirely different ways. Backward utilities are first specified at a given future time T and, they are, subsequently, generated at previous to T times. Forward utilities are defined at present s and are, in turn, generated forward in time. The times T and s, at which the backward and forward utility data are determined, are the backward and forward normalization points. We recall, from Equations (28) and (37), that the BDU and FDU processes U_t^B and U_t^F are $\mathcal{F}_t$-adapted and given, respectively, by

$$U_t^B\left(x; T\right) = -e^{-\gamma x - H(t,T)}$$

and

$$U_t^F\left(x; s\right) = -e^{-\gamma x - h(s,t)},$$

with

$$H\left(t, T\right) = E_{Q^{me}}\left[\int_t^T \frac{1}{2}\left(\lambda_u^2 + \hat{\lambda}_u^2\right) du \,\middle|\, \mathcal{F}_t\right]$$

and

$$h(s, t) = \int_s^t \frac{1}{2}\lambda_u^2 du.$$

Herein, λ and $\hat{\lambda}$ are given in (4) and (18), and Q^{me} is the minimal relative entropy measure.

Both BDU and FDU have an exponential, affine in wealth, structure. However, the backward utility compiles changes in the market environment in an aggregate manner, whereas the forward utility does so in a much finer way.

This is seen by the nature of the processes $H(t,T)$ and $h(s,t)$. It is worth observing that

$$H(t,T) \neq E_{Q^{me}}[h(t,T)|\mathcal{F}_t],$$

and that U_t^F is not affected by $\hat{\lambda}$ (cf. (18) and (19)) that represent the "orthogonal" component of the market price of risk.

Backward and forward utilities generate different optimal investment strategies (see, respectively, Propositions 3 and 7). Under backward dynamic preferences, the investor invests in the risky asset an amount equal to the sum of the myopic portfolio and the excess risky demand. The former investment strategy depends on the risk aversion coefficient γ, but not on the backward normalization point T. The excess risky demand, however, is not affected by γ but depends on the choice of the normalization point, even if investment takes place in a shorter horizons.

Under forward preferences, the investor invests in the risk asset solely within the myopic portfolio. The myopic strategy does not depend on the forward normalization point or the investment horizon.

As a consequence of the above differences, the emerging backward (BIV) and forward (FIV) indifference values $\nu_t^B(C_{\bar{T}};T)$ and $\nu_t^F(C_{\bar{T}};s)$, $0 \leq s \leq t \leq \bar{T} \leq T$ have very distinct characteristics. Concentrating on the class of bounded European claims, we see that $\nu_t^B(C_{\bar{T}};T)$ and $\nu_t^F(C_{\bar{T}};s)$ are constructed via solutions of similar quasilinear PDEs. The nonlinearities in the pricing PDEs are of the same type, however, the associated linear operators $\mathcal{L}^{(S,y),me}$ and $\mathcal{L}^{(S,y),mm}$ differ (see, respectively, (30) and (41)). The former, appearing in the BIV equation, corresponds to the minimal relative entropy measure and the latter, appearing in the FIV equation, to the minimal martingale measure. Denoting the solutions of these PDEs as nonlinear expectations, we may formally represent (with a slight abuse of notation) the two indifference values as

$$\nu_t^B(C_{\bar{T}};T) = \mathcal{E}_{Q^{me}}[C_{\bar{T}};T]$$

and

$$\nu_t^F(C_{\bar{T}};s) = \mathcal{E}_{Q^{mm}}[C_{\bar{T}};s].$$

The FIV is independent on the forward normalization point. The BIV depends, however, on the backward normalization point, even if the claim matures in a shorter horizon.

As the investor becomes risk neutral, $\gamma \to 0$, we obtain

$$\lim_{\gamma \to 0} \nu_t^B(C_{\bar{T}};T) = E_{Q^{me}}[C_{\bar{T}}|\mathcal{F}_t]$$

and

$$\lim_{\gamma \to 0} \nu_t^F(C_{\bar{T}};s) = E_{Q^{mm}}[C_{\bar{T}}|\mathcal{F}_t].$$

However, as the investor becomes infinitely risk averse, $\gamma \to \infty$, both BIV and FIV converge to the same limit given by the super replication value,

$$\lim_{\gamma\to\infty} \nu_t^B\left(C_{\bar{T}};T\right) = \lim_{\gamma\to\infty} \nu_t^F\left(C_{\bar{T}};s\right) = \|C_{\bar{T}}\|_{\mathcal{L}^\infty\{.|\mathcal{F}_t\}}.$$

In the presence of the liability and under backward dynamic utility, the investment in the risky asset consists of the myopic portfolio, the excess risky demand, and the backward indifference risk monitoring strategies (see Proposition 4). With the exception of the myopic portfolio, all other three portfolio components depend on the normalization point T. When, however, the investor uses forward dynamic utility, his optimal integrated policy does not include the excess risky demand (see Proposition 8). The entire policy is independent of the forward normalization point s, and depends exclusively on the maturity of the claim and the changes in the market environment.

When the market becomes complete, the backward and forward pricing measures Q^{me} and Q^{mm} coincide with the unique risk-neutral measure Q^*, and (BIV) and (FIV) reduce to the arbitrage free price.

In general, the backward and forward indifference values do not coincide. The underlying reason is that they are defined via the backward and forward dynamic utilities that process the internal model incompleteness, generated by the stochastic factor Y, in a very different manner. Characterizing the market environments as well as the claims for which the two prices coincide is an open question.

Acknowledgments

Thaleia Zariphopoulou's work is partially supported by NSF grants DMS-0456118 and DMS-0091946. This work was presented at the Workshop in Semimartingale Theory and Practice in Finance, Banff (June 2004), Imperial College (June 2004), Cornell University (November 2004), the Workshop in Mathematical Finance at Carnegie Mellon University (January 2005), the Second Bachelier Colloquium, Metabief (January 2005), London School of Economics (April 2005), the Conference on Stochastic Processes and Their Applications, Ascona (June 2005), the SIAM Conference on Control and Optimization, New Orleans (July 2005), the Stochastic Modeling Workshop on PDEs and their applications to Mathematical Finance at KTH, Stockholm (August 2005), the Bachelier Seminar (May 2006), and the Conference in Honor of S. Sethi at the University of Texas at Dallas (May 2006). The authors would like to thank the participants for fruitful comments and suggestions.

References

1. T. Arai. The relation between the minimal martingale measure and the minimal entropy martingale measure. *Asia-Pacific Financial Markets*, 8(2):167–177, 2001.

2. F. Bellini and M. Frittelli. On the existence of the minmax martingale measures. *Mathematical Finance*, 12:1–21, 2002.

3. F.E. Benth and K.H. Karlsen. A PDE representation of the density of the minimal martingale measure in stochastic volatility markets. *Stochastics and Stochastics Reports*, 77:109–137, 2005.

4. J.Y. Campbell and L.M. Viceira. Consumption and portfolio decisions when expected returns are time-varying. *Quarterly Journal of Economics*, 114:433–495, 1999.

5. J.Y. Campell, J. Cocco, F. Gomes, P.J. Maenhout, and L.M. Viceira. Stock market mean reversion and the optimal equity allocation of a long-lived investor. *European Finance Review*, 5:269–292, 2001.

6. G. Chacko and L. Viceira. Dynamic consumption and portfolio choice with stochastic volatility in incomplete markets. NBER Working papers, No.7377, 1999.

7. M.H.A. Davis and T. Zariphopoulou. American options and transaction fees. *Mathematical Finance, IMA Volumes in Mathematics and its Application*, 65:47–61, Springer-Verlag, 1995.

8. F. Delbaen, P. Grandits, T. Rheinlander, D. Samperi, M. Schweizer, and C. Stricker. Exponential hedging and entropic penalties. *Mathematical Finance*, 12:99–123, 2002.

9. D. Duffie and T. Zariphopoulou. Optimal investment with undiversifiable income risk. *Mathematical Finance*, 3(2):135–148, 1993.

10. H. Foellmer and M. Schweizer. Hedging of contingent claims under incomplete information. *Applied Stochastic Analysis, Stochastic Monographs*, Vol. 5, eds. M.H.A. Davis and R.J. Elliott, Gordon and Breach, 389–414, 1991.

11. M. Frittelli. The minimal entropy martingale measure and the valuation problem in incomplete markets. *Mathematical Finance*, 10:39–52, 2000.

12. P. Grandits and T. Rheinlander. On the minimal entropy measure. *Annals of Probability*, 30(3):1003–1038, 2002.

13. M.R. Grasselli and T. Hurd. Indifference pricing and hedging in stochastic volatility models. preprint, 2004.

14. Y. Kabanov and C. Stricker. On the optimal portfolio for the exponential utility maximization: Remarks to the six-author paper "Exponential hedging and entropic penalties." *Mathematical Finance*, 12(2):125–134, 2002.

15. J. Kallsen and C. Kuehn. Pricing derivatives of American and game type in incomplete markets. *Finance and Stochastics*, 8(2):261–284, 2004.

16. T.S. Kim and E. Omberg. Dynamic nonmyopic portfolio behavior. *Review of Financial Studies*, 9:141–161, 1996.

17. D. Kramkov and W. Schachermayer. The asymptotic elasticity of utility functions and optimal investments in incomplete markets. *Annals of Applied Probability*, 9:904–950, 1999.

18. J. Liu. Portfolio selection in stochastic environments. Working paper, Stanford University, 1999.

19. M. Monoyios. Characterization of optimal dual measures via distortion. *Decisions in Economics and Finance*, 29:95–119, 2006.

20. M. Musiela and T. Zariphopoulou. Indifference prices and related measures. Technical report, The University of Texas at Austin, 2001. http://www.ma.utexas.edu/users/zariphop/.

21. M. Musiela and T. Zariphopoulou. An example of indifference prices under exponential preferences. *Finance and Stochastics*, 8:229–239, 2004.

22. M. Musiela and T. Zariphopoulou. Indifference prices of early exercise claims. *Proceedings of AMS-IMS-SIAM Joint Summer Conference on Mathematical Finance*, eds. G. Yin and Q. Zhang, *Contemporary Mathematics*, AMS, 351:259–272, 2004.

23. M. Musiela and T. Zariphopoulou. The backward and forward dynamic utilities and their associated pricing systems: The case study of the binomial model. *Indifference Pricing*, ed. R. Carmona, Princeton University Press, 2005.

24. M. Musiela and T. Zariphopoulou. Forward dynamic utilities and indifference valuation in incomplete binomial models. Preprint, 2005.

25. A. Oberman and T. Zariphopoulou. Pricing early exercise contracts in incomplete markets. *Computational Management Science*, 1:75–107, 2003.

26. H. Pham. Smooth solutions to optimal investment models with stochastic volatilities and portfolio constraints. *Applied Mathematics and Optimization*, 46:55–78, 2002.

27. T. Rheinlander. An entropy approach to the Stein/Stein model with correlation. *Finance and Stochastics*, 9:399–413, 2003.

28. R. Rouge and N. El Karoui. Pricing via utility maximization and entropy. *Mathematical Finance*, 10:259–276, 2000.

29. M. Schweizer. On the minimal martingale measure and the Foellmer-Schweizer decomposition. *Stochastic Processes and Applications*, 13:573–599, 1995.

30. M. Schweizer. A minimality property of the minimal martingale measure. *Statistics and Probability Letters*, 42:27–31, 1999.

31. R. Sircar and T. Zariphopoulou. Bounds and asymptotic approximations for utility prices when volatility is random. *SIAM Journal on Control and Optimization*, 43(4):1328–1353, 2005.

32. S. Stoikov and T. Zariphopoulou. Optimal investments in the presence of unhedgeable risks and under CARA preferences. *IMA Volumes in Mathematics and its Application*, in print, 2004.

33. N. Touzi. Stochastic control problems, viscosity solutions and application to finance. Technical report, Special research semester on Financial Markets, Pisa, 2002.

34. J. Wachter. Portfolio consumption decisions under mean-reverting returns: An exact solution for complete markets. *Journal of Financial and Quantitative Analysis*, 37(1):63–91, 2002.

35. T. Zariphopoulou. Stochastic control methods in asset pricing. *Handbook of Stochastic Analysis and Applications*, eds. D. Kannan and V. Lakshmikanthan, Statistics, Textbooks, Monographs, 163, Dekker, 679–753, 2002.

Applied and Numerical Harmonic Analysis

J.M. Cooper: *Introduction to Partial Differential Equations with MATLAB* (ISBN 0-8176-3967-5)

C.E. D'Attellis and E.M. Fernández-Berdaguer: *Wavelet Theory and Harmonic Analysis in Applied Sciences* (ISBN 0-8176-3953-5)

H.G. Feichtinger and T. Strohmer: *Gabor Analysis and Algorithms* (ISBN 0-8176-3959-4)

T.M. Peters, J.H.T. Bates, G.B. Pike, P. Munger, and J.C. Williams: *The Fourier Transform in Biomedical Engineering* (ISBN 0-8176-3941-1)

A.I. Saichev and W.A. Woyczyński: *Distributions in the Physical and Engineering Sciences* (ISBN 0-8176-3924-1)

R. Tolimieri and M. An: *Time-Frequency Representations* (ISBN 0-8176-3918-7)

G.T. Herman: *Geometry of Digital Spaces* (ISBN 0-8176-3897-0)

A. Procházka, J. Uhlíř, P.J.W. Rayner, and N.G. Kingsbury: *Signal Analysis and Prediction* (ISBN 0-8176-4042-8)

J. Ramanathan: *Methods of Applied Fourier Analysis* (ISBN 0-8176-3963-2)

A. Teolis: *Computational Signal Processing with Wavelets* (ISBN 0-8176-3909-8)

W.O. Bray and Č.V. Stanojević: *Analysis of Divergence* (ISBN 0-8176-4058-4)

G.T Herman and A. Kuba: *Discrete Tomography* (ISBN 0-8176-4101-7)

J.J. Benedetto and P.J.S.G. Ferreira: *Modern Sampling Theory* (ISBN 0-8176-4023-1)

A. Abbate, C.M. DeCusatis, and P.K. Das: *Wavelets and Subbands* (ISBN 0-8176-4136-X)

L. Debnath: *Wavelet Transforms and Time-Frequency Signal Analysis* (ISBN 0-8176-4104-1)

K. Gröchenig: *Foundations of Time-Frequency Analysis* (ISBN 0-8176-4022-3)

D.F. Walnut: *An Introduction to Wavelet Analysis* (ISBN 0-8176-3962-4)

O. Bratteli and P. Jorgensen: *Wavelets through a Looking Glass* (ISBN 0-8176-4280-3)

H.G. Feichtinger and T. Strohmer: *Advances in Gabor Analysis* (ISBN 0-8176-4239-0)

O. Christensen: *An Introduction to Frames and Riesz Bases* (ISBN 0-8176-4295-1)

L. Debnath: *Wavelets and Signal Processing* (ISBN 0-8176-4235-8)

J. Davis: *Methods of Applied Mathematics with a MATLAB Overview* (ISBN 0-8176-4331-1)

G. Bi and Y. Zeng: *Transforms and Fast Algorithms for Signal Analysis and Representations* (ISBN 0-8176-4279-X)

J.J. Benedetto and A. Zayed: *Sampling, Wavelets, and Tomography* (ISBN 0-8176-4304-4)

E. Prestini: *The Evolution of Applied Harmonic Analysis* (ISBN 0-8176-4125-4)

O. Christensen and K.L. Christensen: *Approximation Theory* (ISBN 0-8176-3600-5)

L. Brandolini, L. Colzani, A. Iosevich, and G. Travaglini: *Fourier Analysis and Convexity* (ISBN 0-8176-3263-8)

W. Freeden and V. Michel: *Multiscale Potential Theory* (ISBN 0-8176-4105-X)

O. Calin and D.-C. Chang: *Geometric Mechanics on Riemannian Manifolds* (ISBN 0-8176-4354-0)

Applied and Numerical Harmonic Analysis (Cont'd)

J.A. Hogan and J.D. Lakey: *Time-Frequency and Time-Scale Methods* (ISBN 0-8176-4276-5)

C. Heil: *Harmonic Analysis and Applications* (ISBN 0-8176-3778-8)

K. Borre, D.M. Akos, N. Bertelsen, P. Rinder, and S.H. Jensen: *A Software-Defined GPS and Galileo Receiver* (ISBN 0-8176-4390-7)

T. Qian, V. Mang I, and Y. Xu: *Wavelet Analysis and Applications* (ISBN 3-7643-7777-1)

G.T. Herman and A. Kuba: *Advances in Discrete Tomography and Its Applications* (ISBN 0-8176-3614-5)

M.C. Fu, R.A. Jarrow, J.-Y. J. Yen, and R.J. Elliott: *Advances in Mathematical Finance* (ISBN 0-8176-4544-6)

Printed in the United States of America